Mathematical Time Capsules

Historical Modules for the Mathematics Classroom

© 2011 by
The Mathematical Association of America (Incorporated)

Library of Congress Control Number 2011926092

Print ISBN 978-0-88385-187-6

Electronic ISBN 978-0-88385-984-1

Printed in the United States of America

Current Printing (last digit):
10 9 8 7 6 5 4 3 2 1

Mathematical Time Capsules

Historical Modules for the Mathematics Classroom

Edited by

Dick Jardine
Keene State College

and

Amy Shell-Gellasch
Beloit College

Published and Distributed by
The Mathematical Association of America

The MAA Notes Series, started in 1982, addresses a broad range of topics and themes of interest to all who are involved with undergraduate mathematics. The volumes in this series are readable, informative, and useful, and help the mathematical community keep up with developments of importance to mathematics.

Committee on Books
Gerald Bryce, *Chair*

Notes Editorial Board
Stephen B Maurer, *Editor*

Deborah J. Bergstrand   Thomas P. Dence
Donna L. Flint   Theresa Jeevanjee
Michael K. May   Judith A. Palagallo
Mark Parker   Susan Pustejovsky
David Rusin   David J. Sprows
Joe Alyn Stickles   Andrius Tamulis

MAA Notes

14. Mathematical Writing, by *Donald E. Knuth, Tracy Larrabee, and Paul M. Roberts.*
16. Using Writing to Teach Mathematics, *Andrew Sterrett,* Editor.
17. Priming the Calculus Pump: Innovations and Resources, Committee on Calculus Reform and the First Two Years, a subcomittee of the Committee on the Undergraduate Program in Mathematics, *Thomas W. Tucker,* Editor.
18. Models for Undergraduate Research in Mathematics, *Lester Senechal,* Editor.
19. Visualization in Teaching and Learning Mathematics, Committee on Computers in Mathematics Education, *Steve Cunningham and Walter S. Zimmermann,* Editors.
20. The Laboratory Approach to Teaching Calculus, *L. Carl Leinbach et al.,* Editors.
21. Perspectives on Contemporary Statistics, *David C. Hoaglin and David S. Moore,* Editors.
22. Heeding the Call for Change: Suggestions for Curricular Action, *Lynn A. Steen,* Editor.
24. Symbolic Computation in Undergraduate Mathematics Education, *Zaven A. Karian,* Editor.
25. The Concept of Function: Aspects of Epistemology and Pedagogy, *Guershon Harel and Ed Dubinsky,* Editors.
26. Statistics for the Twenty-First Century, *Florence and Sheldon Gordon,* Editors.
27. Resources for Calculus Collection, Volume 1: Learning by Discovery: A Lab Manual for Calculus, *Anita E. Solow,* Editor.
28. Resources for Calculus Collection, Volume 2: Calculus Problems for a New Century, *Robert Fraga,* Editor.
29. Resources for Calculus Collection, Volume 3: Applications of Calculus, *Philip Straffin,* Editor.
30. Resources for Calculus Collection, Volume 4: Problems for Student Investigation, *Michael B. Jackson and John R. Ramsay,* Editors.
31. Resources for Calculus Collection, Volume 5: Readings for Calculus, *Underwood Dudley,* Editor.
32. Essays in Humanistic Mathematics, *Alvin White,* Editor.
33. Research Issues in Undergraduate Mathematics Learning: Preliminary Analyses and Results, *James J. Kaput and Ed Dubinsky,* Editors.
34. In Eves Circles, *Joby Milo Anthony,* Editor.
35. Youre the Professor, What Next? Ideas and Resources for Preparing College Teachers, The Committee on Preparation for College Teaching, *Bettye Anne Case,* Editor.
36. Preparing for a New Calculus: Conference Proceedings, *Anita E. Solow,* Editor.
37. A Practical Guide to Cooperative Learning in Collegiate Mathematics, *Nancy L. Hagelgans, Barbara E. Reynolds, SDS, Keith Schwingendorf, Draga Vidakovic, Ed Dubinsky, Mazen Shahin, G. Joseph Wimbish, Jr.*
38. Models That Work: Case Studies in Effective Undergraduate Mathematics Programs, *Alan C. Tucker,* Editor.
39. Calculus: The Dynamics of Change, CUPM Subcommittee on Calculus Reform and the First Two Years, *A. Wayne Roberts,* Editor.
40. Vita Mathematica: Historical Research and Integration with Teaching, *Ronald Calinger,* Editor.
41. Geometry Turned On: Dynamic Software in Learning, Teaching, and Research, *James R. King and Doris Schattschneider,* Editors.
42. Resources for Teaching Linear Algebra, *David Carlson, Charles R. Johnson, David C. Lay, A. Duane Porter, Ann E. Watkins, William Watkins,* Editors.
43. Student Assessment in Calculus: A Report of the NSF Working Group on Assessment in Calculus, *Alan Schoenfeld,* Editor.

44. Readings in Cooperative Learning for Undergraduate Mathematics, *Ed Dubinsky, David Mathews, and Barbara E. Reynolds,* Editors.
45. Confronting the Core Curriculum: Considering Change in the Undergraduate Mathematics Major, *John A. Dossey,* Editor.
46. Women in Mathematics: Scaling the Heights, *Deborah Nolan,* Editor.
47. Exemplary Programs in Introductory College Mathematics: Innovative Programs Using Technology, *Susan Lenker,* Editor.
48. Writing in the Teaching and Learning of Mathematics, *John Meier and Thomas Rishel.*
49. Assessment Practices in Undergraduate Mathematics, *Bonnie Gold,* Editor.
50. Revolutions in Differential Equations: Exploring ODEs with Modern Technology, *Michael J. Kallaher,* Editor.
51. Using History to Teach Mathematics: An International Perspective, *Victor J. Katz,* Editor.
52. Teaching Statistics: Resources for Undergraduate Instructors, *Thomas L. Moore,* Editor.
53. Geometry at Work: Papers in Applied Geometry, *Catherine A. Gorini,* Editor.
54. Teaching First: A Guide for New Mathematicians, *Thomas W. Rishel.*
55. Cooperative Learning in Undergraduate Mathematics: Issues That Matter and Strategies That Work, *Elizabeth C. Rogers, Barbara E. Reynolds, Neil A. Davidson, and Anthony D. Thomas,* Editors.
56. Changing Calculus: A Report on Evaluation Efforts and National Impact from 1988 to 1998, *Susan L. Ganter.*
57. Learning to Teach and Teaching to Learn Mathematics: Resources for Professional Development, *Matthew Delong and Dale Winter.*
58. Fractals, Graphics, and Mathematics Education, Benoit Mandelbrot and Michael Frame, Editors.
59. Linear Algebra Gems: Assets for Undergraduate Mathematics, *David Carlson, Charles R. Johnson, David C. Lay, and A. Duane Porter,* Editors.
60. Innovations in Teaching Abstract Algebra, *Allen C. Hibbard and Ellen J. Maycock,* Editors.
61. Changing Core Mathematics, *Chris Arney and Donald Small,* Editors.
62. Achieving Quantitative Literacy: An Urgent Challenge for Higher Education, *Lynn Arthur Steen.*
64. Leading the Mathematical Sciences Department: A Resource for Chairs, *Tina H. Straley, Marcia P. Sward, and Jon W. Scott,* Editors.
65. Innovations in Teaching Statistics, *Joan B. Garfield,* Editor.
66. Mathematics in Service to the Community: Concepts and models for service-learning in the mathematical sciences, *Charles R. Hadlock,* Editor.
67. Innovative Approaches to Undergraduate Mathematics Courses Beyond Calculus, *Richard J. Maher,* Editor.
68. From Calculus to Computers: Using the last 200 years of mathematics history in the classroom, Amy Shell-Gellasch and Dick Jardine, Editors.
69. A Fresh Start for Collegiate Mathematics: Rethinking the Courses below Calculus, *Nancy Baxter Hastings,* Editor.
70. Current Practices in Quantitative Literacy, *Rick Gillman,* Editor.
71. War Stories from Applied Math: Undergraduate Consultancy Projects, *Robert Fraga,* Editor.
72. Hands On History: A Resource for Teaching Mathematics, *Amy Shell-Gellasch,* Editor.
73. Making the Connection: Research and Teaching in Undergraduate Mathematics Education, *Marilyn P. Carlson and Chris Rasmussen,* Editors.
74. Resources for Teaching Discrete Mathematics: Classroom Projects, History Modules, and Articles, *Brian Hopkins,* Editor.
75. The Moore Method: A Pathway to Learner-Centered Instruction, *Charles A. Coppin, W. Ted Mahavier, E. Lee May, and G. Edgar Parker.*
76. The Beauty of Fractals: Six Different Views, *Denny Gulick and Jon Scott,* Editors.
77. Mathematical Time Capsules: Historical Modules for the Mathematics Classroom, *Dick Jardine and Amy Shell-Gellasch,* Editors.

MAA Service Center
P.O. Box 91112
Washington, DC 20090-1112
1-800-331-1MAA   FAX: 1-301-206-9789

Preface

Mathematical Time Capsules offers teachers historical modules for immediate use in the mathematics classroom. Relevant history-based activities for a wide range of undergraduate and secondary mathematics courses are included. The genesis of this volume was a Contributed Papers Session on *Using History of Mathematics in Your Mathematics Courses*, organized by the editors at the Joint Mathematics Meetings, San Antonio, Texas, in January of 2006. That session was very well attended, which prompted Andrew Sterrett from MAA publications to suggest that we put together our second volume for the MAA Notes series.

Purpose

For a wide variety of reasons, instructors are looking for ways to include the history of mathematics in their courses. It is not uncommon to see requests for "how to" posted to the History of Mathematics Special Interest Group of the MAA (www.homsigmaa.org) email list, such as this 2008 posting:

> ...I am a newcomer to HOM. Where and how should a newcomer begin? Right now, I would liketo include HOM in a meaningful way in the courses that we teach. Weteach courses from college arithmetic to linear algebra.

In response to such inquiries, we hope to serve the broader mathematical community by offering practical suggestions on how to use the history of mathematics quickly and easily in the mathematics classroom.

A *time capsule* can be defined as a container preserving articles and records from the past for scholars of the future. Of course our volume does not fit that precise definition, but readers who open this book will find articles and activities from mathematics history that enhance the learning of topics typically associated with undergraduate or secondary mathematics curricula. Each capsule presents one topic or perhaps a few related topics, or a historical thread that can be used throughout a course. The capsules were written by experienced practitioners to provide other teachers with the historical background, suggested classroom activities, and further references and resources on the chapter subject. An instructor reading a capsule will have increased confidence in engaging students with at least one activity rich in the history of mathematics that will enhance student learning of the mathematical content of the course. Most of the historical topics contained in a capsule can be implemented in one class period with minimal additional preparation on the part of the teacher.

How to use Mathematical Time Capsules

Teaching styles have been categorized along a spectrum from lecture-oriented practices at one extreme to student-centered approaches in which the teacher guides student work in the classroom. *Mathematical Time Capsules* respects the diversity of teaching styles which individual teachers adopt. Some of the capsules are, in some sense, ready-made lectures the instructor can adopt and adapt as appropriate. Examples of those include Victor Katz's "Copernican Trigonometry," Roger Cooke's "Numerical Solution of Equations," or Jim Tattersall's "Finding the Greatest Common Divisor and More... ." Other capsules clearly engage the students more actively, such as Vicky Klima's "A Different Sort of Calculus Debate." But the capsules should not be categorized as appropriate for one pedagogical approach or the other. For example, "Finding the Greatest Common Divisor and More..." could be adapted for use as a student project to be presented by the student(s) in class after the Euclidean algorithm is covered.

The reader may note that the authors of the capsules demonstrate a variety of approaches to integrating history. The differences are consistent with the nature of this volume, created with respect for the diversity of our authors and our readers. We acknowledge that many teachers prefer to develop their own course materials, and we encourage readers to modify the offerings of the authors.

Mathematical Time Capsules is organized in three sections. The first capsules have as their target mathematical topics that are usually addressed in courses taught in secondary school, at two-year colleges, or during the first two years of the undergraduate mathematics curriculum. These courses are not often taken by mathematics majors, and include, for example, algebra, geometry, mathematics for elementary teachers, trigonometry, or precalculus.

The third section of capsules address topics included in courses traditionally taken by mathematics majors, such as calculus, differential equations, number theory, abstract algebra, differential equations, and analysis.

As an interlude between the first and third, we offer some ideas that can be applied to a wide variety of courses throughout the undergraduate (including two-year college) or secondary curriculum. These interlude capsules (15, 16, and 17) are of a general pedagogical nature, not mathematical, and could be adapted for use in any course.

We could have arranged the capsules differently, and we ask that the reader not limit investigating the offerings in this book thinking that lower-level material is in the front and upper-level material at the back. For example, a teacher interested in historical ideas for a numerical methods course should consider Roger Cooke's "Numerical Solution of Equations," Randy Schwartz's "Rule of Double False Position," Clemency Montelle's "Roots, Rocks, and Newton-Raphson Algorithms for Approximating $\sqrt{2}$ 3000 Years Apart," and Dick Jardine's "Euler's Method in Euler's Words." Some teachers are interested in having students read original sources in the history of mathematics, and Montelle's "Amo, Amas, Amat! What's the Sum of That?" and Jardine's "Euler's Method in Euler's Words" provide opportunities to do just that with brief excerpts in the words of the originators.

Where possible, we grouped the capsules within the sections according to mathematical subject area. As an example, there are three capsules that address Pythagorean triples, but each of those capsules approaches the topic in a very different way and from the perspective of a different era of the history of mathematics. That latter notion is significant, as it is important for students to see the evolution of a mathematical idea and how it is viewed through a different lens depending on how mathematics was done at various times in history. One of the goals of *Mathematical Time Capsules* is to provide a vehicle for teachers to help their students learn the historical context for a mathematical development while they are learning the specific mathematical concept. Learning the history of an idea promotes deeper understanding of the idea.

Additional purposes—assessment and teacher certification

Beyond a teacher's personal interest in using the history of mathematics in teaching, accrediting agencies and state certifying agencies now require pre-service teachers to be well-versed in the history of mathematics in specific content areas. The requirement that school teachers demonstrate understanding of the connection between a mathematics topic and the history of that specific topic is explicitly documented in state and national standards. It is becoming an imperative that many college teachers introduce the history of mathematic in their teaching of mathematics in order to pass muster for state and national certification of teacher education programs. The National Council for Accreditation of Teacher Education (NCATE) and the National Council of Teachers of Mathematics (NCTM) have published standards which list *within the content areas* a requirement that candidates for teacher certification demonstrate their knowledge of the historical development of that content area. For example, the current content Standard 10 is on the subject of algebra. For certification, in addition to demonstrating expertise in the usual algebra concepts, candidates must "Demonstrate knowledge of the historical development of algebra including contributions from diverse cultures" [2].

In this era of outcomes-based assessment, to satisfy state and national evaluators it is not sufficient to show that history was included in the syllabus or to claim that history was included in a lecture. *Actual student work* (interestingly for us interested in history, the student work is called an *artifact* in current assessment jargon) must be presented to the accrediting agency or state department of education evaluators. The material presented herein will provide teachers with actual activities that students can do. The products of these activities become the artifacts necessary to validate that students are engaged in learning the historical development of the mathematical topic. *Mathematical Time Capsules* provides materials enhancing student understanding and interest in mathematics, always keeping the learning of mathematics clearly at the center of each capsule's focus.

Acknowledgements

We could not have taken on this project without the interest and enthusiasm of the many contributors from campuses around the world. We are indebted to the successful leadership efforts of V. Frederick Rickey and Victor Katz in bringing together a community of educators and encouraging our passion for improving student learning through the use of the history of mathematics. The editors of this volume met as participants in the NSF funded Institute for the History of Mathematics and its use in Teaching, organized by Fred and Victor, which provided a sound foundation for our work and a network of like-minded colleagues and friends. We also offer special thanks to Chris Arney, who led the way as Department Head of the Department of Mathematical Sciences at the United States Military Academy at West Point in not only documenting that the history of mathematics is a significant learning outcome at the program level in undergraduate mathematics departments but also provided an exemplar of effective implementation. We both worked for Chris and he made a real difference by encouraging and supporting our early use of the history of mathematics to deepen student learning of our discipline. Finally, and perhaps most importantly, we also acknowledge the interest and enthusiasm of our students, who let us know each semester that we teach that learning the history of mathematics is important to them.

References

1. Amy Shell-Gellasch and D. Jardine, ed., *From Calculus to Computers: Using the Last 200 Years of Mathematics History in the Classroom*, Mathematical Association of America, Washington, 2005.

2. *NCATE/NCTM Program Standards (2003), Programs for Initial Preparation of Mathematics Teachers, Standards for Secondary Mathematics Teachers*

    `www.nctm.org/uploadedFiles/Math_Standards/`,

accessed May 12, 2009.

Contents

Preface   vii

1   The Sources of Algebra, Roger Cooke   1
    1.1   Introduction   1
    1.2   Egyptian problems   1
    1.3   Mesopotamian problems   2
    1.4   "Algebra" in Euclid's geometry   2
    1.5   Chinese problems   4
    1.6   An Arabic problem   4
    1.7   A Japanese problem   5
    1.8   Teaching note   5
    1.9   Problems and Questions   5
    1.10   Further reading   6

2   How to Measure the Earth, Lawrence D'Antonio   7
    2.1   Introduction   7
    2.2   Historical Introduction   8
    2.3   In the Classroom   13
    2.4   Taking it Further   15
    2.5   Conclusion   16
    Bibliography   16

3   Numerical solution of equations, Roger Cooke   17
    3.1   Introduction   17
    3.2   The ancient Chinese method of solving a polynomial equation   18
    3.3   Non-integer solutions   19
    3.4   The cubic equation   20
    3.5   Problems and questions   20
    3.6   Further reading   21

4   Completing the Square through the Millennia, Dick Jardine   23
    4.1   Introduction   23
    4.2   Historical preliminaries   23
    4.3   Student activities   27
    4.4   Summary and conclusion   27
    Bibliography   27
    Appendix: Student activities   28

5 Adapting the Medieval "Rule of Double False Position" to the Modern Classroom, Randy K. Schwartz   29
    5.1   Introduction 29
    5.2   Historical Background 29
    5.3   In the Classroom 31
    5.4   Taking It Further 35
    5.5   Conclusion 37
    Bibliography 37

6 Complex Numbers, Cubic Equations, and Sixteenth-Century Italy, Daniel J. Curtin   39
    6.1   Introduction 39
    6.2   Historical Background 39
    6.3   In the Classroom 40
    6.4   Rafael Bombelli 41
    6.5   Conclusion 43
    Bibliography 43

7 Shearing with Euclid, Davida Fischman and Shawnee McMurran   45
    7.1   Introduction 45
    7.2   Historical Background 46
    7.3   In the Classroom 50
    7.4   Conclusion 51
    Appendix 52
    Bibliography 53

8 The Mathematics of Measuring Time, Kim Plofker   55
    8.1   Introduction 55
    8.2   Historical Background 55
    8.3   In the Classroom 57
    8.4   Taking It Further 60
    8.5   Conclusion 61
    Bibliography 61

9 Clear Sailing with Trigonometry, Glen Van Brummelen   63
    9.1   Introduction 63
    9.2   Historical Background 65
    9.3   Navigating with Trigonometry 65
    9.4   In the Classroom 67
    9.5   Conclusion 68
    Bibliography 69
    Appendix: Michael of Rhodes: Did He Know the Law of Sines? 70

10 Copernican Trigonometry, Victor J. Katz   73
    10.1 Introduction 73
    10.2 Historical Background 73
    10.3 In the Classroom 74
    10.4 Conclusion 88
    Bibliography 88

11 Cusps: Horns and Beaks, Robert E. Bradley — 89
- 11.1 Introduction — 89
- 11.2 Historical Background — 89
- 11.3 In the Classroom — 90
- 11.4 Conclusion — 98
- 11.5 Notes on Classroom Use — 98
- Bibliography — 99

12 The Latitude of Forms, Area, and Velocity, Daniel J. Curtin — 101
- 12.1 Introduction — 101
- 12.2 Historical Background — 101
- 12.3 In the Classroom — 102
- 12.4 Taking It Further — 104
- 12.5 Conclusion — 104
- 12.6 Comments — 104
- Bibliography — 105

13 Descartes' Approach to Tangents, Daniel J. Curtin — 107
- 13.1 Introduction — 107
- 13.2 Historical Background — 107
- 13.3 In the Classroom — 107
- 13.4 Conclusion — 110
- Bibliography — 110

14 Integration à la Fermat, Amy Shell-Gellasch — 111
- 14.1 Introduction — 111
- 14.2 Historical Background — 111
- 14.3 In the Classroom — 113
- 14.4 Taking it Further — 114
- 14.5 Conclusion — 116
- Bibliography — 116

15 Sharing the Fun: Student Presentations, Amy Shell-Gellasch and Dick Jardine — 117
- 15.1 Introduction — 117
- 15.2 Getting Started — 117
- 15.3 Presentations — 118
- 15.4 Assessment — 119
- 15.5 Conclusion — 119
- Bibliography — 119

16 Digging up History on the Internet: Discovery Worksheets, Betty Mayfield — 123
- 16.1 Introduction — 123
- 16.2 In the Classroom — 124
- 16.3 Learning about History — and about the Web — 124
- 16.4 Conclusion — 124
- Bibliography — 124
- Appendix 1: Who Was Gauss? — 125
- Appendix 2: History of Matrices and Determinants — 125
- Appendix 3: Stephen Smale, a contemporary mathematician — 126
- Appendix 4: Nancy Kopell, a female mathematician — 126

17 Newton vs. Leibniz in One Hour!, Betty Mayfield — 127
- 17.1 Introduction — 127
- 17.2 Historical Background — 128
- 17.3 In the Classroom — 128
- 17.4 Taking it Further — 129
- 17.5 Conclusion — 129
- Bibliography — 130
- Appendix 1: Instructions to the class — 130

18 Connections between Newton, Leibniz, and Calculus I, Andrew B. Perry — 133
- 18.1 Introduction — 133
- 18.2 Historical Background — 133
- 18.3 Newton's Work — 134
- 18.4 Leibniz's Work — 135
- 18.5 Berkeley's Critique — 136
- 18.6 Later Developments — 137
- 18.7 In The Classroom — 137
- 18.8 Conclusion — 137
- Bibliography — 137

19 A Different Sort of Calculus Debate, Vicky Williams Klima — 139
- 19.1 Introduction — 139
- 19.2 Historical Background — 139
- 19.3 In the Classroom — 141
- 19.4 Conclusions — 142
- Bibliography — 143
- Appendix A: Fermat's Method Worksheet — 144
- Appendix B: Barrow's Theorem Worksheet — 146
- Appendix C: The Debates: Roles, Structure, Hints — 148

20 A 'Symbolic' History of the Derivative, Clemency Montelle — 151
- 20.1 Introduction — 151
- 20.2 The Derivative — 151
- 20.3 Isaac Newton (1643–1727): — 152
- 20.4 Gottfried Wilhelm von Leibniz (1646–1716): — 153
- 20.5 Joseph-Louis Lagrange (1736–1813) — 154
- 20.6 Louis François Antoine Arbogast (1759-1803) — 155
- 20.7 Conclusion — 157
- 20.8 Reflective Questions — 157
- Bibliography — 158

21 Leibniz's Calculus (Real Retro Calc.), Robert Rogers — 159
- 21.1 Introduction — 159
- 21.2 Differential Calculus (Rules of Differences) — 160
- 21.3 Conclusion — 165
- Appendix — 165
- Bibliography — 167

22 An "Impossible" Problem, Courtesy of Leonhard Euler, Homer S. White — 169
- 22.1 Introduction … 169
- 22.2 Historical Setting … 169
- 22.3 In the Classroom … 171
- 22.4 Conclusion … 176
- Appendix: Remarks on Selected Exercises … 176
- Bibliography … 177

23 Multiple Representations of Functions in the History of Mathematics, Robert Rogers — 179
- 23.1 Introduction (A Funny Thing Happened on the Way to Calculus) … 179
- 23.2 Area of a Circle … 180
- 23.3 Ptolemy's Table … 181
- 23.4 From Geometry to Analysis to Set Theory … 184
- 23.5 Conclusion … 187
- Bibliography … 187

24 The Unity of all Science: Karl Pearson, the Mean and the Standard Deviation, Joe Albree — 189
- 24.1 Introduction … 189
- 24.2 Karl Pearson: Historical preliminaries … 189
- 24.3 A data set … 190
- 24.4 The Mean and the First Moment … 191
- 24.5 The Second Moment and the Standard Deviation … 193
- 24.6 In the Classroom … 195
- 24.7 Conclusions … 195
- 24.8 Activities and Questions … 195
- Bibliography … 197

25 Finding the Greatest Common Divisor, J.J. Tattersall — 199
- 25.1 Introduction … 199
- 25.2 Historical Background … 199
- 25.3 More Historical Background … 201
- 25.4 In The classroom … 201
- 25.5 Conclusion … 202
- Bibliography … 202

26 Two-Way Numbers and an Alternate Technique for Multiplying Two Numbers, J.J. Tattersall — 203
- 26.1 Introduction … 203
- 26.2 Historical Background … 203
- 26.3 In The Classroom … 205
- 26.4 Taking It Further … 205
- 26.5 Conclusion … 206
- Bibliography … 207

27 The Origins of Integrating Factors, Dick Jardine — 209
- 27.1 Introduction … 209
- 27.2 Historical preliminaries … 209
- 27.3 Mathematical preliminaries: Integrating factors … 211
- 27.4 Bernoulli's and Euler's use of integrating factors … 212
- 27.5 Student activities … 213
- 27.6 Summary and conclusion … 213
- Bibliography … 213
- Appendix: Student activities … 214

28 Euler's Method in Euler's Words, Dick Jardine — 215
- 28.1 Introduction — 215
- 28.2 Historical preliminaries — 215
- 28.3 Euler's description of the method — 217
- 28.4 Student Activities — 218
- 28.5 Summary and conclusion — 218
- Bibliography — 219
- Appendix A: Student Assignments — 220
- Appendix B: Original source translation — 221

29 Newton's Differential Equation $\frac{\dot{y}}{\dot{x}} = 1 - 3x + y + xx + xy$, Hüseyin Koçak — 223
- 29.1 Introduction — 223
- 29.2 Newton's differential equation — 223
- 29.3 Newton's solution — 224
- 29.4 Phaser simulations — 225
- 29.5 Remarks: Newton, Leibniz, and Euler — 226
- 29.6 Suggested Explorations — 227
- Bibliography — 228

30 Roots, Rocks, and Newton-Raphson Algorithms for Approximating $\sqrt{2}$ 3000 Years Apart, Clemency Montelle — 229
- 30.1 Introduction — 229
- 30.2 The Problem — 230
- 30.3 Solving $\sqrt{2}$ — Second Millennium C.E. Style — 230
- 30.4 Time Warp: Solving $\sqrt{2}$ — Second Millennium B.C.E. Style — 232
- 30.5 Taking it Further: Final Reflections — 237
- 30.6 Conclusion — 238
- Bibliography — 239

31 Plimpton 322: The Pythagorean Theorem, More than a Thousand Years before Pythagoras, Daniel E. Otero — 241
- 31.1 Introduction — 241
- 31.2 Historical Background — 241
- 31.3 Reading the Tablet — 242
- 31.4 Sexagesimal numeration — 243
- 31.5 So What Does It All Mean? — 245
- 31.6 Why Tabulate These Numbers? — 248
- 31.7 Plimpton 322 in the Classroom — 248
- 31.8 Conclusion — 250
- Bibliography — 250

32 Thomas Harriot's Pythagorean Triples: Could He List Them All?, Janet L. Beery — 251
- 32.1 Introduction — 251
- 32.2 Mathematical Background — 251
- 32.3 Historical Background — 252
- 32.4 In the Classroom — 252
- 32.5 Conclusion — 258
- Bibliography — 258
- Appendix: Who Was Thomas Harriot? — 259

33 Amo, Amas, Amat! What's the sum of that?, Clemency Montelle — **261**
- 33.1 Introduction — 261
- 33.2 The Harmonic Series — 262
- 33.3 Bernoulli and the Harmonic Series — 262
- 33.4 The Mathematical Explanation — 265
- 33.5 Conclusion — 267
- Bibliography — 267

34 The Harmonic Series: A Primer, Adrian Rice — **269**
- 34.1 Introduction — 269
- 34.2 Historical preliminaries — 269
- 34.3 Introducing the harmonic series — 271
- 34.4 A "prime" piece of mathematics — 273
- 34.5 Conclusion — 275
- Bibliography — 276

35 Learning to Move with Dedekind, Fernando Q. Gouvêa — **277**
- 35.1 Historical Background: What Dedekind Did — 277
- 35.2 In the Classroom 1: Transition to Proofs — 280
- 35.3 In the Classroom 2: Abstract Algebra — 282
- 35.4 Moving with Dedekind — 283
- Bibliography — 283

About the Editors — **285**

1
The Sources of Algebra

Roger Cooke
University of Vermont

1.1 Introduction

Nowadays we recognize written algebra by the presence of letters (called *variables*) standing for unspecified numbers, and especially by the presence of equations involving those letters. These two features—letters and equations—reveal the *techniques* of algebra, but algebra itself is *not* these techniques. Rather, algebra consists of problems in which the goal is to find a number knowing certain indirect information about it. If you were told to multiply 7 by 3, then add 26 to the product, you would be doing arithmetic, that is, you would be given not only the data, but also told which operations you must perform (multiplication followed by addition). But if you were asked for a number having the property that if it is multiplied by 3 and 26 is added to the product, the result is 57, you would be facing an algebra problem. In an algebra problem, the operations and some of the data are given to you, but these operations are not for you to perform. Rather, you assume someone else has performed them, and you need to find the number(s) on which they were performed. Using this definition, we can recognize algebra problems in very ancient texts that contain no equations at all.

But how do such problems arise? Why were people interested in solving them? Those are questions that any student who looks beyond the horizon of tomorrow's homework assignment is bound to ask. In the following paragraphs, we shall look at some examples and see if we can answer such questions. In this article, we are going to state a number of problems taken from "classic" textbooks, paraphrased and reformulated in plain English and in a style reflective of contemporary textbooks. Students who have had at least one year of algebra can study these problems in the language of xs and ys and solve them using modern methods. At the same time, students are encouraged to use their imaginations in order to think of the motivation that led each author to believe it was worthwhile to write about problems of this sort.

1.2 Egyptian problems

Several problems from the *Rhind Mathematical Papyrus*, which was written some 3500 years ago, make use of the notion of *pesu*, which measures the amount to which grain is "stretched" or diluted in making bread or beer. If you get three (standard-size) loaves of bread per *hekat* of grain, the *pesu* of that bread is 3. Obviously a high *pesu* means very thin or light bread (or a very small loaf) or very weak beer. Here is Problem 73 from the Rhind Papyrus: *One hundred loaves of* pesu 10 *are to be traded for loaves of* pesu 15. *How many of the latter will there be?*

Implicit in this problem is a "conservation of grain" principle. The hundred loaves of *pesu*-10 bread represent ten hekats of grain (at ten loaves per *hekat*). So, if we "expand" the grain by stretching it so as to get 15 loaves per *hekat*, we will obviously get 150 loaves of the weaker bread. The scribe who wrote the papyrus had no difficulty figuring this out.

The concept of *pesu* is analogous to our modern concepts of specific gravity or density. Problems involving these concepts nearly always lead to equations of first degree (linear equations). They are often so simple that it is not even necessary to write down the equation in order to solve them. Others, however, are more complicated. For example, Problem 40 asks how to distribute 100 loaves to five people so that each one (except the last) receives a fixed amount more than the next and so that the first three together receive seven times as much as the last two together. Although this is a "linear" problem, it contains two unknowns, as we would analyze it, namely the amount received by the first person (x) and the amount received by the second person (y). (The amounts received by the last three are then determined as $2y - x$, $3y - 2x$, and $4y - 3x$.) The equations to be satisfied are $10y - 5x = 100$ and $3y = 7(7y - 5x)$.

1.2.1 Quadratic equations

The Egyptians seem to have considered algebra problems that lead to equations more complicated than linear ones. The Berlin Papyrus (from about 3800 years ago) contains a problem that (with some modern conjectural restorations of lost parts) asks for the sides of two squares in the ratio of 3 to 4 with the total area of the two squares to be 100. That is, to solve the simultaneous equations $3x = 4y$ and $x^2 + y^2 = 100$. As an algebra problem, this is not the most general type of quadratic equation, but the geometry of the problem suggests that the Egyptians may have known that a triangle with sides of lengths 3, 4 and 5 is a right triangle.

1.3 Mesopotamian problems

While the surviving Egyptian papyri contain mostly linear algebra problems and a very few quadratic ones, the more durable clay tablets from Mesopotamia, dating to the same period, about 3500 years ago, contain many problems leading to quadratic equations and even a few that seem to call for cubic equations.

One Mesopotamian problem that creates a quadratic equation is the following: *The area of a square less its side is 870. What is the side of this square?*

We would write this problem as the equation $x^2 - x = 870$. Since the Mesopotamian number system was based on 60 and was written in a place-value notation analogous to our decimal system, the number 870 was regarded as $14 \times 60 + 30$ and the right-hand side of this equation was expressed as 14, 30. Needless to say, no equation with the letter x or its Mesopotamian equivalent was written. Instead, the instructions for finding the unknown were given as a recipe: *Take half of* 1, *which is* 30 *[sixtieths]. Multiply it by itself to get* 15 *[sixtieths]. Add this to* 14, 30 *to get* 14, 30; 15 *[870.25]. Take the square root to get* 29; 30 *[29.5]. Now add* 30 *[sixtieths] to this number to get* 30 *[units], which is the side of the square.*

The recipe by itself is nearly incomprehensible, even if you know the quadratic formula for solving this equation. To see how the author was guided, look at the corresponding geometric figure (Fig. 1.1).

The author imagines a strip half a unit wide being peeled off from the right-hand side and then another from the bottom. The total amount removed would be numerically equal to the side of the square, except that the small square of side $\frac{1}{2}$ at the bottom right was already missing after the strip on the right was removed, and hence didn't get removed when the strip was taken off the bottom. Hence what is left is the given amount (870) plus $\frac{1}{4}$. The rest is a matter of taking the square root of 870.25 to get 29.5 as the side of the smaller square, then adding 0.5 to that side to get the side of the original square.

1.4 "Algebra" in Euclid's geometry

Using our definition of an algebra problem as one that requires finding an unknown quantity from certain information about it, we can find examples in Euclid's *Elements*, especially Book 6, that might count as algebra. Most historians of mathematics are convinced that these problems were actually not intended by Euclid to be algebra problems disguised as geometry. In that sense the re-interpretation of those problems in algebraic language that we are about to perform is

1.4. "Algebra" in Euclid's geometry

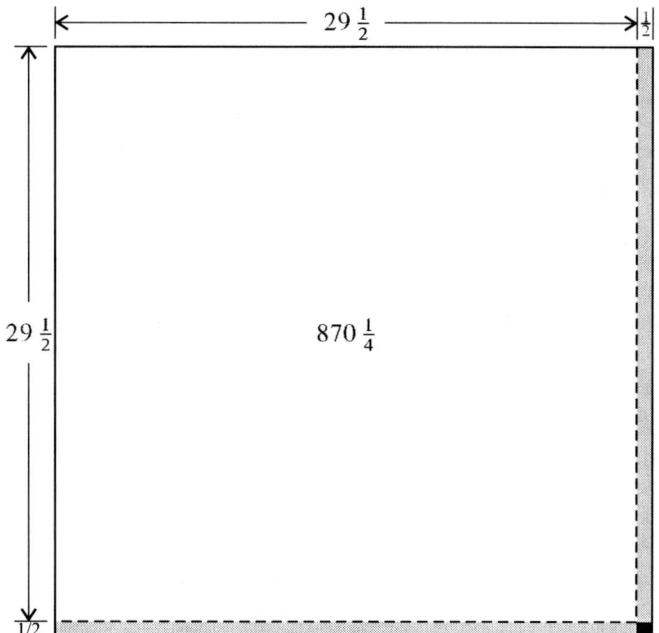

Figure 1.1. An ancient Mesopotamian algebra/geometry problem

not historical. Euclid never gave numerical data. Instead, he posed the problem of finding the point at which a certain line segment is to be divided so as to get a certain figure. It looks as if Euclid deliberately suppressed the notion of *length* of a line segment, which fairly leaps off the page at the modern reader. He had a good reason for doing so. He knew as well as we do that some line segments, such as the diagonal of a square whose side has length 1, have lengths that we would call *irrational* numbers (in this example, $\sqrt{2}$), which are numbers that Euclid would not recognize as numbers at all, since he had no arithmetic procedure for adding, subtracting, multiplying, or dividing them exactly. These operations could be represented geometrically, but any numerical translation of them would be only approximate. Thus our inclusion of a problem from Euclid is really a backward projection of algebra onto a problem that was originally solved using geometry alone.

The problem we choose is a simplified version of Proposition 28 of Book 6. This problem presents a line segment OL and an area A as data and asks for the point P on OL such that a rectangle built on OP with height PL will have area A. Since the maximum area that can be attained in this way is $\frac{\overline{OL}^2}{4}$, A must not be larger than this quantity. If l is the length of OL and x the length of OP, this problem presents us with the equation

$$x(l - x) = A.$$

which is a quadratic equation that can be rewritten as $x^2 + A = lx$. There is obviously some symmetry here: If x satisfies this equation, so does $l - x$. The equation shows that the full rectangle on OL consists of a rectangle on OP equal to A, plus the square on PL. That square is called the *defect* in the construction, and this problem is known as *application with defect*. An illustration is given in Fig. 1.2 with an area A equal to 35 and a line segment OL of length 12.

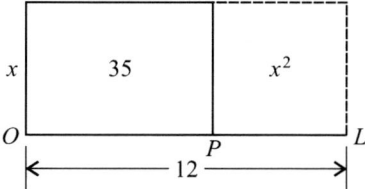

Figure 1.2. A Greek "geometric algebra" problem: $x^2 + 35 = 12x$

1.5 Chinese problems

Chinese mathematicians solved linear and quadratic problems at an early stage, then developed numerical procedures for solving equations of any degree approximately. An early linear problem dating to the Han Dynasty, about 1900 years ago, is found in Chapter 7 of the *Nine Chapters on the Mathematical Art*: *An unknown number of people are buying hens to be held in common. If each gives* 9 *units of money, they will receive* 11 *units of money in change. If each gives* 6 *units, they will be short of the amount they need by* 16 *units.* The problem is to determine how many people P are making this joint purchase and how much money M they need. Without writing down the equations $9P - M = 11$ and $6P - M = -16$, you can do the subtraction in your head. A difference of 3 units per person makes a difference of 27 (that is, $11 - (-16)$) units of available money. Hence there must be 9 people involved, and therefore $M = 70$. The Chinese recipe was more formal: First, the sum $9 \cdot 16 + 6 \cdot 11 = 210$ was taken; then it was divided by $9 - 6$, to get 70 as the purchase price M.

1.5.1 Quadratic equations

Quadratic equations arise in the last chapter of the *Nine Chapters* in the form of right-triangle problems. For example, a square town with a gate in each of its walls (which are aligned along the four cardinal compass points) is such that a tree 20 paces north of the north gate becomes visible to a person who walks 14 paces south from the south gate and then 1775 paces west. The problem is to find the size of the town, that is, the length of each of its four walls. As the similar triangles OAD and EAB in Fig. 1.3 show, we have the proportion

$$\frac{20}{x/2} = \frac{AB}{EB} = \frac{AD}{OD} = \frac{34+x}{1775},$$

leading to the equation

$$x^2 + 34x = 71000,$$

whose only positive solution is $x = 250$.

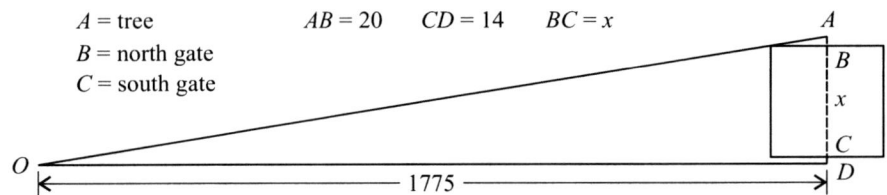

Figure 1.3. A Chinese algebra/geometry problem

1.6 An Arabic problem

The word *algebra* itself, and the central idea of studying equations as an object of interest and developing general methods of solving them comes to us from the medieval Muslim world. The first Arabic text on the subject, written by Muhammed ibn-Musa al-Khwarizmi (ca. 790–ca. 840) about 1200 years ago, contains a number of problems involving the division of an estate, problems that lead to linear equations. It also contains some geometric problems that lead to quadratic equations. In a commentary on this work written about 1100 years ago by Abu Kamil (ca. 850–930), we find the following problem: *We divide* 50 *by a certain number and get a quotient. If the divisor is increased by* 3, *the quotient decreases by* $3\frac{3}{4}$. *What is the divisor?* These conditions lead to the relation

$$\frac{50}{x+3} = \frac{50}{x} - \frac{15}{4},$$

which is equivalent to

$$x^2 + 3x = 40.$$

From this, you can conclude that $x = 5$. (The solution $x = -8$ would not have been recognized by Abu Kamil, since negative numbers were not used.)

1.7 A Japanese problem

Japanese mathematicians began to work independently in the seventeenth century, after mastering the work of Chinese algebraists, which included the numerical solution of higher-degree equations. For some 250 years, from 1600 to about 1850, they posed challenge problems to one another. One such challenge problem, from the 1627 *Treatise on Large and Small Numbers* of Yoshida Koyu (1598–1672), asks where to make cross cuts so as to divide a log into three equal volumes. The log is described as a frustum of a cone, 18 feet long with a circumference of 2.5 feet at the smaller end and 5 feet at the larger end, as in Fig. 1.4.

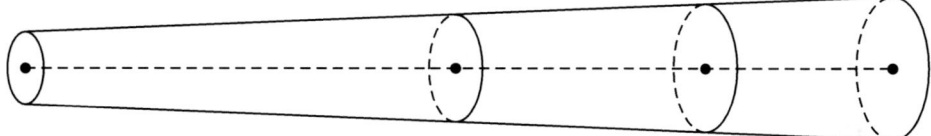

Figure 1.4. Cutting a log into three equal pieces requires solving two cubic equations.

The volume of this log from its smaller end to a plane parallel to that end at a distance of x feet is

$$V(x) = \frac{25}{15552\pi}(x^3 + 54x^2 + 972x).$$

The total volume is $V(18) = \frac{525}{8\pi}$. For if the frustum is capped so as to make a full cone, its apex will be another 18 feet beyond the smaller end, since the circumference decreases by half over the 18-foot length of the cone. The radius is directly proportional to the height, and at a height of 18 feet from the apex it is $2.5/2\pi$ feet, so that $r = \frac{2.5}{36\pi}h$. When $h = x + 18$, this gives $r = \frac{2.5}{36\pi}(x + 18)$. As a result, the volume between the plane at a distance of $18 + x$ feet from the apex and the smaller end is

$$V(x) = \frac{\pi}{3}\left(\frac{2.5}{36\pi}(x+18)\right)^2(x+18) - \frac{\pi}{3}\left(\frac{2.5}{36\pi}18^2\right)(18) = \frac{25}{15552\pi}\left((x+18)^3 - 18^3\right).$$

Hence, this problem leads to the two equations

$$x^3 + 54x^2 + 972x = 13608,$$
$$x^3 + 54x^2 + 972x = 27216$$

for the two distances from the smaller end at which the cuts should be made. Their solutions are $x = 6(-3 + \sqrt[3]{90}) \approx 8.88843$ and $x = 6(-3 + \sqrt[3]{153}) \approx 14.0909$.

1.8 Teaching note

This material is self-explanatory. Probably it would be best to pose each problem except the Japanese problem, which involves a cubic equation, as a challenge at the end of a class, with no background at all provided. Ask the students to set the problem up as an equation or equations, choosing the unknown suitably. At the next class meeting, students can present their algebraic formulation of the problem, and then methods of solving the equations can be discussed and later on compared with the solutions given above. After all the problems have been used in this way, the material in the following two sections can be added to consolidate learning.

1.9 Problems and Questions

Problem 1. *Verify that the equations $10y - 5x = 100$ and $3y = 7(7y - 5x)$ satisfy the conditions of Problem 40 in the Rhind Mathematical Papyrus, and find the number of loaves each person receives.*

Problem 2. *Show how to solve the quadratic equation $x^2 - x = 870$ using the quadratic formula*

$$x = \frac{b}{2} + \sqrt{\left(\frac{b}{2}\right)^2 + c}$$

for solving $x^2 - bx = c$. How is each term in the quadratic formula represented in Fig. 1.1?

Problem 3. *In the first of the Chinese problems explain why "cross-multiplication" of each per-person amount with the surplus or shortage from the* other *per-person amount yields the total price times the difference of two per-person amounts. (Hint: If you have solved linear systems of equations using determinants, you will be able to see this easily from the two equations* $9P - M = 11$ *and* $6P - M = -16$.)

Question 1. *Of what practical use might the* pesu *problems in the Rhind Mathematical Papyrus be? Could there be any practical use for the loaf-distribution problem?*

Question 2. *What purpose could anyone have for solving an "application with defect" problem?*

Question 3. *How does each of these six examples fit the definition of algebra as finding unknown numbers given the result of performing operations on them? In each case, what are the operations performed, and what is the result?*

1.10 Further reading

Walter Eugene Clark, ed., *The Aryabhatiya of Aryabhata*, University of Chicago Press, Chicago, 1930.

Henry Thomas Colebrooke, *Algebra, with Arithmetic and Mensuration from the Sanscrit of Brahmegupta and Bhascara*, J. Murray, London, 1817.

John N. Crossley and Alan S. Henry, "Thus spake al-Khwārizmī: A translation of the text of Cambridge University Library Ms. ii.vi.5," *Historia Mathematica*, 17 (1990), 103–131.

Tobias Dantzig, *Number, the Language of Science*, Fourth edition, The Free Press, New York, 1967. (See especially, Part Two, Chapter C.)

Richard J. Gillings, *Mathematics in the Time of the Pharaohs*, MIT Press, Cambridge, MA, 1972.

Lancelot Hogben, *Mathematics for the Million*, Third Edition, W. W. Norton, New York, 1952. (See especially Chapter VII. Be warned, however, that Hogben erroneously calls the *Aryabhatiya* of the fifth-century Hindu astronomer Aryabhata I by the name *Lilavati*, which is the name of a work by the twelfth-century mathematician Bhaskara II.))

Lay-Yong Lam, "Jiu Zhang Suanshu (Nine Chapters on the Mathematical Art): An Overview," *Archive for History of Exact Sciences*, 47 (1994) 1–51.

Yoshio Mikami, *The Development of Mathematics in China and Japan*, Chelsea, New York, 1961. (Reprint of 1913 edition.)

Otto Neugebauer, *The Exact Sciences in Antiquity*, Princeton University Press, Princeton, NJ, 1952.

Eleanor Robson, *Mesopotamian Mathematics, 2100–1600 BC: Technical Constants in Bureaucracy and Education*, Clarendon Press, Oxford, 1999.

Frederic Rosen, *The Algebra of Mohammed ben Musa*, Oriental Translation Fund, London, 1831.

V. S. Varadarajan, *Algebra in ancient and modern times*, American Mathematical Society, Providence, RI, 1998.

B. L. van der Waerden, *Science Awakening*, Wiley, New York, 1963.

——, *A History of Algebra from al-Khwārizmī to Emmy Noether*, Springer-Verlag, New York, 1985.

Yan Li and Shiran Du, *Chinese Mathematics: A Concise History*, translated by John N. Crossley and Anthony W.-C. Lun, Clarendon Press, Oxford, 1987.

2
How to Measure the Earth

Lawrence D'Antonio
Ramapo College of New Jersey

2.1 Introduction

Who first determined the size of the Earth? How did they do it? These fundamental questions arise in studying early Greek, Indian and Islamic mathematical astronomy. In this article we look at the attempts of Eratosthenes, Posidonius, and al-Bīrūnī to determine the circumference of the Earth and ways to use this topic in the classroom. These calculations use only basic knowledge of geometry and trigonometry, so that instructors in many different courses can include this topic in their syllabus. It would be appropriate to discuss the problem in a high school or college geometry class, in a precalculus class, a history of mathematics class, or in a freshman mathematics survey class.

There are three primary methods for determining the circumference of the Earth: using the lengths of shadows, the elevation of stars, or the altitude of a mountain. Explaining these methods can be done in roughly two hours of class time. If an instructor wants to assign students a project to carry out one of these calculations then one or two more hours may be needed to complete the topic (assuming that the students do measurements during class time).

There are certain geographical and astronomical terms that are frequently used in this topic and should be defined for students. The position of a point on the Earth's surface is given by two coordinates, its latitude and longitude. The *latitude* of a point is measured by how far it is north or south of the equator, so that points of equal latitude form a circle parallel to the equator. Latitude is measured in degrees from the equator (0°) to the poles (90°). *Longitude* measures how far east or west the point lies. Points on a semicircle passing between the poles form what is called a meridian. Today we measure longitude relative to the meridian passing through Greenwich, England called the prime meridian. Longitude is measured in degrees from 0° at the prime meridian to 180° east or west.

The apparent path of the Sun through the sky is called the *ecliptic*. The plane in which the Earth orbits around the Sun is called the ecliptic plane, although the early astronomers that we will discuss generally believed that the Sun orbited the Earth. The Earth's axis of rotation is tilted with respect to the ecliptic, see Figure 2.1. In that figure, the angle α represents the angle between the Earth's equator and the plane of the ecliptic. This tilt is called the *obliquity* of the ecliptic and measures approximately 23°26′. Because of the Earth's tilt, the place in the northern hemisphere where the Sun is directly overhead at noon on the summer solstice, is not on the equator but at a latitude called the Tropic of Cancer (the winter solstice and the Tropic of Capricorn play an equivalent role for the southern hemisphere). This implies that the obliquity of the ecliptic is equal to the latitude of the Tropic of Cancer.

Eratosthenes determined that the arc between the Tropic of Cancer and the equator was 11/83 of a meridian. This translates into a tilt of 23°51′20″, a very accurate result. Further discussion of the obliquity calculation can be found in [10, 11]. A general reference for the history of ancient astronomy is the treatise of Evans [6].

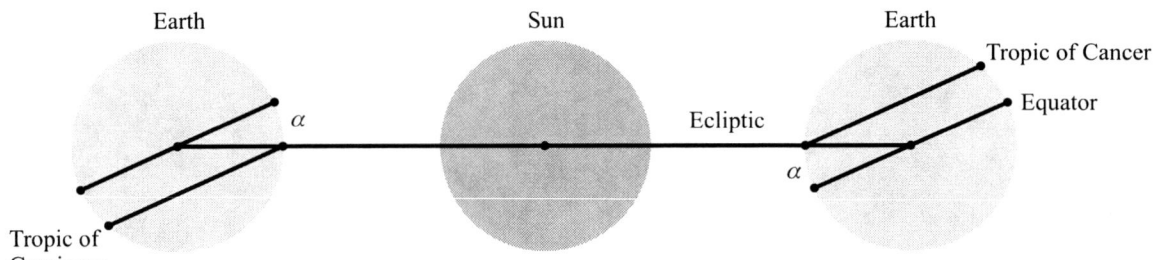

Figure 2.1. The obliquity of the ecliptic

2.2 Historical Introduction

In this section we consider the problem of how we know that the Earth is a sphere and then present the methods of Eratosthenes, Posidonius, and al-Bīrūnī to measure the circumference of the Earth. Note that the methods of Eratosthenes and Posidonius compute the polar circumference (i.e., along a longitude) while the method of al-Bīrūnī measures the circumference of some great circle (although this method can usually be arranged to compute a polar circumference).

2.2.1 Why is the Earth a sphere?

Before one can determine the circumference of the Earth there must be a prior assumption that the Earth is a sphere. Asking students to give reasons why the Earth is spherical makes for an interesting class discussion (the instructor should be ready for surprising responses).

Greek, Indian and Islamic astronomy accepted the hypothesis of a spherical Earth, while Chinese astronomy ascribed to a flat-world theory. How did this idea of a spherical Earth arise? Perhaps from the Pythagoreans who believed that the sphere was the most perfect shape, hence all celestial bodies were spheres. Aristotle gave three empirical arguments why the Earth is round.

- Matter is drawn to the center of the Earth.
- As you move north to south, new constellations are seen in the sky.
- During a lunar eclipse, the Earth's shadow on the Moon is round.

Aristotle stated in *De Caelo* Book II, "Also, those mathematicians who try to calculate the size of the Earth's circumference arrive at the figure 400,000 stades [1] This indicates not only that the Earth's mass is spherical in shape, but also that as compared with the stars it is not of great size." This clearly indicates that the Greeks had a basic understanding of geographical and astronomical distances. They had a rough idea of the size and shape of the Earth, of the distances between Earth, Moon, and Sun, and the fact that the stars are far away. Another interesting topic for classroom discussion is the problem of how we know the stars are very far away (compared to distances within the Solar System).

Of course the Earth is not a perfect sphere, but instead is an oblate spheroid, being slightly flattened at the poles and bulging at the equator. The circumference of the equator (24,902.4 miles) is greater than the polar circumference (24,860.2 miles). This fact was reported by Pierre de Maupertuis in 1737 [13].

As we will discuss in the next section, the circumference of 400,000 stades reported by Aristotle is far too large. A more accurate computation was done in the century after Aristotle by Eratosthenes.

2.2.2 How to measure the Earth using shadows

The earliest method for computing the circumference of the Earth for which we have a detailed account is that of Eratosthenes (276 BCE–194 BCE). Born in the northern African city of Cyrene, Eratosthenes was known for his accomplishments in a wide variety of disciplines. He was a noted astronomer, mathematician, geographer, librarian, poet, and philosopher. Eratosthenes followed his teacher Callimachus in becoming the librarian at the famed library of Alexandria.

---

[1] The stade is the standard Greek unit of distance, representing the length of a foot race. As discussed below, the actual length of a stade is highly disputed.

2.2. Historical Introduction

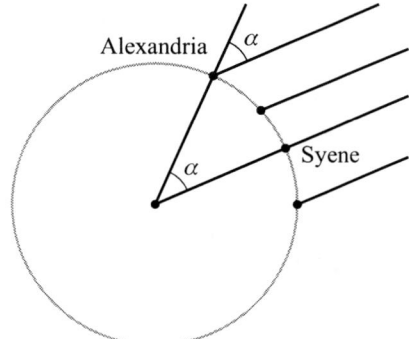

Figure 2.2. The method of Eratosthenes

The calculation of the Earth's circumference appeared in the now lost work, *On the measurement of the Earth*. Later descriptions of this work appear in commentaries by Cleomedes, Strabo and Pliny. The method of Eratosthenes consists of the following steps.

1. Choose two locations a known distance apart, with the same longitude, but different latitudes. Eratosthenes used the Egyptian cities of Alexandria and Syene (the present-day Aswan).

2. Next, simultaneously measure the angle of inclination of the Sun at both locations. Here it is assumed that the Sun is so distant that all rays of light hitting the Earth are parallel. This implies that the difference in the angle of the Sun at these two locations will equal the difference in their latitude. Eratosthenes believed that Syene was on the Tropic of Cancer, so that at noon on the summer solstice the Sun is directly overhead at Syene. Hence the angle of the Sun measured at noon in Alexandria on the summer solstice will equal the difference in latitude of the two cities, see Figure 2.2.

3. Then use the following formula to determine the circumference of the Earth,

$$\frac{\text{circumference}}{\text{distance between locations}} = \frac{\text{angular measure of a circle}}{\text{angular difference of latitudes}}. \qquad (2.1)$$

Thus, in order to compute the circumference of the Earth, Eratosthenes only needed to know two data values: the angle of the Sun in Alexandria at noon on the summer solstice and the distance between Alexandria and Syene.

How did Eratosthenes measure the angle of the Sun? Presumably Eratosthenes used a gnomon (a vertical rod) to measure the shadow's length. In Figure 2.3, h is the height of the gnomon, s the length of the shadow. The angle of the shadow α can be computed by the formula

$$\alpha = \arctan \frac{s}{h}.$$

Of all the current methods to measure the inclination of the Sun, the gnomon is the simplest to use in the classroom. Students can easily construct a gnomon, measure the length of the shadow at noon and compute the above arctangent. But how did Eratosthenes compute the inverse tangent? The ancient Greeks did not have trigonometry as we know it today. Hipparchus in the next century developed a table of chords, but not a triangle based theory. It seems more

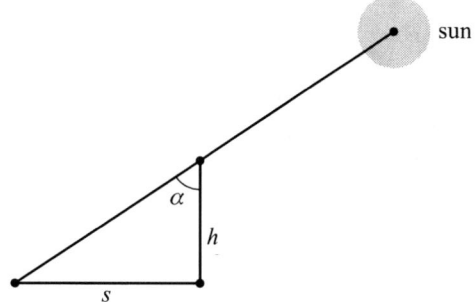

Figure 2.3. Angle of Sun using a gnomon

likely that Eratosthenes used a type of graduated sundial known as a *skaphe*. This is a hemisphere with a gnomon in its center and markings on the inside of the bowl which could be used to determine the angle of the Sun.

Eratosthenes found that the noon-day shadow at Alexandria traversed 1/50 of a circle, which translates to an angle of 7°12′. There are several sources of error in this calculation. In reality Syene is 21′ north of the Tropic of Cancer. Also, Syene is 3° east of Alexandria. The correct difference in latitudes is 7°7′. Despite these errors, the value given by Eratosthenes, 7°12′, is only slightly more than 1% in excess of the true value.

The other piece of data that Eratosthenes needed for the circumference calculation is the distance between Syene and Alexandria. He used a value of 5000 stades for this distance. It's not clear how this distance was determined. Perhaps it was an accepted figure based on existing data of traders sailing on the Nile between the cities. Another possibility is that the distance was calculated by *bematists*, specialists trained to measure distance by counting their paces as they walked.

Assuming the figure of 5000 stades and the ratio of 1/50 for the arc of the Earth's surface between Alexandria and Syene, Eratosthenes would have computed the Earth's circumference to be $5000 \cdot 50 = 250,000$ stades. Cleomedes gives this result [14, vol. 2, pp. 267–273], while Strabo and Pliny give 252,000 stades as the figure computed by Eratosthenes. It has been speculated that the figure of 252,000 was used because it is divisible by 360, thus giving a figure of 700 stades per degree of meridian.

How long is a stade? The stade is a unit of distance related to the length of a sprint in ancient Greek athletic competitions. In different locations, different distances were used for the race. One scholar determines that the stade used by Eratosthenes is 166.7 m [8], while another, citing the testimony of Pliny that 40 stades equaled 5 Roman miles, gives a stade as 184.98 m [5].

If we use the larger figure of 184.98 m. then Eratosthenes' 250,000 stades translates to 46,245 km (28,735.31 miles), which is nearly 16% in excess compared to the true polar circumference of 40,008.6 km (24,860.2 miles). Whereas, if we use the figure of 166.7 m for a stade then 250,000 stades equals 41,675 km (25,895.64 miles), which is only 4% in excess of the correct value. We may never know the precise value of Eratosthenes' stade, but in any case it is clear that the accuracy of his calculation of the Earth's circumference is extremely impressive.

2.2.3 How to measure the Earth using stars

The Stoic philosopher and mathematician, Posidonius of Rhodes (135 BCE–51 BCE), used a method that is slightly different from that of Eratosthenes [4]. Consider the cities of Rhodes and Alexandria, which are on the same meridian (more or less). In Rhodes, the star Canopus just appears on the horizon (actually it reaches a height of 1° above the horizon). In Figure 2.4 this is shown by Canopus being on the tangent to Rhodes. As one sails south from Rhodes, Canopus appears higher and higher in the sky. At Alexandria, the star appears at a maximum altitude of 7°30′ above the horizon.

Posidonius takes the angle of elevation of Canopus at Alexandria to be α in Figure 2.4. This angle also corresponds to the difference in latitudes of the cities. This may seem puzzling, for it would appear by looking at the figure that the angle of elevation of Canopus at Alexandria is in fact larger than α. But Posidonius is assuming that the distance from the Earth to Canopus is far greater than the distance between Rhodes and Alexandria, which is of course true. In this case, the angle of elevation at Alexandria will be practically equal to α.

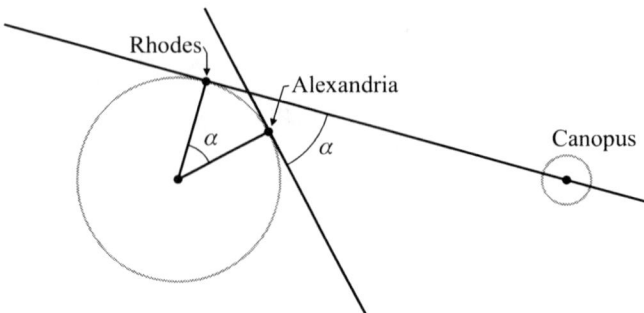

Figure 2.4. The method of Posidonius

Posidonius used 5000 stades as the distance between Rhodes and Alexandria (leading to the hypothesis that all Greek cities at the time were 5000 stades apart). This gives a computed circumference of

$$5000 \cdot \frac{360}{7.5} = 240,000 \text{ stades}.$$

Strabo instead stated that Posidonius found the circumference to be 180,000 stades, making it the smallest such measure from antiquity. Some scholars believe that this shorter distance was used by Columbus as an argument to sponsor his journeys, i.e, India is not really all that far away.

The discrepancy seems to result from an error in Posidonius' calculation. The distance of 5000 stades between Rhodes and Alexandria is too large. Strabo says that Eratosthenes determined the distance to actually be 3750 stades. Using this distance instead of 5000 stades leads to the circumference of $3750 \cdot 360/7.5 = 180,000$ stades, as given by Strabo.

If we accept the value of 240,000 stades, this converts to 40,008 km (24,859.82 miles), which is only 0.001% smaller than the current value of 40,076.5 km. Given the errors in Posidonius' calculations, this is a rather amazing coincidence! The article of Fischer [7] has a good discussion of the methods of Eratosthenes and Posidonius.

2.2.4 How to measure the Earth using mountains

The Islamic mathematician Abu Arrayhan Muhammad ibn Ahmad al-Bīrūnī (973 - 1048) developed a method for calculating the circumference of the Earth by sighting the horizon from the top of a mountain of known height. He was a noted Islamic scholar, mathematician, astronomer, physician and astrologer. Al-Bīrūnī was born in the region of Khwarazm, in present day Uzbekistan. He studied and later collaborated with the mathematician and astronomer Abu Nasr Mansur. He and Mansur carried on scientific studies under the patronage of the brothers Abu'l Abbas and Ali ibn Ma'mun. The Ma'muns were later overthrown by Sultan Mahmud who then became al-Bīrūnī's patron, or perhaps captor; the relationship isn't clear.

Mahmud led a military campaign to India and took al-Bīrūnī with him. While in India, al-Bīrūnī studied the language, literature, customs, religion and scientific achievements of that country. As a result of his studies he wrote the treatise *India - Containing an Explanation of the Doctrines of the Indians* [3].

Our discussion of al-Bīrūnī's calculation of the circumference of the Earth is drawn from his treatise *The Determination of the Coordinates of Positions for the Correction of Distances between Cities* ("Kitāb Taḥdīd Nihāyāt al-Amākin Litaṣḥīḥ Masāfāt al-Masākin") [2]. Another source for this material is Berggren [1, pp. 141–143].

Al-Bīrūnī presents three methods for determining the circumference of the Earth, [2, pp. 183–189]. We will study the second method, which uses a mountain of given height. Before introducing his method, al-Bīrūnī states, seemingly as a comment on the method of Eratosthenes,

> Here is another method for the determination of the circumference of the Earth. It does not require walking in deserts.
> [2, p. 183]

Assume the existence of a mountain of known height, represented by the segment EL in Figure 2.5. As you stand at the top of the mountain, at point E, you sight the horizon along line ET using an instrument, such as a theodolite, that can measure angles. In its simplest form a theodolite can be made by attaching a protractor to a sighting tube. In the figure, circle ABZ may be thought of as representing the protractor, held vertically. The actual radius of the protractor, admittedly rather large in the figure, is irrelevant to the calculation. The angle $\alpha = \angle BET$, called the dip angle, will be found by this observation. Draw line MZ perpendicular to EL. We wish to find KT, the radius of the Earth.

Note that $\alpha = \angle EZM = \angle EKT$ and triangles EZM, EKT are similar. Hence

$$\frac{EZ}{ZM} = \frac{EK}{KT}. \tag{2.2}$$

The ratio on the left side is known, since

$$\frac{ZM}{EZ} = \cos \alpha.$$

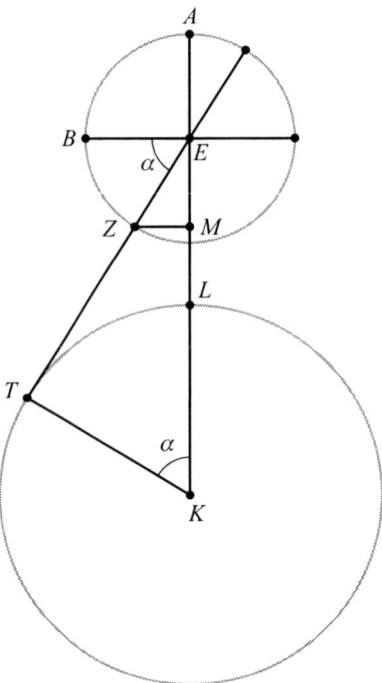

Figure 2.5. The method of al-Bīrūnī

It then follows, since $KT = KL$, that

$$\frac{EZ}{EZ - ZM} = \frac{EK}{EK - KT} = \frac{EK}{EK - KL} = \frac{EK}{EL}. \tag{2.3}$$

But the ratio on the left is known since

$$\frac{EZ}{EZ - ZM} = \frac{1}{1 - \frac{ZM}{EZ}} = \frac{1}{1 - \cos\alpha}. \tag{2.4}$$

Since EL, which is the mountain's height, is known, this implies that Equation (2.3) can be solved for EK, namely,

$$EK = EL \times \frac{EZ}{EZ - ZM} = EL \times \frac{1}{1 - \cos\alpha}. \tag{2.5}$$

Then the Earth's radius KT can be computed from Equation (2.2). To summarize, the radius of the Earth, KT, is given by

$$KT = EK \times \frac{ZM}{EZ} = EL \times \frac{\cos\alpha}{1 - \cos\alpha}. \tag{2.6}$$

The presence of trigonometric functions in this formula did not cause al-Bīrūnī any problem at all. He made highly accurate tables of the sine, cosine, and tangent functions.

This method has an advantage over that of Eratosthenes in that it doesn't require measurements at two different locations. But on the other hand, in practice $\angle EZM$ will be extremely small unless the mountain is very high. For example, for a mountain of height 1000 ft, $\angle EZM = 0°34'$. Even for a mountain of 10,000 ft the dip angle is still quite small, $\angle EZM = 1°46'$. And if al-Bīrūnī were an expert mountain climber and measured the angle from some Himalayan peak 20,000 ft high, the angle would only be $2°30'$.

The above calculation requires that we know the height of a mountain. How did al-Bīrūnī calculate this height? He gives more than one method for computing the height of mountains. We consider a method which involves a very simple calculation but requires two sightings of the mountain, see [12]. For another procedure that only requires one sighting, but a somewhat more difficult calculation, see [2, pp. 187–188] or [1, pp. 141–142]. In Figure 2.6 the height of the mountain which we wish to determine is $h = AB$. Make one sighting of the mountain top from point C, so that the angle $\alpha = \angle ACB$ is determined. Make another sighting from point D, computing the angle $\beta = \angle ADB$. It is

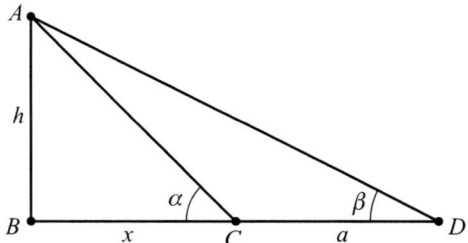

Figure 2.6. Computing the height of a mountain

necessary to assume that the two locations C, D are themselves at the same height. Al-Bīrūnī computes h by

$$h = \frac{a}{\cot \beta - \cot \alpha}. \tag{2.7}$$

This follows since

$$\cot \beta - \cot \alpha = \frac{x+a}{h} - \frac{x}{h} = \frac{a}{h}.$$

While this formula is very simple, if the points C and D are not far apart then the angles α, β will be practically indistinguishable and the calculation of h may involve considerable round-off error.

Al-Bīrūnī actually used his method to compute the Earth's circumference. On one of his journeys to India with Sultan Mahmud, al-Bīrūnī stayed at Nandana Fort, which is in the Punjab region of present-day Pakistan. According to some accounts al-Bīrūnī was being held in detention. He found a nearby peak that was convenient for his calculations. It had an open view on the plains to the south. The peak is 1795 ft above sea level, but al-Bīrūnī measured the height from the surrounding plains. He found the height above the plains to be 1055.18 ft. The unit of measurement that al-Bīrūnī used was the cubit. The height is equal to 652.055 cubits (using the conversion of 1 cubit = 1.61825 ft found in [12]). Al-Bīrūnī measured the dip angle to be $0°34'$. Using Equation (2.4), al-Bīrūnī finds a radius of 12803337;2,9 cubits, al-Bīrūnī uses sexagesimal notation for the fractional part. The radius translates into 3924.05 miles, which is only 0.6% smaller than the true value. This leads to a circumference of 25,044.99 miles, which is 0.7% larger than the true value of 24,860.2 miles. This calculation seems suspiciously accurate, perhaps due to the conversion from cubit to feet.

2.3 In the Classroom

There are several projects that may be assigned on this topic. Each project requires students to not only understand the underlying mathematics but also deal with the problems of applying mathematics in the real world.

2.3.1 Sample Projects

- Measure the circumference of the Earth using Eratosthenes' method. In order to use the method of Eratosthenes in the classroom there are several practical issues to consider.

  It is recommended that the students use a gnomon to measure the length of shadows. Gnomons are simple enough for almost any student to construct. A potential problem in using a gnomon is to ensure that it is vertical. This can be accomplished by using a level while constructing and using the gnomon.

  Since my school isn't in session on the summer solstice (or winter solstice for those in the Southern Hemisphere), when do I take measurements? There are a couple of solutions to this problem.

    - Using one measurement of the Sun: The easy way to do this is to first look up the distance between your location and the Tropic of Cancer. This then requires only one measurement of the Sun's inclination. Instead of using noon at the summer solstice one can use noon on the vernal or autumnal equinoxes. On either of these dates, the Sun is directly overhead at noon on the Equator. So if the students measure the Sun's inclination at noon on an equinox (conveniently each semester has an equinox) and if the distance between your present location and the Equator is known then Equation (2.1) can be used to compute the circumference. Of course having to look up the distance to the Equator may seem like cheating. So the next method is more in the spirit of Eratosthenes.

- Using two measurements of the Sun: This method is a lot more flexible in terms of scheduling the project (this method can be used at any time of year during daylight), but one needs to measure the Sun's inclination from two locations. The locations have to be on the same longitude, have to be a known distance apart, and must be far enough apart for the difference in angle of inclination to be detectable. All that needs to be done is to schedule two observations of the Sun at the same time on the same day in different locations on the same meridian. In order for this method to be practical one either needs to be conveniently located near a road that runs due north or south, or one must find a companion school on the same latitude to collaborate with.

If measurements are supposed to be taken when the Sun is the highest in the sky (namely, solar noon) then when exactly should the observations be done? The simplest way to compute solar noon is to take the midpoint of the local times for sunrise and sunset (which can be found in any newspaper).

- Measure the circumference using the method of Posidonius. How would one use this method in the classroom? Here are some of the issues that need to be addressed.

  - One obvious problem is that it involves night observations. There is no way to get around this problem.
  - How should one measure the elevation of stars in the night sky? The simplest way to measure this angle is to use a sextant. These are reasonably easy to construct; for further discussion see the Web site:

    http://www.tecepe.com.br/nav/XTantProject.htm.

  - A major issue is that the instructor is probably not as fortunate as Posidonius. In order to do the calculation making only one measurement, one needs to find a star that rises only up to the horizon in some city that is on the same longitude and a known distance from one's current location.
  - The previous point means that the calculation will likely require two simultaneous measurements from two locations on the same meridian and a known distance apart. How far apart should the measurements be? A distance of 300 miles corresponds to a difference in the angle measured of approximately $3°46'$. Using a distance much smaller than this requires highly accurate measurements.

- Measure the circumference using a variation of Posidonius' method. The earliest known attempt to measure the Earth was by Archytas, who was a contemporary of Plato. The method used by Archytas was to find two cities on the same meridian, each city having a star directly over their locations at the same time of night. The angular distance between the two cities is then equal to the angle between the stars along a celestial meridian.

- Measure height of an object using the method of al-Bīrūnī discussed above. For example, measure the height of a hill, mountain, tall building, tethered balloon or other tall object. A word of caution is necessary. It is preferable to use an object that has a well-defined peak (a hill can have a too rounded top to be useful). The two points from which to make your sightings have to be fairly far apart (the farther the better). If possible, take the first sighting close to the base of the object. In my history of mathematics class, I used the four story classroom building in which the class is taught.

- Measure the circumference of the Earth using the method of al-Bīrūnī. This presents some difficulties when used in the classroom.

  - The main difficulty may be the need for a height that not only can be measured but that can also be climbed in order to measure the dip angle. If the instructor is hindered by the lack of a convenient mountain nearby the classroom, a sufficiently tall building may be substituted for the mountain.
  - Even accounting for the availability of a nearby height, another obstacle is how to accurately measure the very small dip angle. One strategy to alleviate this problem is to make several sightings and compute the average dip angle.
  - In any case, one needs an accurate instrument for measuring angles. Either a sextant or theodolite may be constructed for this purpose. It is to be expected that a homemade instrument will be rather inaccurate and impact the final value for the circumference.

- Determine how to compute height using only one sighting. This is described in detail in Berggren [1, pp. 141–142].

- Make your own surveying instrument (such as sextant, theodolite, astrolabe, or sundial) and then use this instrument in making real world measurements. There is a variety of interesting mathematics involved in constructing these instruments. Here are some Web resources for making such instruments.

  Making your own sextant:
  http://www.tecepe.com.br/nav/XTantProject.htm

  Making your own theodolite:
  http://www.dar.csiro.au/airwatch/awballoon.html
  http://www.apogeerockets.com/education/downloads/newsletter93.pdf

  Making your own astrolabe:
  http://www.astrolabes.org/
  http://celebrating200years.noaa.gov/edufun/book/MakeyourownAstrolabe.pdf

  Making your own sundial:
  http://www.lmsal.com/YPOP/Classroom/Lessons/Sundials/sundials.html

  Making your own vertical sundial:
  http://www.mysundial.ca/sdu/graphical_vertical_sundial.html

- Determine the latitude or longitude of your location. One way to compute latitudes in the northern hemisphere is to use the elevation of the North Star. This star is more or less directly overhead at the North Pole, so that at any location in the northern hemisphere the angle of the North Star's elevation will equal that point's latitude. Computing longitude is more difficult. The relative longitude of two locations can be determined by comparing the time at which a lunar eclipse occurs at both locations. The ratio of the difference in longitude to $360°$ equals the ratio of the difference in time of the eclipse to the length of a sidereal day, which is the time it takes for stars to return to their highest point (a little less than 24 hours).

- For high school students an interesting organized activity is the online Eratosthenes Experiment,

  http://www.youth.net/eratosthenes/.

This is a world-wide project in which students from around the world measure the angle of the noon-day Sun. Using this information together with the school's latitude and longitude allows the students to calculate the Earth's circumference. The experiment takes place twice a year, on the spring and autumn equinoxes.

2.4 Taking it Further

In this article we have examined various methods for measuring terrestial distances. This discussion may be extended to the measurement of distances between the Earth and other heavenly bodies. This uses the concept of parallax, which is the apparent angular displacement of an object, due to the motion of the observer. There are three common uses of parallax: to measure the distance between the Earth and the Moon, between the Earth and the Sun, and between the Earth and nearby stars. The use of parallax in measuring astronomical distances can be found in the work of Greek mathematicians such as Aristarchus, Hipparchus and Ptolemy. A good history of the use of parallax in astronomy can be found in the survey of Hirshfeld [9].

Both the solar and stellar parallax are too small to measure without refined instruments, but it is feasible to measure the lunar parallax as a class project and hence to determine the distance to the Moon. Lunar parallax involves the different position of the Moon relative to the background of fixed stars when the Moon is viewed simultaneously from two different locations on Earth. In fact, the method of comparing sightings of the Moon from two different locations on Earth is very reminiscent of the procedure of Eratosthenes.

2.5 Conclusion

Let us conclude with an object lesson. Given the lack of technology available to these early astronomers, it is remarkable how truly accurate their calculations were. But such accomplishments also come with a classical admonition against hubris. Archytas, famed for his achievements in mathematics, astronomy and philosophy was the subject of a renowned ode of Horace (Odes I.28). The beginning of the poem states the theme that all of the triumphs of Archytas are made trivial by the great leveler, death.

> Archytas, you who measured the Earth and the sea and the numberless sands,
> are now confined in a small mound of dirt near the Matine shore,
> what does it avail you that you once
> explored the mansions of the skies and that you traversed
> the round celestial vault — you with a soul born to die?

Bibliography

[1] J. L. Berggren, *Episodes in the Mathematics of Medieval Islam*, Springer Verlag, New York, 1986.

[2] al-Bīrūnī, *The Determination of the Coordinates of Positions for the Correction of Distances between Cities*, tr. Jamil Ali, Centennial Publications, American University of Beirut, 1967.

[3] ———, *India — Containing an Explanation of the Doctrines of the Indians*, tr. as *alBeruni's India* by E. Sachau, Kegan Paul, Trench, Trübner & Co., London, 1910.

[4] I. E. Drabkin, "Posidonius and the Circumference of the Earth", *Isis*, **34**(6) (1943), 509–512.

[5] D. Engels, "The Length of Eratosthenes' Stade", *Amer. J. Philology*, **106**(3) (1985), 298–311.

[6] J. Evans, *The Theory and Practice of Ancient Astronomy*, Oxford University Press, Oxford, 1998.

[7] I. Fischer, "Another Look at Eratosthenes' and Posidonius' Determinations of the Earth's Circumference", *Quart. Jnl. Roy Astro. Soc.*, **16**(1975), 152–167.

[8] E. Gulbekian, "The Origin and Value of the Stadion Unit used by Eratosthenes in the Third Century B.C.", *Arch. Hist. Exact. Sci.*, **37**(4) (1987), 359–363.

[9] A. W. Hirshfeld, *Parallax: The Race to Measure the Cosmos*, W.H. Freeman, New York, 2001.

[10] A. Jones, "Eratosthenes, Hipparchus, and the Obliquity of the Ecliptic", *J. Hist. Astron.*, **33**(2002), 15-19.

[11] D. Rawlins, "Eratosthenes' Geodesy Unraveled: Was There a High-Accuracy Hellenistic Astronomy?", *Isis*, **73**(2) (1982), 259–264.

[12] S. Rizvi, "A Newly Discovered Book of Al-Bīrūnī: "Ghurrat-uz-Zījāt", and al-Bīrūnī's Measurements of Earth's Dimensions", *Al-Bīrūnī Commemorative Volume*, ed. H. Said, Hamdard National Foundation, Karachi, Pakistan, 1979, 605–680.

[13] M. Terrall, *The man who flattened the Earth : Maupertuis and the sciences in the enlightenment*, University of Chicago Press, Chicago, 2002.

[14] I. Thomas, *Selections Illustrating the History of Greek Mathematics*, Loeb Classical Library, Harvard University Press, 1941.

3

Numerical solution of equations

Roger Cooke
University of Vermont

3.1 Introduction

Methods of solving polynomial equations lie at the heart of classical algebra. There are two interpretations of the problem of solving an equation, leading to two different approaches to its solution. In most courses, the emphasis is on the structure of the equation and finding a way to express the roots as a *formula* in terms of the coefficients. The simplest example of such a formula is the quadratic formula, which gives the solution of the equation $ax^2+bx+c=0$ as

$$x = \frac{-b \pm \sqrt{b^2 - 4ac}}{2a}.$$

This approach is elegant and leads to some exceedingly profound mathematics. However, for one who actually needs to know a *number* that satisfies the equation, this approach leaves something to be desired. It works with maximum efficiency in the case of the quadratic equation, but even in that case, if the quantity under the radical is not the square of a rational number, one is forced to resort to approximations in order to get a usable number. For cubic and quartic equations, there are formulas, but they work even less well, since they often involve taking the cube root of a complex number, which is a problem just as complicated as the original equation was, if not more so. Once again, one is forced to resort to numerical approximations. Beyond the fourth degree, the only formulas involve non-algebraic expressions, and are of little practical use. Higher-degree equations are the realm of numerical methods. To understand how numerical methods work, it is useful to begin with the simplest cases and the simplest methods. That is what we are about to do.

For the past few decades we have had access to calculators that can solve equations lightning fast. Some of these calculators can use the formulas that are taught in algebra to solve low-degree equations. In addition, nearly all calculators nowadays can proceed directly to find numerical approximations to the roots of an equation, without using an "exact" formula for those roots. In this article, we assume that the student knows how to use such a calculator. What we are going to do is try to imagine what is going on *inside* the calculator. We shall do that by examining an early method of doing these computations on a counting board. The insight thereby gained into the behavior of the polynomials that make up the equation is worth acquiring, even though the student may never be stranded without a calculator and forced to find the roots by hand.

If we look for numerical approximations from the beginning, we don't have to search for abstract relations between the coefficients and the roots. We can proceed directly from the equation to the roots. A procedure for doing so was developed in China over a thousand years ago. It was rediscovered in Britain in the nineteenth century and taught to students under the name *Horner's method*. This method works very well on a counting board, where it is possible to keep rows and columns of numbers in good order. In China, the numbers might have been represented by sticks placed on the squares of a large board, and the arithmetical operations involved in using the method would have been carried out by moving the sticks around. Needless to say, in China, as elsewhere in the world, the modern method of solving equations is to use a calculator. The method we are about to discuss is therefore mostly of historical interest, but also of interest to a person who wishes to construct a new algorithm for solving equations, since certain properties of polynomials, which will become apparent in the course of the discussion, are essential in all such algorithms.

3.2 The ancient Chinese method of solving a polynomial equation

We are now going to see how these counting-board techniques can be used to solve an equation. For simplicity, we start with a quadratic equation, say $x^2 - 229x - 462 = 0$. In order to get started, we need to guess some interval in which the solution lies. Direct inspection shows us that $p(200) < 0$ and $p(300) > 0$, so we know there is a root between 200 and 300. Putting that fact in language that fits the Chinese method better, the root has at least three digits left of the decimal point, and first digit of the root is 2.

To get the second digit, we take 200 as a "base value" and let $x = 200 + y$, where now we know that $0 < y < 100$. We need to rewrite the equation in terms of y. The Chinese found a very simple way to do this on a counting board, by filling in the blanks in the following array

$$\begin{array}{cccc} 1 & 1 & 1 & 1 \\ -229 & & & 0\,. \\ -462 & & 0 & 0 \end{array}$$

Before giving the rule for completing this array, we note two things. First, each entry in the top row is equal to the leading coefficient of the equation, while the left-hand column is simply the full set of coefficients. Second, the zeros here would be merely empty squares on the counting board. We inserted them as "stop signs" for the procedure about to be described, but they have an additional advantage that will appear shortly, in that they can be included in the data for the next step so that we are always looking for a digit between 1 and 9.

The rule for filling in the array is simple. Work from left to right and top to bottom. To find what goes in an empty space, multiply the entry immediately above the space by the current "base value" (200) and add the adjacent number on the left. The result is

$$\begin{array}{cccc} 1 & 1 & 1 & 1 \\ -229 & -29 & 171 & 0\,. \\ -462 & -6262 & 0 & 0 \end{array}$$

The coefficients of the equation that y has to satisfy can now be read *diagonally downward* from right to left, that is, $p_1(y) = y^2 + 171y - 6262 = 0$. We will not take the time to explain in full why this procedure always works, although it is not difficult to analyze. You can verify that it has given the correct result in this case, since $0 = p(x) = (y + 200)^2 - 229(y + 200) - 462 = y^2 + 400y + 40000 - 229y - 45800 - 462 = y^2 + 171y - 6262$.

Now we know that $p_1(y)$ has a zero between 0 and 100. Calculation shows that $p_1(30) = -232 < 0$ and $p_1(40) = 2178 > 0$. Hence the zero is between 30 and 40, and so the second digit of the root is 3.

To get the third digit, we repeat the process, writing $y = 30 + z$ (using 30 as the current "base value") and filling in the array to get

$$\begin{array}{cccc} 1 & 1 & 1 & 1 \\ 171 & 201 & 231 & 0\,. \\ -6262 & -232 & 0 & 0 \end{array}$$

Thus z satisfies $p_2(z) = z^2 + 231z - 232 = 0$, and we know that z is between 0 and 10. We then find very quickly that $z = 1$ gives an exact root, so that $x = 231$ is the root of the original polynomial.

Although the equation is now solved, we might continue to experiment with this method. What would happen if we

3.3. Non-integer solutions

continued, letting, say $z = 1 + w$? What would the equation for w look like? The method would yield

$$\begin{array}{cccc} 1 & 1 & 1 & 1 \\ 231 & 232 & 233 & 0 \\ -232 & 0 & 0 & 0 \end{array}.$$

In other words, w would satisfy $p_3(w) = w^2 + 233w = 0$, so that $w(w + 233) = 0$. What this tells us is that w might be *either* 0 *or* -233, so that z might have been either 1 or -232, y might have been either 31 or -202, and x might have been either 231 (as we found) or -2.

Explanation of the method

Why does this method work? In order to understand that, we need to look at these equations using the algebraic notation that occurs in modern textbooks. Suppose we are trying to solve the equation

$$ax^2 + bx + c = 0.$$

According to our instructions, we start with the array

$$\begin{array}{cccc} a & a & a & a \\ b & & & 0 \\ c & & 0 & 0 \end{array}.$$

Suppose we have guessed a first approximation to the root. Let it be denoted u. Our exact solution is then $x = u + y$, where we have chosen u, and we need to find y. The assertion is that we can find an equation for y by filling in the array according to the rule "multiply u by the number above the space, and add the number to the left." That rule leads us to the array

$$\begin{array}{cccc} a & a & a & a \\ b & au + b & 2au + b & 0 \\ c & au^2 + bu + c & 0 & 0 \end{array}.$$

Our rule says that y will then satisfy the equation

$$ay^2 + (2au + b)y + (au^2 + bu + c) = 0.$$

Why is this true? Well, the equation transforms as follows, given that $x = y + u$:

$$\begin{aligned} 0 &= ax^2 + bx + c \\ &= a(y + u)^2 + b(y + u) + c \\ &= a(y^2 + 2uy + u^2) + by + bu + c \\ &= ay^2 + (2au + b)y + (au^2 + bu + c). \end{aligned}$$

Here we see the usefulness of the technique of representing unknown or unspecified numbers by symbols. It takes some imagination to think of a square on a counting board as the representative of a number having a certain relation to another number. The symbols are definitely easier. That is one reason they are used everywhere nowadays. But once the counting-board technique has been mastered, it can be applied very rapidly.

3.3 Non-integer solutions

Before considering cubic equations, we need to work one more example of this procedure to introduce a small complication that arises when the solutions are not integers. We illustrate it by finding the zeros of the polynomial $p(x) = 8x^2 - 18x - 11$. We start as usual by noting that $p(2) = -15$ and $p(3) = 7$, so that there is a root between 2 and 3. As before, we let $x = 2 + y$ and get the equation for y from the array

$$\begin{array}{cccc} 8 & 8 & 8 & 8 \\ -18 & -2 & 14 & 0 \\ -11 & -15 & 0 & 0 \end{array}.$$

Thus, y satisfies $p_1(y) = 8y^2 + 14y - 15 = 0$, and y is between 0 and 1. Since we want the next digit of the solution, we should try the numbers .1, .2, .3, and so on, as values of y until we find the point where $p_1(y)$ changes sign. It is a tiny bit simpler, however, to do a decimal shift and consider $10y$ instead of y. That is, we let $z = 10y$, so $y = z/10$. It is quite simple to see that z satisfies $q_1(z) = 8z^2 + 140z - 1500 = 0$, and this is easy to remember, since all we have to do is adjoin the zeros already in the array to the coefficients. By trial, we find that $q_1(7) = -128 < 0$ and $q_1(8) = 132 > 0$, so the next digit will be 7. We then write $z = 7 + u$ and continue.

Again, since u is between 0 and 1, it is simpler to multiply it by 10 and write $v = 10u$, $u = v/10$. The array

	8	8	8	8
	140	196	252	0
	−1500	−128	0	0

tells us that v satisfies $q_2(v) = 8v^2 + 2520v - 12800 = 0$, and v is between 0 and 10. This time, we find that $v = 5$ gives an exact solution. Therefore the solution of the equation is $x = 2.75$.

If we wanted to know the other solution, we could continue the procedure one more step, as we did above. The array would be

	8	8	8	8
	2520	2560	2600	0
	−12800	0	0	0

In other words, if $v = 5 + w$, then w satisfies $8w^2 + 2600w = 0$, so $w = 0$ (as already found) or $w = -\frac{2600}{8} = -325$. Then $x = 2 + y = 2 + z/10 = 2.7 + v/100 = 2.75 + w/100 = 2.75 - 3.25 = -0.5$.

3.4 The cubic equation

To show that this procedure is perfectly general, we shall solve a cubic equation by the same method. To do this, we need one extra row and one extra column. The polynomial for which we shall find a zero is $p(z) = x^3 - 2x^2 + x - 3 = 0$. Since $p(2) = -1$ and $p(3) = 9$, there is a root between 2 and 3. We then write the solution as $x = 2 + y$ and rewrite the equation in terms of y. The array that gives the equation for y is

	1	1	1	1	1
	−2	0	2	4	0
	1	1	5	0	0
	−3	−1	0	0	0

We find that y must satisfy the equation $p_1(y) = y^3 + 4y^2 + 5y - 1 = 0$. Since we are moving into fractions at this point, however, let us once again multiply by 10 and write the equation for $z = 10y$, namely $q_1(z) = z^3 + 40z^2 + 500z - 1000 = 0$. Since z is between 0 and 10, we find it by locating the integer where $q_1(z)$ changes sign. Since $q_1(1) = -459$ and $q_1(2) = 168$, we take $z = 1 + u = 1 + v/10$ and continue the procedure. At this point we know that $x = 2.1 \ldots$.

In this way (working with sufficient patience and accuracy), it is possible to find any number of decimal digits of a root of any equation with real coefficients, no matter its degree. The Japanese mathematician Seki Kowa (Seki Takakazu, 1642–1708) is said to have solved an equation of degree 1458, over a period of several days, on the floor of a large room ruled into squares.

In 1819, a technique essentially the same as this ancient Chinese method, except that it applied to infinite series as well as polynomials, was developed by the British scholar William George Horner (1787–1837). It was taught for about a century in American high-school algebra books under the name *Horner's method*, with the computations simplified using "synthetic division."

3.5 Problems and questions

Problem 1. *Using the examples given above as a model, solve the equation $x^2 - 5 = 0$ to two decimal places. When you finish, you should have the first two digits of $\sqrt{5}$, truncated rather than rounded off. In other words, you should*

know that the root lies between 2.23 and 2.24. *The computations should be easy, at least at the first stage, because of the zero coefficients.*

Problem 2. *Find an approximation to $\sqrt[3]{2}$ by solving the equation $x^3 - 2 = 0$. Get at least four decimal places.*

Problem 3. *We have been vague about the way to find the initial approximation to a root. Verify that the largest a root can be in absolute value is the sum of the absolute values of the non-leading coefficients divided by the absolute value of the leading coefficient, thus providing an upper bound on the size of a root.*

Question 1. *What search algorithm would you use to minimize the number of trials that need to be made when looking for the next digit of a root? Show that you could get by with at most four trials before finding the place where the polynomial changes value.*

3.6 Further reading

Lay-Yong Lam and Tian-Se Ang, *Fleeting Footsteps. Tracing the Conception of Arithmetic and Algebra in Ancient China*, World Scientific, River Edge, NJ, 1992.

Ulrich Libbrecht, *Chinese Mathematics in the Thirteenth Century*, MIT Press, Cambridge, MA, 1973.

Jean-Claude Martzloff, "Li Shanlan (1811–1882) and Chinese traditional mathematics," *The Mathematical Intelligencer*, 14 (1982), 32–37.

Margaret McGuire, "Horner's Method," in *A Source Book in Mathematics*, David Eugene Smith, ed., Dover, New York, 1959.

4
Completing the Square through the Millennia

Dick Jardine
Keene State College

4.1 Introduction

Solving quadratic equations is a topic relevant to modern mathematics instruction, as it has been for thousands of years. As we start the 21st century, more often than not students will use calculators and computer algebra systems to solve quadratics. Today, we associate solving quadratics with curves (parabolas) rather than rectangles and squares (even though the word quadratic is from the Latin *quadratum*, a four-sided figure). A centuries old method which hopefully will survive in classrooms in this millennium is the method of completing the square. Understanding the process of completing the square is important for our students, for a wide range of reasons including that it provides arguably the best approach to deriving the quadratic formula. In the examples below, we outline the use of completing the square as it was done in four previous millennia.

Over the years, the method has had various representations. Understanding the historical, geometric representation may help students internalize the method when algorithmic or algebraic representations alone may not. Multiple ways of learning and knowing are offered by including the historical perspective. The examples given in this capsule are actual problems solved in the past, and your students are invited to solve them today using the methods of antiquity as well as current techniques. In my courses, I present the information as an interactive lecture that extensively involves students, as described below.

4.2 Historical preliminaries

About 4000 years ago, Mesopotamian scribes pressed the method of completing the square into clay tablets, the technology used to record information in that time. Just over a millennium later, during the centuries when Hellenistic Greek culture flourished, Euclid included a proof for completing the square in his most famous work, the *Elements* (ca. 300 BCE). Over the next thousand years, Arabic mathematicians not only continued but extended the work of the Greeks, completing the square to solve quadratics in their own way, as will be seen in reviewing the work of al-Khwarizmi (ca. 780–850). In translating and extending Arab works, Renaissance Europeans, among them Girolamo Cardano (1501–1576), included the method of completing the square to solve quadratic equations. These four examples of the method, passed down in history from the Mesopotamian scribes through Euclid, al-Khwarizmi, and Cardano, will be used to demonstrate the historical transition from a geometric representation to what we now think of as an algebraic process.

Babylonian mathematicians, like their Greek and Arab successors, associated number with length. The multiplication of one length by another was done for applications, such as computing the area of land to be planted with barley or calculating the area of an enemy army's encampment for the purpose of estimating the size of the opposing force. There were sufficiently many of these applications that the scribes developed and recorded on clay tablets specific procedures for solving the resulting mathematical problems.

One such tablet is BM13901, written between 2000 and 1800 BCE. The solution technique on BM13901 was described in geometric terms, and the method used was a literal completing of a square, preserved in solidified clay to this day after almost four thousand years. The method described below was for a *specific* problem, as was all the mathematics written by the Mesopotamian scribes.

Finding the unknown area of a square was well known in ancient Mesopotamia. The inverse problem of finding the lengths of the sides given the area of a square was also a known process, although slightly more involved. Finding the lengths of the sides of a rectangle of a given area proved to be even more involved. On BM13901, the scribe converted a problem involving a rectangle to the easier problem involving a square.

In one problem the scribe had a rectangle with total area $\frac{3}{4}$, paraphrasing Neugebauer's and Hoyrup's translations described by Robson [6] and converting from sexagesimal numbers. He wrote: *I totalled the area and the side of my square which is $\frac{3}{4}$*. Using our convenient notation, we would write $x^2 + x = \frac{3}{4}$. Although there were no drawings on the surviving tablet, the scribe presented a geometrical approach to find the length of the unknown side. In Figure 4.1 is the first construction described by the scribe, the splitting of the unknown rectangle into a known rectangular

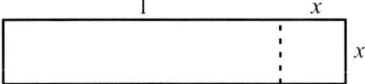

Figure 4.1. Rectangle and square with total area $\frac{3}{4}$

projection of length one and a square with side of unknown length:

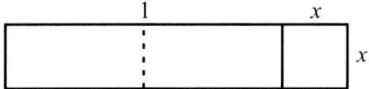

Figure 4.2. Halving the projection

He then broke the projection in half, and attached one of the pieces below the unknown square, then added to the

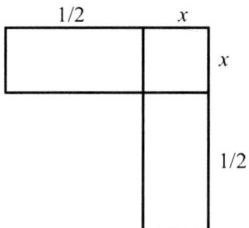

Figure 4.3. Splitting the halved rectangle

resulting figure (a *gnomon*) with area $\frac{3}{4}$ a square with area $\frac{1}{2} \cdot \frac{1}{2} = \frac{1}{4}$, which "completed the square" of area $\frac{3}{4} + \frac{1}{4} = 1$.

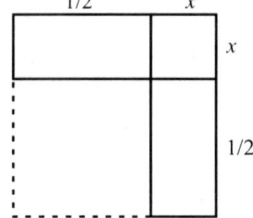

Figure 4.4. Completing the new square

All that remained for the scribe to do was to subtract $\frac{1}{2}$, the known side of a square with area $\frac{1}{4}$, from 1 to obtain the

4.2. Historical preliminaries

length of the unknown side as $\frac{1}{2}$. Building on knowledge of solving for the sides of squares, the scribe solved the problem involving the rectangle by geometrically constructing a square and then solving the resulting square.

The ancient Greeks were familiar with this technique developed by the Mesopotamian scribes. Greek mathematicians, however, went beyond the Babylonian "recipe" approach to solving specific problems. They developed *general* theorems, and provided proofs of those results. Euclid's *Elements*, mostly known for geometric content but which also contains significant algebra and number theory, was compiled around 2300 years ago. The *Elements* contains the method of completing the square in Book II, Proposition 6:

> If a straight line be bisected and produced to any point, the rectangle contained by the whole line thus produced and the part of it produced, together with the square on half the line bisected, is equal to the square on the straight line which is made up of the half and the part produced.[3]

Figure 4.5 is a depiction of an outline of Euclid's proof. Note that the figure and description by Euclid is similar to that of the Mesopotamian scribe. Euclid started by bisecting a line of length a, so that each half has length $\frac{a}{2}$. He extended the line an arbitrary amount x. A rectangle was created with width x and length $x + a$. That rectangle was divided into two rectangles with length $\frac{a}{2}$ and width x and a square with sides of length x. A rectangle was constructed below the square with sides of length x, a construction equivalent to moving the left-most of the original rectangles with side $\frac{a}{2}$ to that position. That construction formed a square with sides of length $\frac{a}{2}$, and also completed a larger square having sides of length $x + \frac{a}{2}$. That square was completed by summing the square with sides of length x, the two rectangles with area x by $\frac{a}{2}$, and the square with sides of length $\frac{a}{2}$. In this proposition, Euclid documented a proof of the general process of completing the square.

The wording from Euclid's *Elements* may seem awkward to us, but remember that mathematics was done differently 2300 years ago. A geometric approach to problem solving prevailed then. The key idea in the evolution of completing the square demonstrated by Euclid is that he was not solving a specific problem as the Mesopotamian scribes did, but was instead proving a general mathematical result. A parallel approach to solving quadratics can be found in the mathematical writings of Arabs, who learned from and continued the mathematical traditions of the ancient Greeks.

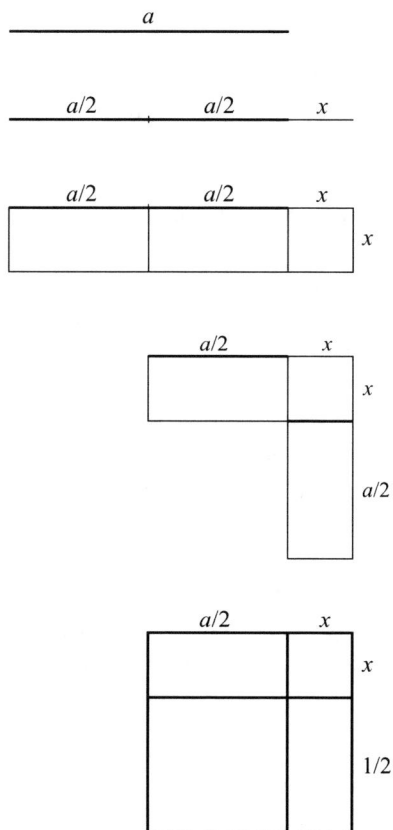

Figure 4.5. Completing the square in Euclid's *Elements*

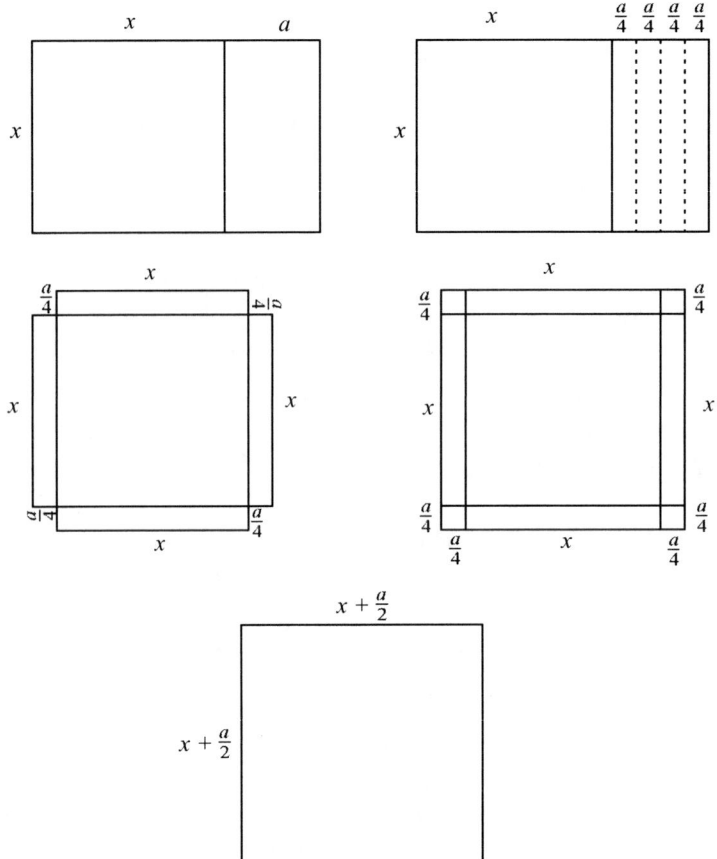

Figure 4.6. A representation of al-Khwarizmi's completing the square

Al-Khwarizmi was a 9th century Islamic mathematician who was encouraged by his Arab sponsor to compile a text on the applications of mathematics. Al-Khwarizmi produced *Kitab al-jabr waal-muquabala*[1], in which he described well-defined procedures for solving equations. We often use the word *algorithm* to describe procedural methods for solving problems, and al-Khwarizmi's name is the source of the word algorithm.

One application described in *Kitab al-jabr waal-muquabala* demonstrated the method of completing the square. The method was used to solve a financial problem involving the relationship, *a square and 10 roots equal 39 dirhems*. A *dirhem* was a unit of money, and in our notation, al-Khwarizmi found the roots of the quadratic $x^2 + 10x = 39$. Al-Khwarizmi did not have the benefit of our concise notation, and resorted to long verbal explanations to express the relationship between quantities. Figure 4.6 depicts an interpretation of the process al-Khwarizmi used to prove the process of completing the square [7].

The square in the middle corresponds to the x^2-term of the quadratic. The rectangle (representing $10x$) to the right of the square is divided into four parts. Those rectangles are then positioned one to each of the sides of the square. The square is completed by adding the four corner squares, which have a total area of 4 times $(10/4)^2$, or 25. So 25 must be added to both sides of the original equation. The completed square has area $39 + 25 = 64$ and sides with length 8. The length of x, then, is $x = 8 - 5 = 3$.

Leonardo of Pisa (1170–1250), also known as Fibonacci, translated and included Arab mathematics in his text *Liber Abaci* (1202). Girolamo Cardano's *Ars Magna* (1545) was influenced by Fibonacci's compilation. Cardano was a true Renaissance man, having written on such diverse subjects as medicine, astronomy, philosophy, gambling, and mathematics, among other topics. In chapter 5 of the *Ars Magna*, problem 2 is:

> There were two leaders each of whom divided 48 aurei among his soldiers. One of these had two more soldiers than the other. The one who had two soldiers fewer had 4 aurei more [than the other] for each soldier.

---

[1] It is from the second word in the title that the word algebra is derived. To al-Khwarizmi, *al-jabr* meant taking the subtracted quantity of a mathematical expression and adding it to the other side.

What is to be found is how many soldiers each had [1].

Using proportional reasoning, Cardano converted the word problem to the equation $\frac{48}{x} = \frac{48}{x+2} + 4$ in our notation, where x is the smaller number of soldiers. Simplifying, the equation is written $x^2 + 2x = 24$. Cardano solved the quadratic using completing the square, but relied on geometric methods (and citation of Euclid's *Elements*) to prove the general process.

This brief historical presentation of completing the square ends with Cardano. Descartes (1596-1650) was among the first to label line segments by letters representing the length of the segments, and then to multiply the two line segments to obtain a third line segment with length corresponding to their product. This new approach enabled Descartes to use algebraic equations to describe a wide variety of curves. This helps explain why we associate quadratics with parabolas, not with the squares and rectangles used by those who first conceived and later used the method of completing the squares for the first 3500 years of its application.

4.3 Student activities

In the Appendix are activities for students to practice completing the square on quadratic equations from the past. I present a lecture on the use of completing the square as part of my History of Mathematics and Applied Algebra and Trigonometry classes that I teach at our college. I have also used parts of the presentation when the topic of completing the square comes up in calculus class. To keep the students engaged in the lecture and actually doing mathematics, not just watching me do mathematics, I have the students do exercises in small groups at the appropriate points of the lecture. For example, after I introduce BM13901 and the resulting quadratic, I stop the lecture and have students do the exercise practicing completing the square to solve another quadratic from that era. Additionally, students take part in making 5-minute presentations on the named mathematicians as part of the classroom activities, as described in Capsule 15. My classroom lecture provides the introductory material, fills gaps that may exist after the student presentations, and makes the transitions necessary as we move from millennium to millennium. The extent to which students participate varies depending on the time I make available in class meetings for this topic in each course. I use the papers students write for both course grades and as artifacts for program assessment, as the student work demonstrates that they are learning the historical development of mathematics.

4.4 Summary and conclusion

The historical evolution of the method of completing the square outlined here can be readily incorporated in high school and college algebra classes. Additionally, concrete representations of the process of completing the square can add to the experience through the use of algebra tiles, mimicking the different geometric techniques used in the past to solve quadratics. This would be particularly useful in mathematics classes for pre-service teachers. Some of the problems presented here are taken from translations of original sources identified in the References below, and there are more to be found there if you wish to expose students to working with original sources in the history of mathematics. In solving the exercises done in the past using various representations and a historical approach, students can gain deeper insight into the methods, breadth and evolution of mathematics.

Bibliography

[1] Girolamo Cardano, *The Great Art or The Rules of Algebra,* translated and edited by T. Richard Witmer, Cambridge, MA: The MIT Press, 1968.

[2] Roger Cooke, *The History of Mathematics: A Brief Course*, New York: Wiley-Interscience, 1997.

[3] Euclid's *Elements,* as translated by Sir Thomas Heath, Dana Densmore, ed., Santa Fe, New Mexico: Green Lion Press, 2002.

[4] Jens Høyrup, "The Four Sides and the Area: Oblique Light on the Prehistory of Algebra," in *Vita Mathematica: Historical Research and Integration with Teaching*, Ronald Calinger, ed., Washington: The Mathematical Association of American, 1996.

[5] Victor Katz. *A History of Mathematics: An Introduction*, 2nd ed., Reading, MA: Addison-Wesley, 1998.

[6] Eleanor Robson, review of Jens Hoyrup's *Lengths, Widths, Surfaces: A portrait of Old Babylonian algebra and its kin*, MAA Online book review column, www.maa.org/reviews/lsahoyrup.html, accessed June 3 2008.

[7] Muhammad ibn Musa al-Khwarizmi, as translated in Dirk Struik, *A Source Book in Mathematics, 1200–1800*, Harvard University Press, Cambridge, MA, 1969.

[8] Muhammad ibn Musa al-Khwarizmi, as translated in Frederick Rosen, "The Algebra of Mohammed Ben Musa", in *The Treasury of Mathematics*, Henrietta O. Midonick, ed., Philosophical Library, New York, 1965.

[9] John J. O'Connor and Edmund F. Robertson *The MacTutor History of Mathematics Archive*, www-groups.dcs.st-and.ac.uk/~history/Biographies/Al-Khwarizmi.html, accessed June 26, 2007.

Appendix
Student activities

I have created worksheets using the exercises below, enabling students to solve quadratics and practice completing the square using the actual quadratic problems solved by the students' mathematical predecessors. I have used combinations of the exercises in a college-level algebra course, in a calculus course, and in a history of mathematics course. As our students are not accustomed to the proportional reasoning of Cardano, they will need some help constructing the quadratic from the quote given.

Historical completing the square exercises

1. A cuneiform tablet at the British Museum is BM13901, written between 2000 BCE and 1800 BCE by a Mesopotamian scribe. On this tablet the scribe wrote mathematical problems, presumably derived from an applications. One of the problems: *the length of a side of a square was to be found given that the area of the square added to 4/3 of a side of the square was 11/12.*

    a. Write the quadratic equation that follows from the problem described on table BM13901.
    b. Solve the problem by completing the square graphically as a Babylonian scribe would.
    c. Solve the problem using the completing the square algorithm that we use today.
    d. Check your solution by substituting your answer into the original equation.

2. The Arab mathematician al-Khwarizmi wrote about financial problems that led to relationships such as, "two squares and ten roots are equal to forty-eight dirhems" in the 9th Century. A dirhem was a unit of money, and in al-Khwarizmi sought the roots of the quadratic that arose from that problem statement.

    a. Write the quadratic equation that follows from the problem described by al-Khwarizmi.
    b. Solve the problem by completing the square graphically as al-Khwarizmi would.
    c. Solve the problem using the completing the square algorithm that we use today.
    d. Check your solution by substituting your answer into the original equation.

3. Girolamo Cardano included the following problem in his 1545 book, *Ars Magna*:

    There were two leaders each of whom divided 48 aurei among his soldiers. One of these had two more soldiers than the other. The one who had two soldiers fewer had 4 aurei more [than the other] for each soldier. What is to be found is how many soldiers each had.

    a. Write the quadratic equation that follows from the problem described by Cardano.
    b. Solve the problem by completing the square.
    c. Check your solution by substituting your answer into the original equation.

4. Use completing the square to derive the quadratic formula, beginning with the standard form of a quadratic, $ax^2 + bx + c = 0$. Use the quadratic formula to solve each of the quadratics above.

5

Adapting the Medieval "Rule of Double False Position" to the Modern Classroom

Randy K. Schwartz
Schoolcraft College

5.1 Introduction

The rule of double false position is an arithmetical procedure for evaluating linearly-related quantities. The method does not rely on variables or equations, but is based instead on interpolating between, or extrapolating from, two guesses, or suppositions. Although the technique is seldom mentioned today in North American curricula, it was routinely used in much of Europe, Asia, and North Africa from medieval times to the 19th Century, and is still taught in many classrooms there today. Historically, the approach was especially convenient for practical tradesmen whose knowledge did not normally extend to a mastery of algebra; they could pull the algorithm from their mathematical toolkits whenever needed and deploy it as a rote arithmetical procedure.

I have adapted for instructional use a North African version of the rule of double false position. The topic is well suited to college or high school courses in College Algebra, Precalculus, Calculus, Applied Calculus, and Linear Algebra. In my experience, only 30–50 minutes of class time needs to be devoted to teaching the method in order for students to grasp the mechanics, justification, and various applications. Instruction can take any of various forms, ranging from a traditional lecture to a self-guided instructional module for individual or group work. I describe such a module below, in the section "In the Classroom".

Covering a technique that students will find handy in solving certain problems helps round out their technical skills. In addition, it helps introduce them to the contributions of a variety of cultures, and provides some historical perspective on mathematics. Learning about double false position highlights the fact that practical algorithms were being used many centuries before the modern era and without recourse to algebra.

5.2 Historical Background

In 1202, Leonardo Fibonacci of Pisa, Italy, devoted Chapter 13 of his famous treatise *Liber Abaci* to this technique. His nearly boundless enthusiasm for the method comes through from the moment he introduces the chapter, writing that "the Arabic *elchataym* by which the solutions to nearly all problems are found is translated as the method of double false position." [8, p. 447].

To illustrate the technique, consider this sample problem from Fibonacci:

A certain worker received 7 *bezants* per month if he worked, and if he did not work he had to pay 4 *bezants* per month to the foreman; [one month,] for whatever he worked or did not work he received at the end of the month 1 *bezant* from the foreman; it is sought how many days of the month he worked. [8, p. 453].

Suppose that the worker had labored, say, 20 of the 30 days of the month, and been idle the remaining one-third of the month. Then he would have received

$$\frac{2}{3}(7) - \frac{1}{3}(4) = \frac{10}{3} \text{ bezants (gain)},$$

which is too high by 7/3 *bezant*, since we are told that he actually earned 1 *bezant*. Thus, the supposition of 20 days of work is too high. Suppose, instead, that he had labored only 10 days and been idle the remaining two-thirds of the month. Then he would have paid the foreman

$$\frac{2}{3}(4) - \frac{1}{3}(7) = \frac{1}{3} \text{ bezant (loss)},$$

which, again compared to the actual gain of 1 *bezant*, is too low by 4/3 *bezant*. Thus, the supposition of only 10 days of work is too low. The correct number of days that the worker labored during the month must, therefore, be somewhere between 10 and 20. It can now be found as a weighted average of these "input" values, weighting them with the resulting "errors" (the excess of 7/3 and the deficit of 4/3). We must give the second input (10 days) the heavier weight, since it came closer to the actual gain of 1 *bezant*:

$$20\left(\frac{4/3}{4/3 + 7/3}\right) + 10\left(\frac{7/3}{4/3 + 7/3}\right) = \frac{20(4/3) + 10(7/3)}{4/3 + 7/3}.$$

With routine arithmetic we can now complete the calculation, arriving at the answer of $13\frac{7}{11}$ days worked during the month.

Notice that the two inputs and the resulting errors ended up being "cross-multiplied", i.e., each input is multiplied by the error associated with the other input. In modern symbolism, the resulting algorithm can be represented as a formula,

$$x = \frac{x_1 e_2 + x_2 e_1}{e_2 + e_1},$$

where e_1 and e_2 denote the errors (one excess and one deficit) resulting from the suppositions x_1 and x_2, respectively.

In a case in which both of the errors are deficits, we could adapt the above formula by treating one of the deficits as a negative excess, i.e., replace e_1 with $-e_1$ or else e_2 with $-e_2$. Likewise, if both of the errors are excesses, then we could treat one of the excesses as a negative deficit, again replacing e_1 with $-e_1$ or else e_2 with $-e_2$. We find that all of these cases reduce to

$$x = \frac{x_1 e_2 - x_2 e_1}{e_2 - e_1}.$$

In medieval times, signed numbers were generally not used, so a discrepancy, whether excess or deficit, was always signified with an unsigned number. The above formulaic process thus had to be rendered in different forms corresponding to the different cases. In my course activities (see below), I simplify the situation by allowing signed numbers; as a result, a single common procedure can be used in all cases.

So long as the relationship underlying a given problem is affine linear, any two inputs x_1 and x_2, even wild guesses, will yield the same final result x, since any two points determine a line.[1] Historically, the input values were thought of as suppositions, or erroneous guesses, and the algorithm was conceived as a process of discerning the truth from two falsehoods. In Arabic the algorithm is known as *hisab al-khata'ayn*, which can be translated as "reckoning from two falsehoods." Fibonacci translated this into Latin as *regulis elchatayn* [2, p. 318] or *elchataym* [8, p. 447]. Eventually in Latin Europe, *regula falsi positionis* ("the rule of false position") and *regula duorum falsorum* ("the rule of two falsehoods") became the most common terms for the technique.

The tradition of double false position was especially strong in the Maghreb (Northwest Africa). It was often used there, for example, by legal specialists in the complex Qur'anic rules for division of legacies. North African scholars

---

[1] If the relationship is nonlinear, then the result will be only an approximation, and will vary with the choice of input values.

developed mnemonics, notably poems and diagrams, to aid such nonscientists in recalling the steps of the algorithm. A diagrammatic "method of scales" was widely used in the Maghreb by the 12th Century and persisted for centuries. Fibonacci, who had studied mathematics in this region as a boy, used a simplified version of the scales diagram in his *Liber Abaci*.

To illustrate the method of scales, consider again Fibonacci's problem of the worker and the foreman. This time, I will use Fibonacci's own suppositions of 20 and 15 days (instead of 20 and 10). We substitute these guesses directly into the statement of the problem:

$$\text{For 20 days of work: } \frac{2}{3}(7) - \frac{1}{3}(4) = \frac{10}{3} \text{ or } 3\frac{1}{3} \text{ bezants}$$

$$\text{For 15 days of work: } \frac{1}{2}(7) - \frac{1}{2}(4) = \frac{3}{2} \text{ or } 1\frac{1}{2} \text{ bezant.}$$

Comparing these results with the target value of 1 *bezant*, we find that they represent excesses of 2 1/3 *bezants* and 1/2 *bezant*, respectively. These data can be placed in a scales (balance) diagram, drawn here in the style associated with ibn al-Banna' of Marrakech (1256–1321) [3, pp. 101–103]:

The target value, 1, is placed on the "dome" of the balance, and the two guesses are placed in the "pans", with the corresponding excesses above them. The lines in the diagram are a guide in carrying out the cross-multiplication and other steps of the algorithm:

$$\frac{15 \times 2\frac{1}{3} - \frac{1}{2} \times 20}{2\frac{1}{3} - \frac{1}{2}} = \frac{25}{1\frac{5}{6}} = 13\frac{7}{11} \text{ days.}$$

For more details, including for the cases of two deficits or one excess and one deficit, see [3, p. 102].

5.3 In the Classroom

I have had great success incorporating an adaptation of the rule of double false position into self-paced modules that I developed for courses in Linear Algebra and in Calculus for Business and Social Sciences. These take the form of written activities, each about 10 pages in length, and rich with graphics. The modules themselves are available online in my article, "Combining Strands of Many Colors: Episodes from Medieval Islam for the Mathematics Classroom," mathdl.maa.org/mathDL/46/?pa=content&sa=viewDocument&nodeId=3546&bodyId=3913.

Parts of the written modules are expository, describing the mechanics of the technique or its cultural and historical context. The exercises, on the other hand, guide the student in discovering why the technique works, in exploring its relation to other methods, or in applying it to solve various types of story problems. The students begin the activity in class, working for about 30 minutes either individually or in groups of two or three, as they prefer. Each student takes the module home to complete it, and later submits it to me for grading and comments.

The version of double false position that I teach my students allows them to more easily grasp, recall, and carry out the technique. I created this version of the method of scales by making the following modifications:

1. The dome and levers are replaced by a circuit of arrows.

2. The diagram is turned 90 degrees to match how pairs of coordinates are usually tabulated in our courses.

3. Signed numbers are used (with deficits being represented by negative numbers), so that all cases can be treated in a unified way.

To teach students the mechanics of the technique, the instructor can include an example such as Fibonacci's problem of the worker and the foreman. I also include a brief exposition of the historical context for this method, by summarizing the forms and uses of mathematics that arose in the Middle East and how these were transmitted to Europe by Fibonacci and others. (More information on this can be found below, in the section "Taking It Further".)

Here is how Fibonacci's problem of the worker and the foreman, described earlier, would be solved in this style. The student might want to first tabulate the inputs and outputs:

days worked	excess pay
20	2 1/3
15	1/2
?	0

If these are thought of as coordinates, then the problem amounts to finding the missing coordinate on the left. The steps are as follows:

(1) Write the four values in a square arrangement:

$$20 \qquad 2\ 1/3$$

$$15 \qquad 1/2$$

(2) Draw the two diagonals of the square.

(3) Draw one of the two vertical sides of the square; the "missing" vertical side needs to align with the "missing" coordinate in the table:

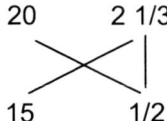

(Note how the drawn vertical side corresponds to the "dome" in the scales diagram.)

(4) Place arrowheads on the three line segments so that they connect head-to-tail, either clockwise (as below) or counterclockwise:

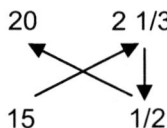

(5) Imagine walking along the path in the direction of the arrows, successively from tail to head to tail, etc. To recall the order of operations, multiply along the diagonals and subtract along the vertical. Write the complete path in the numerator, and the vertical portion alone in the denominator:

$$\frac{15 \times 2\frac{1}{3} - \frac{1}{2} \times 20}{2\frac{1}{3} - \frac{1}{2}} = \frac{25}{1\frac{5}{6}} = 13\frac{7}{11} \text{ days.}$$

Note that reversing the direction of the arrows results in negating both numerator and denominator, so that the final answer is the same. Students will be familiar with this same invariance property in the rise-over-run formula for slope.

To help the students discover that the technique is based on the concept of ratio and proportion, the instructor can make use of coordinate axes and symbolic algebra. Note, first, that the triangles in the figure below are geometrically similar.

Students can be asked to set up a corresponding proportion, such as:

$$\frac{b}{x-a} = \frac{d}{x-c}.$$

They can then be asked to solve the relation algebraically for x, giving a "formula" for double false position:

$$x = \frac{ad - bc}{d - b}.$$

5.3. In the Classroom

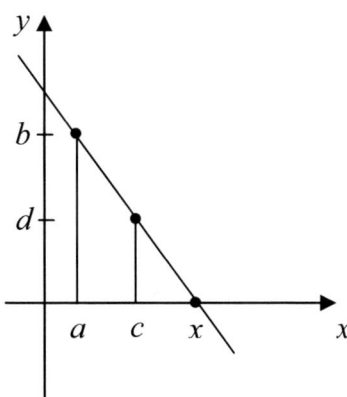

By being challenged with many types of practice problems, students will become quite adept at using the method of scales. I like to include some interesting historical examples from different cultures, such as these drawn from Italy and China:

Example 5.1. Two birds fly from the tops of two towers whose bases are 50 yards apart, one 40 yards high, the other 30, starting at the same time and flying at the same rate, and reaching the center of a fountain between the two towers at the same moment. How far is the fountain from the base of each tower? [paraphrased from [8, pp. 462–463]]

(a) Make two wild guesses as to the distance from the fountain to the base of the tall tower. For each guess, use the Pythagorean Theorem and the fact that the towers are 50 yards apart to calculate the discrepancy between the *squares* of the distances flown by the two birds. An example is given in the chart below; record your own results in the next two rows.

distance from fountain to base of tall tower (yards)	distance from fountain to base of short tower (yards)	squared distance from fountain to top of tall tower (square yards)	squared distance from fountain to top of short tower (square yards)	discrepancy (column 4 minus column 3)
example: 15	$50 - 15 = 35$	$15^2 + 40^2 = 1825$	$35^2 + 30^2 = 2125$	$2125 - 1825 = 300$
x				

(b) Use your two input and two output values (in columns 1 and 5 of the chart) along with the rule of double false position to solve the problem. Record your answer in the fourth row of the chart, and verify that it solves the problem.

(c) Using algebra, complete the last row of the chart to verify that— despite the squaring that is involved— there is a linear relationship between the input variable in column 1 and the output variable in column 5. Set the resulting discrepancy in column 5 to zero, and show that this leads to the same answer that you got in part (b).

Example 5.2. A tub of full capacity 10 *dou* contains a certain quantity of husked rice. Unhusked rice is added to fill up the tub. When the rice is all husked, it is found that the tub contains 7 *dou* of husked rice altogether. Assume each *dou* of unhusked rice produces 3/5 *dou* of husked rice. Find the original amount of husked rice in the tub. [paraphrased from [7, p. 365]]

Husked rice (*dou*)	Unhusked rice (*dou*)	Total amount of rice when all is husked (*dou*)	Actual amount of rice when all is husked (*dou*)	Discrepancy (column 4 minus column 3)
example: 2	$10 - 2 = 8$	$2 + \frac{3}{5}(8) = 6\frac{4}{5}$	7	$7 - 6\frac{4}{5} = \frac{1}{5}$
			7	
			7	

Example 5.3. Now an item is purchased jointly; everyone contributes 8 [coins], the excess is 3; everyone contributes 7, the deficit is 4. Tell: the number of people, the item price, what is each? [7, p. 358].

Instructors can also formulate their own interesting story problems, based, for example, on the linear relationship between temperatures in the Celsius and Fahrenheit scales, or between years in the Christian and Muslim calendars.

I inform my business students that those who conduct affairs both in predominantly Christian and predominantly Muslim countries will need to understand the two calendars and be able to translate dates back and forth between them. The dates differ for two main reasons. First, the accounting is pegged to two different events: the traditional year of Christ's birth in one case, and the year of the Hijra, or emigration of Muhammed from Mecca to Medina, in the other case. Second, the Christian calendar is based on a solar year, while the Islamic calendar is based on a lunar year. In the Middle East, the use of lunar and solar calendars (especially by nomadic and sedentary peoples, respectively) both flourished. Of course, because these calendars are so rooted in culture their use persists today, even in regions where few people are nomads or farmers.

Example 5.4. This exercise explores the relationship between the "Christian" year used in the West, and the "Islamic" year used among Muslims. For example, the Christian year 1492 roughly corresponded to the Muslim year 897, while the Christian year 1990 roughly corresponded to the Muslim year 1410. The relation between Muslim and Christian years is very close to being linear, $y = mx + b$.

(a) Use the sample years given above and the definition of slope to estimate m with a high degree of precision.

(b) Use *al-khata'ayn* (double false position) to estimate b with a high degree of precision.

(c) Use your model $y = mx + b$ to complete this table; round your answers to the nearest year.

Muslim year, x	Christian year, y
1	
1000	
	2020

Based on the top row, in which Christian year do you estimate that the Hijra took place?

(d) Use your model $y = mx + b$ to estimate when the Muslim and Christian years will be the same (round your answer to the nearest year).

For business students, I also devised problems such as the following.

Example 5.5. A management consultant charges a base fee for each consultation, plus an hourly rate. Her records for two different consultations show a charge of $281.50 for 3 hrs 20 mins of work, and a charge of $249.30 for 2 hrs 45 mins.

(a) Use double false position to determine her base fee, in dollars (round to the nearest cent).

(b) Use the rise-over-run procedure to determine her hourly rate, in dollars per hour (round to the nearest cent).

Example 5.6. To stimulate sales during a recession, General Motors Corp. decided to temporarily lower its financing rate for new vehicle purchases. When the rate was lowered to 8 3/4 %, sales jumped by 12% compared to those that were being recorded under the standard financing terms. When the rate was lowered all the way to 3 1/2 %, sales rose 19% higher than those under standard financing. Assume that the trend is linear.

(a) What sales increase can be expected under zero-percent financing?

(b) What standard interest rate is offered by GM to its customers?

Most textbooks for Business Calculus include a whole series of optimization exercises in which a linear constraint must first be determined. Students can be encouraged to use the method of double false position to determine the intercept, and the rise-over-run procedure to determine the slope. An example follows.

5.4. Taking It Further

Example 5.7. A certain toll road averages 36,000 cars per day when charging $1 per car. A survey concludes that increasing the toll will result in 300 fewer cars for each cent of increase. What toll should be charged in order to maximize the revenue? [4, p. 191].

(a) Use double false position to determine the intercept that is missing in the table below.

toll (dollars)	average daily traffic (number of vehicles)
1.00	36,000
1.01	35,700
0	

(b) Use the rise-over-run procedure to determine the slope.

(c) Use your answers to parts (a) and (b) to write the linear relation between toll and traffic.

(d) Use the optimization technique to determine how to maximize revenue.

For Linear Algebra, I devised the following surveying problem.

Example 5.8. A buoy B was positioned in a water channel many years ago. Now, it's desired to know how far it sits from the sides of the channel. When rangefinders are placed at opposite points A and C, it's found that the channel width there is 97.61 meters. When one of the rangefinders is moved 30 meters due north of A, it's found that the other rangefinder must be moved 42.66 meters due south of C to remain aligned with the buoy.

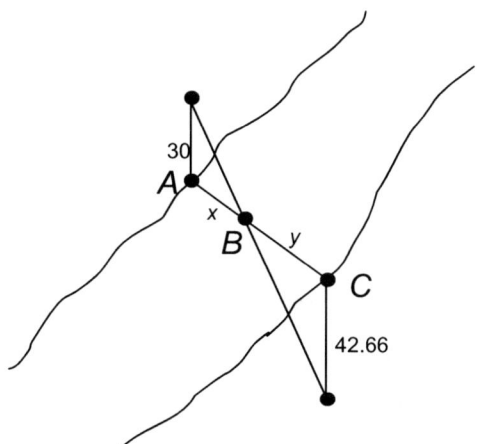

(a) Use the similarity of the two triangles to show the relation $42.66x - 30y = 0$.

(b) Make two wild guesses for x. For each guess, use the fact that $AC = 97.61$ to calculate $42.66x - 30y$, which should be zero.

(c) Use double false position and your data from part (b) to solve the problem.

5.4 Taking It Further

When I teach an historical method in the classroom, I am always interested in having students see its relation to techniques used in other cultures and other historical periods.

Students who have studied matrix methods will recognize that in the "formula" for double false position, the numerator resembles a determinant. Those in Linear Algebra, in particular, will be interested to see relationships between this method and Cramer's Rule of Determinants, which came much later in history. Students can, for instance, be guided to derive the formula for double false position from Cramer's Rule as follows. Consider a relation of the form $ax + b = c$, which we might solve algebraically as

$$x = \frac{c-b}{a}.$$

As an alternative to that method, suppose there are two suppositions x_1 and x_2, and corresponding excesses e_1 and e_2:

$$ax_1 + b = c + e_1$$
$$ax_2 + b = c + e_2.$$

$$x_1 a + 1(b-c) = e_1$$
$$x_2 a + 1(b-c) = e_2.$$

Applying Cramer's Rule to this system gives

$$a = \frac{\begin{vmatrix} e_1 & 1 \\ e_2 & 1 \end{vmatrix}}{\begin{vmatrix} x_1 & 1 \\ x_2 & 1 \end{vmatrix}} = \frac{e_1 - e_2}{x_1 - x_2}$$

$$b - c = \frac{\begin{vmatrix} x_1 & e_1 \\ x_2 & e_2 \end{vmatrix}}{\begin{vmatrix} x_1 & 1 \\ x_2 & 1 \end{vmatrix}} = \frac{x_1 e_2 - e_1 x_2}{x_1 - x_2},$$

and thus the familiar false-position formula,

$$x = \frac{c-b}{a} = \frac{b-c}{-a} = \frac{x_1 e_2 - e_1 x_2}{e_2 - e_1}.$$

Fibonacci's naming of this method with a term *elchatayn* derived from Arabic is not surprising. His travels in, and borrowings from, the Muslim-influenced lands surrounding the Mediterranean are well known. During medieval times, scholars living under Muslim rule became preeminent in mathematics and many other sciences, and their work disseminated across various networks of trade and scholarship. They recovered and synthesized much of the classical mathematics of Greece, Byzantium, and India, including the "Hindu reckoning" made possible by decimal place-value numeration. Then they greatly extended these, making major breakthroughs in plane and solid geometry, plane and spherical trigonometry, root extraction and other arithmetical algorithms, algebra, the analysis of polynomials, number theory, and combinatorics. In fact, the very word *algorithm* ultimately derives from the name of al-Khuwarizmi, a leading mathematician working in Baghdad in the 9th Century. Details about many of these contributions are available in [1] and [5] as well as more general works on the history of mathematics.

Although the movement of *hisab al-khata'ayn* from the Middle East to Europe is evident, the actual origins of the method are unclear. Somewhat similar techniques were used in ancient China bearing names such as *ying bu tsu shu* ("the rule of too much and not enough"), possibly before 100 BCE [7, pp. 349–385]. However, I have shown [6] that it is very unlikely that Arabs borrowed their technique from Chinese literature, based on differences in the respective algorithms, their justifications, terminologies, and applications. In the same article, I argued that *hisab al-khata'ayn* might well have first found its way into Arabic treatises from a prior tradition of practical usage by merchants, jurists, surveyors, builders, and the like.

The earliest discussions of this technique exemplify how medieval scientists built on the ancient mathematics that came before them. Greek geometers themselves do not seem to have known of double false position, but mathematicians in the early Islamic world justified the technique by devising formal geometric proofs in the Greek style. This can be seen in the oldest surviving Arabic writing on *hisab al-khata'ayn*, that of Qusta ibn Luqa (late 9th C.), a Christian

Arab mathematician from Baalbek on the coast of Lebanon. He represented the linearity of the problem as a right triangle whose height and hypotenuse increase proportionally with its base. His meticulous three-part proof relies on Euclid I:43. A translation of the first part of the proof (the case of two deficits) is provided in [3, pp. 98–101], while [9] includes a German translation of the entire treatise.

5.5 Conclusion

Teaching the medieval rule of double false position "fills in a gap" in people's understanding of linear relationships. We certainly drill into our students how to calculate a slope (rate of change) purely from four pieces of data: two inputs and two outputs. Shouldn't we be able to quickly compute the intercepts from the same four numbers? As we have seen, a wide range of practical problems can be reduced to finding such intercepts. Double false position is an efficient, purely arithmetical procedure for solving such problems.

Instruction in this topic also "fills in a gap" in another sense. In the West, there is a startling lack of knowledge of the historical contributions to mathematics and science made by non-European peoples. Especially in industrialized nations, many also have the mistaken impression that mathematics is purely a product of abstract thought, not realizing the extent to which practical problems have been a major force in its development. Learning about double false position and the various ways in which it was conceived, used, and justified broadens our students' perspective on how mathematics and science develop. It provides a case study in how science has been driven by the curiosity, hard work, and creativity of people in many different cultures.

Bibliography

[1] J. Lennart Berggren, *Episodes in the Mathematics of Medieval Islam*, Springer-Verlag, New York, 1986.

[2] Baldassarre Boncompagni, *Scritti di Leonardo Pisano*, vol. 1, Tipografia delle Scienze Matematiche e Fisiche, Rome, Italy, 1857.

[3] Jean-Luc Chabert, et al., *A History of Algorithms: From the Pebble to the Microchip*, Springer-Verlag, Berlin, Germany, 1999.

[4] Larry J. Goldstein and David C. Lay, David I. Schneider, and Nakhl H. Asmar, *Brief Calculus & Its Applications*, 12th ed., Pearson Prentice Hall, Upper Saddle River, NJ, 2010.

[5] Roshdi Rashed, *The Development of Arabic Mathematics: Between Arithmetic and Algebra*, Kluwer Academic Publishers, Dordrecht, The Netherlands, 1994.

[6] Randy K. Schwartz, "Issues in the Origin and Development of *Hisab al-Khata'ayn* (Calculation by Double False Position)," in *Actes du Huitime Colloque Maghrébin sur l'Histoire des Mathématiques Arabes, Tunis, les 18-19-20 Dcembre 2004*, Tunisian Association of Mathematical Sciences, Tunis, Tunisia, 2006. Also available at http://facstaff.uindy.edu/~oaks/Biblio/COMHISMA8paper.doc.

[7] Shen Kangshen, John N. Crossley and Anthony W.-C. Lun, *The Nine Chapters on the Mathematical Art: Companion and Commentary*, Oxford University Press, Oxford, UK, 1999.

[8] Laurence E. Sigler, *Fibonacci's Liber Abaci: A Translation into Modern English of Leonardo Pisano's Book of Calculation*, Springer-Verlag, New York, 2002.

[9] Heinrich Suter, "Die Abhandlung Qosta ben Luqas und zwei andere anonyme ber die Rechnung mit zwei Fehlern und mit der angenommenen Zahl," *Bibliotheca Mathematica*, Third Series, 9 (1908) 111–122.

6

Complex Numbers, Cubic Equations, and Sixteenth-Century Italy

Daniel J. Curtin
Northern Kentucky University

6.1 Introduction

The complex numbers are important in modern mathematics and science, yet they receive almost no attention in the modern curriculum, which is heavily weighted towards preparation for the Calculus. Most pre-calculus treatments of the complex numbers give no insight into where they came from. They are mainly seen as supplying a full set of roots for polynomials that do not have all real roots. In fact, they first arose because they were needed to find real roots for cubic equations, precisely in the case where all three roots are real. The material in this article can be used anywhere complex numbers are introduced. It mixes geometric and algebraic ideas, in a way that should be particularly useful in a functional approach to pre-calculus. Technology can be used or not, as seems appropriate to the instructor.

6.2 Historical Background

The idea of complex numbers, at least in the sense of calculations involving square roots of negative numbers, first arose in Italy around 1540, as mathematicians solved cubic and quartic equations. In modern courses, complex numbers often first arise to discuss solutions of quadratic equations where $ax^2 + bx + c = 0$ has solutions

$$x = \frac{-b \pm \sqrt{b^2 - 4ac}}{2a},$$

and the question of what happens when $b^2 - 4ac$ is a negative number arises. Solutions to quadratic problems, analogous to the quadratic formula, have been known since at least 2000 BC, yet no one appears to have shown any interest in the possibility of negatives under the square root. Most likely this is because they only occur when both roots are complex, so from the point of view of real numbers the equation has no solutions.

For the cubic and higher equations, mathematicians had sought in vain for formulas similar to the quadratic formula. Success came at last in Italy just after 1500. The intrigue and feuding surrounding this discovery might be worthy of operatic treatment. We haven't time to pursue it, but the interested reader should seek out [3; 4, pp. 7–25; 5, pp. 358–367]. We will see below that precisely in the case where the cubic equation has three real roots the relevant formula must involve square roots of negative numbers. It was to resolve this difficulty that mathematicians began to calculate with such quantities. In fairly short order, Rafael Bombelli produced his book, *L'Algebra* [1], which includes a careful treatment of such calculations and thus provided a remarkably complete treatment of the arithmetic of what we now call complex numbers.

6.3 In the Classroom

6.3.1 Quadratic Equations

If the students have not investigated the discriminant for quadratics, it may be worthwhile to first have them look at examples. To tie in with what we do later consider the graph of $f(x) = x^2 + 2x$, either drawing it by hand or using technology. Then examine the intersection of this graph with $y = d$ for various specific values of d and study how the number of intersections corresponds with the values of $b^2 - 4ac$. See Figure 6.1.

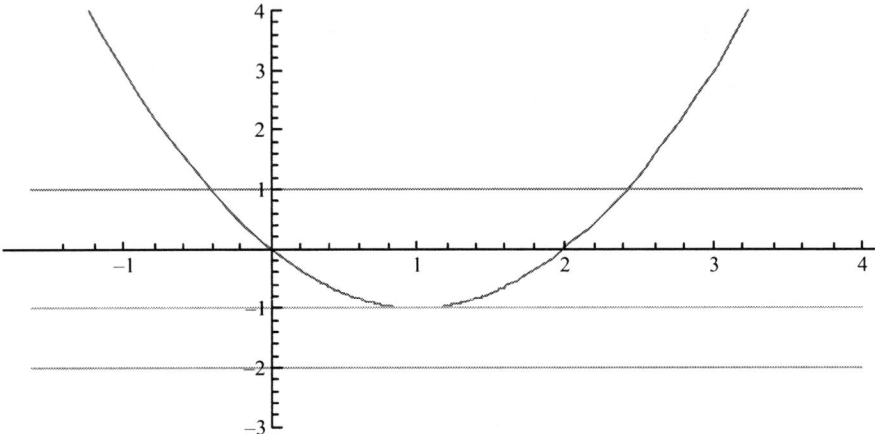

Figure 6.1. Various values of the discriminant. For $y = 1$, $b^2 - 4ac = 8$, $y = -1$, $b^2 - 4ac = 0$, and $y = -2$, $b^2 - 4ac = -4$.

6.3.2 Cubic Equations

The general cubic equation would have the form $ax^3 + bx^2 + cx + d = 0$, with $a \neq 0$. It is convenient to assume the equation has been divided by a, so the form is $x^3 + bx^2 + cx + d = 0$. Substituting $x = y - (b/3)$ and expanding eliminates the square term, so we may take the general form as $x^3 + px + q = 0$.

The first complete solution to the cubic equation (and the fourth-degree, or quartic, equation) is given in the *Ars Magna* (*Great Art*) of Girolamo Cardano, published in 1545 [2]. The elimination of the square term is his idea. It should be noted that although negative numbers were used in calculations they were not really accepted as coefficients. Even as solutions they were regarded as somewhat fictitious. Cardano explained the negative (he called it "false") solution $x = -3$ of the equation $x^3 + 21 = 2x$ as really being the positive solution $x = 3$ of $x^3 = 2x + 21$.

In fact, since negative coefficients were not allowed, Cardano would view the equations in the last sentence as being of different forms $x^3 + q = px$ and $x^3 = px + q$, respectively. The remaining cubic without a square term would be $x^3 + px = q$. Our standard version $x^3 + px + q = 0$ would seem absurd to Cardano since with x, p, and q all positive it is obvious no solution is possible!

This division of cases is somewhat reasonable if we consider the number of possible real solutions. For the case $x^3 + px = q$, consider a typical example: $x^3 + 6x = 20$. The graph of $f(x) = x^3 + 6x$ is increasing (and it is for $f(x) = x^3 + px$ whenever $p > 0$). Thus a line $y = d$ hits the graph exactly once, and in fact this type of equation has one real root. For $x^3 + 6x = 20$, it is $x = 2$.

We will concentrate on the case $x^3 = px + q$; the case $x^3 + q = px$ is similar. To visualize we will rewrite it as $x^3 - px = q$, though Cardano might not approve. Look at the graph of $f(x) = x^3 - px$. See Figure 6.2. The functions in this family are very similar for any particular positive value of p. Since $x^3 - px = x(x^2 - p)$, the graph crosses the x-axis at three points: $x = -\sqrt{p}, 0, \sqrt{p}$, with a local maximum between $x = -\sqrt{p}$ and $x = 0$ and a local minimum between $x = 0$ and $x = \sqrt{p}$. For a large enough value of q there is clearly one intersection with $y = q$

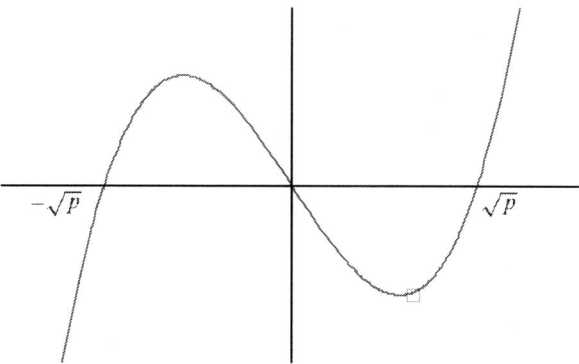

Figure 6.2. The graph of the cubic $f(x) = x^3 - px$.

and in fact exactly one real root. For small (positive) values there will be three intersections, thus three real roots. For one positive value there are exactly two intersections. If q is allowed to be negative we are effectively in the case $x^3 + q = px$.

For the equation $x^3 = px + q$, Cardano gives the solution

$$x = \sqrt[3]{\frac{q}{2} + \sqrt{\frac{q^2}{4} - \frac{p^3}{27}}} + \sqrt[3]{\frac{q}{2} - \sqrt{\frac{q^2}{4} - \frac{p^3}{27}}}.$$

See [5, pp. 362–363] for details. This gives only one root, but the others can be found by reducing to a quadratic.

Try this on $x^3 = 6x + 40$. You get $\sqrt[3]{20 + \sqrt{392}} + \sqrt[3]{20 - \sqrt{392}}$, a positive number (note $\sqrt{392} < 20$.) Look at the graphs of $y = x^3 - 6x$ and $y = 40$. It appears there is exactly one solution. In fact, try $x = 4$. That's the solution.

Now consider $x^3 = 15x + 4$. The Cardano formula gives $\sqrt[3]{2 + \sqrt{-121}} + \sqrt[3]{2 - \sqrt{-121}}$. Hmmm! A look at the graphs shows there is one positive solution, but this time there are two negative solutions. If you try $x = 4$ it works again.

A calculator will tell us that $\sqrt[3]{20 + \sqrt{392}} + \sqrt[3]{20 - \sqrt{392}} = 4$, so that's probably OK. If it can handle complex numbers it will tell us $\sqrt[3]{2 + \sqrt{-121}} + \sqrt[3]{2 - \sqrt{-121}} = 4$ also.

This raises two questions. What does the second solution mean? In both cases how do we simplify?

6.3.3 Complex Numbers

A small amount of complex arithmetic is needed. Most algebraic rules remain the same. The main new ingredient is that we include square roots of negative numbers. That is we include numbers such as $\sqrt{-121}$. Now $\sqrt{2}$ is defined to be the (positive real) number whose square is 2. So we assert that $\sqrt{-121}$ is some sort of number whose square is -121. Now assuming the rules of algebra work, $\sqrt{-121} = \sqrt{121(-1)} = \sqrt{121}\sqrt{-1}$. In fact, $\sqrt{-b} = \sqrt{b}\sqrt{-1}$ in general, so we really only need to come up with $\sqrt{-1}$ (often called i) to make this work.

A complex number is simply one of the form $a + b\sqrt{-1}$, where a and b are ordinary real numbers. The operations of addition and multiplication are

$$\left(a + b\sqrt{-1}\right) + \left(c + d\sqrt{-1}\right) = (a + c) + (b + d)\sqrt{-1},$$
$$\left(a + b\sqrt{-1}\right) \times \left(c + d\sqrt{-1}\right) = ac + ad\sqrt{-1} + bc\sqrt{-1} + bd(\sqrt{-1})^2 = (ac - bd) + (ad + bc)\sqrt{-1}.$$

6.4 Rafael Bombelli

Cardano's work was brilliant but difficult to read. Rafael Bombelli, a civil engineer by trade, decided to put together a book that would lay out the background needed to understand the *Ars Magna* and give detailed proofs of its conclusion. The result was *l'Algebra*, from which later mathematicians including Newton and Leibniz drew inspiration. It was finally eclipsed when modern algebraic notation, beginning with Descartes, made Bombelli's own notation — though not his ideas — obsolete.

Bombelli attacks $\sqrt[3]{2+\sqrt{-121}} + \sqrt[3]{2-\sqrt{-121}}$ as follows.

It seems reasonable to guess that something of the form $\sqrt[3]{a+\sqrt{-b}}$ can be written $x + \sqrt{-y}$. We would say a cube root of a complex number should be a complex number. Of course then it is equally reasonable to assume $\sqrt[3]{a-\sqrt{-b}}$ is $x - \sqrt{y}$.

If so, then
$$a + \sqrt{-b} = \left(x + \sqrt{-y}\right)^3 = x^3 + 3x^2\sqrt{-y} - 3xy^2 - y\sqrt{-y},$$

since $\left(\sqrt{-y}\right)^2 = -y$. Putting like with like,
$$a + \sqrt{-b} = (x^3 - 3xy^2) + (3x^2 - y)\sqrt{-y},$$

so $x^3 - 3xy^2 = a$.

Now if $\sqrt[3]{a - \sqrt{-b}} = x - \sqrt{y}$, then $a - \sqrt{-b} = \left(x - \sqrt{-y}\right)^3$. We multiply $a - \sqrt{-b}$ by $a + \sqrt{-b}$ to get
$$a^2 + b^2 = \left(x - \sqrt{-y}\right)^3 \left(x + \sqrt{-y}\right)^3 = \left(x^2 + y\right)^3.$$

So we have
$$x^2 + y = \sqrt[3]{a^2 + b^2} \quad \text{and} \quad x^3 - 3xy^2 = a.$$

We want to find x and y. If we just plow ahead this leads us back to another cubic equation, which is no help. Instead Bombelli proceeds by what he calls "*trovare al tentone*," that is, "to find by groping along." Of course, he doesn't grope blindly.

For $\sqrt[3]{2 + \sqrt{-121}}$ he is solving
$$x^2 + y = \sqrt[3]{125} = 5 \tag{6.1}$$

and
$$x^3 - 3xy^2 = 2, \quad \text{with } x \text{ and } y \text{ positive.} \tag{6.2}$$

From (6.1) he knows $x^2 < 5$, from (6.2) $x^3 > 2$, so he tries $x = 2$, in which case $y = 1$, so that
$$\sqrt[3]{2 - \sqrt{-121}} = 2 - \sqrt{-1} \quad \text{and} \quad \sqrt[3]{2 + \sqrt{-121}} = 2 + \sqrt{-1}.$$

This can be checked by multiplying out $\left(2 + \sqrt{-1}\right)^3$.

Then the solution to $x^3 = 15x + 4$ is
$$x = \sqrt[3]{2 + \sqrt{-121}} + \sqrt[3]{2 - \sqrt{-121}} = \left(2 + \sqrt{-1}\right) + \left(2 - \sqrt{-1}\right) = 4,$$

as claimed. Notice how the square roots of the negatives, so essential to the solution, slip demurely off-stage at the very last moment.

In fact, in his text Bombelli devotes many pages to the discussion of how to perform such calculations, as well as other important matters such as extraction of fifth roots. It really is a complete book of algebra for its time!

Try your hand at Bombelli's second example of simplifying complex cube roots [1, p. 141]: $\sqrt[3]{52 + \sqrt{-2209}}$. It turns out to be $4 + \sqrt{-1}$.

Are you bold enough for his third example? See [1, p. 142]: $\sqrt[3]{8 + \sqrt{-232\frac{8}{27}}}$. Bombelli shows that it is $\left(\sqrt{2} + 1\right) + \sqrt{-\left(3\frac{2}{3} - \sqrt{8}\right)}$. Note how his *tentoni* stretches to guesses like $x = \sqrt{2} + 1$.

Bombelli's method does not lead to a method for reducing all such radicals, though it pointed the way for later work.

What about $x^3 = 6x + 40$, whose solution $x = \sqrt[3]{20 + \sqrt{394}} + \sqrt[3]{20 - \sqrt{394}}$ is also 4? Similar ideas can be applied, assuming $\sqrt[3]{a \pm \sqrt{b}} = x \pm \sqrt{y}$ and equating the terms that involve roots and also the terms that don't. For a modern look at this question, see [6].

6.5 Conclusion

It is perhaps unfortunate that the solution to the cubic equation is no longer taught in algebra classes. In complete generality it is probably too complicated for most purposes. Still a treatment of one case — I'd recommend $x^3 = ax + b$ — could be very useful for students to meet the complex numbers and sharpen their algebraic skills. Today's graphical devices allow them to do this while firmly anchoring their calculations to a lovely picture.

Bibliography

[1] Rafael Bombelli, *l'Algebra*, E. Bortolotti, ed., Feltrinelli, Milan, Italy, 1966.

[2] Girolamo Cardano, *The Rules of Algebra (Ars Magna)*, translated by Richard Witmer, Dover Publications, Inc., Mineola, New York, 2007.

[3] Daniel J. Curtin, "The Solution of the Cubic Equation: Renaissance Genius and Strife," *Cubo Matematica Educacional*, 4:2 (2002) 29–42.

[4] Hal Hellman, *Great Feuds in Mathematics*, John Wiley, New York, 2006.

[5] Victor J. Katz, *A History of Mathematics: An Introduction*, Second Edition, Addison-Wesley, Reading, Massachusetts, 1998.

[6] Chi-Kwong Li and David Lutzer, "The Arithmetic of Algebraic Numbers: An Elementary Approach," *The College Mathematics Journal*, 35:4 (2004), 307–309.

7
Shearing with Euclid

Davida Fischman and Shawnee McMurran
California State University, San Bernardino

7.1 Introduction

The Pythagorean Theorem is one of those intriguing geometric concepts that provide a never-ending source of ideas at all levels. Proofs of this theorem abound in print[1], and one wonders whether humans will ever stop looking for yet another. Indeed, it would be unusual for a student who has taken algebra or geometry not to have been exposed to at least one proof of the theorem, but how many have had occasion to explore the proof appearing in Euclid's *Elements*? In this proof, Euclid introduces a clever and elegant application of the concept of *shearing*. It is a proof that provides a golden opportunity not only to bring some history into the classroom, but that also provides us a natural venue to highlight connections between algebraic and geometric concepts. Moreover, the proof presented by Euclid has the useful property that it provides for generalizations of the theorem in a number of different directions. For example, by using shearing one may prove Pappus' theorem, which is a Pythagorean-like theorem for arbitrary triangles. The concept of shearing itself can then be generalized in the form of Cavalieri's principle to determine the volume of more general solids.[2]

In Book XII, Euclid again applies a technique that is connected to the concept of shearing, this time in three dimensions. The problem is seemingly unrelated: determining the relationship between the volumes of pyramids and prisms that share the same base and height. This application provides contemporary teachers an opportunity to motivate and illuminate the ostensibly nonintuitive formula for the volume of a pyramid. We will offer some techniques for providing our students with a hands-on activity that will allow them to explore Euclid's proof while developing three-dimensional visualization skills.

Throughout, we use Heath's English translation of Euclid's *Elements* [7]. In addition, we will refer to each proposition by the book in which it is found followed by its number. For example, "Proposition XII-5" would denote Proposition 5 of Book XII.

The activities in this chapter are appropriate for teaching the Pythagorean Theorem and the volume of pyramids in foundational algebra, geometry, and trigonometry classes. They are designed to be completed in a one-hour lesson. Additional activities may be added as desired.

---

[1] Elisha Loomis has collected well over 300 proofs in *The Pythagorean Proposition* [10].
[2] See [6] for some history and applications of Cavalieri's principle.

7.2 Historical Background

In his 1925 preface to the second edition of his English translation of Euclid's *Elements* from the text of Heiberg, Sir Thomas L. Heath claimed,

> So long as mathematics is studied, mathematicians will find it necessary and worth while to come back again and again, for one purpose or another, to the twenty-two-centuries-old book which, notwithstanding its imperfections, remains the greatest elementary textbook in mathematics that the world is privileged to possess. [7, p. ix]

To be sure, mathematicians, philosophers, statesmen, and scientists of all disicplines are among the multitudes who have studied the *Elements*, admired the logic therein, and waxed poetic about the beauty of the mathematics found there, despite the fact that we have no original copies of Euclid's work, nor do we know much about Euclid himself. Much of what we know about Euclid comes from Proclus [11], who lived about seven centuries later. Nevertheless, Euclid's work has been translated from Greek into any number of languages, has gone through hundreds of editions, and has been annotated and commented on ad infinitum – fortunately so for many students of geometry. It is only in the last century or so that the dominance of Euclid's rendition has waned in geometry classrooms, having been supplanted by contemporary texts on Euclidean geometry which sometimes fail to capture the allure of the original.

The *Elements* consists of thirteen books. Within the first four lies the development of plane geometry, including most of the ideas introduced in a secondary-level geometry course. Books five through ten deal with ratios, proportions and elementary number theory. Finally, the fundamentals of spatial geometry are developed in books eleven through thirteen. Euclid's *Elements* is undeniably one of the greatest textbooks ever written, and we hope that this glimpse of the wisdom and beauty contained within its pages will capture the imagination of some of our own students just as it has inspired so many others throughout the ages.[3]

7.2.1 What is a Shear Transformation?

We will connect the proofs of two very different propositions from Euclid via a concept referred to as shearing. A *shear transformation* is a type of affine transformation that preserves area or volume. A *two-dimensional shear* is defined in [12] as,

> A transformation of a figure in which all points along a given line L remain fixed while other points are shifted parallel to L by a distance proportional to their perpendicular distance from L.

The generalization to three dimensions would be that all points along a given plane P remain fixed while other points are shifted in a given direction on a plane parallel to P by a distance proportional to their orthogonal distance from P.

To demonstrate a shear transformation in two dimensions, we might imagine a rectangle with its base fixed in place and its top placed on a rail. Now slide the top along the rail, allowing the sides to grow as needed. The result, as illustrated in Figure 7.1, is a non-rectangular parallelogram with the same base and height as the original rectangle.

A nice activity for beginning students is to dissect the resulting parallelogram in such a way as to recreate the original rectangle, and thus illustrate that the area is preserved.

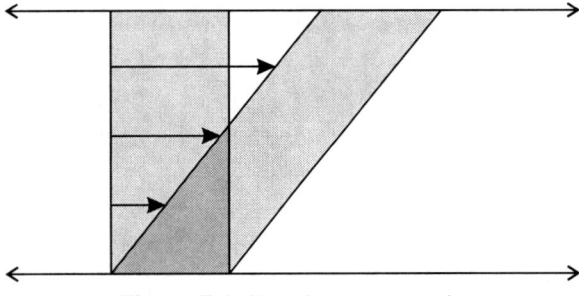

Figure 7.1. Shearing a rectangle

---

[3] Those interested in a little more historical background on Euclid and the *Elements* that can easily be brought to the classroom might consider starting with [5], [3], or [9].

7.2. Historical Background

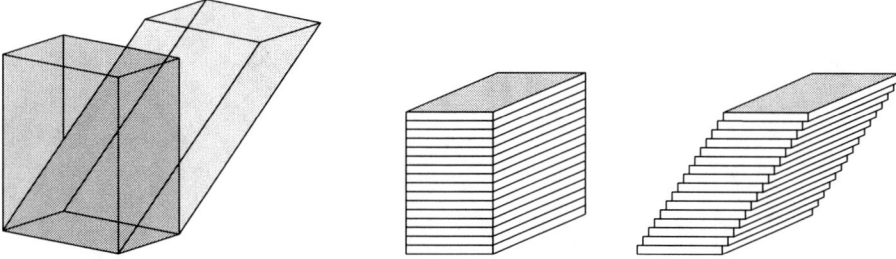

Figure 7.2. Shearing a prism

How can we demonstrate a shear transformation in three dimensions with simple tools? One option is to stack one or more decks of cards to create a right rectangular prism; take a flat surface and use it to push a face or an edge of the stack to create a slanted prism with the bottom card still in place (See Figure 7.2). This new prism, also known as a parallelepiped, still has the same base and height as the original rectangular prism. Moreover, since the two prisms both share the same number of cards, this demonstration helps clarify why the two prisms should have the same volume.

Shearing is more often seen in its algebraic form as a linear transformation expressed by matrix multiplication. This form of shearing is given in the appendix.

7.2.2 Euclid's Proof of the Pythagorean Theorem

The Pythagorean relationship between the side lengths of a right triangle is quite amazing and intriguing. In the words of Charles Dodgson, whom some of us may better recognize under the pen name Lewis Carroll,

> But neither thirty years, nor thirty centuries, affect the clearness, or the charm, of Geometrical truths. Such a theorem as "the square of the hypotenuse of a right-angled triangle is equal to the sum of the squares of the sides" is as dazzlingly beautiful now as it was in the day when Pythagoras first discovered it. [4, p. xvi]

The Pythagorean Theorem appears as the penultimate proposition in Book I of Euclid's *Elements*. The theorem, along with its converse, Propostion I-48, present a climactic scene in the first book of Euclid's ingenious geometric triskaidecalogy. The *Elements* states the Pythagorean Theorem as follows.

Proposition I-47. *In right-angled triangles the square on the side subtending the right angle is equal to the squares on the sides containing the right angle.*

Note that the implied inference in this context is that "equal" refers to equal area or content.

We begin our discussion with Proposition I-35 in which Euclid introduced a relationship between parallelograms of the same base and height, and Proposition I-37 which gives the analogous result for triangles.

Proposition I-35. *Parallelograms which are on the same base and in the same parallels are equal to one another.*

Proposition I-37. *Triangles which are on the same base and in the same parallels are equal to one another.*

Although Euclid's view of the relationships stated in these propositions was probably static, his wording "in the same parallels" evokes a dynamic view of the relationship that connects naturally to the shearing concept. In modern terms, Proposition I-35 states that parallelograms with the same base and equal heights have the same area. In particular, a parallelogram will have the same area as a rectangle with the same base and height, i.e., the product of the base length and the height. Thus, as illustrated in Figure 7.1, transforming a rectangle, or a parallelogram, via shearing does not change its area. Proposition I-37 follows in a straightforward way from Proposition I-35 when we view a triangle as half a parallelogram with the same base and height.

Now let's examine the role played by the shearing concept in Euclid's proof of Pythagoras' Theorem. We start with a right triangle $\triangle ABC$ set on its hypotenuse $\overline{BC}$ and construct the altitude $\overline{AI}$ by dropping a perpendicular from vertex A to the hypotenuse $\overline{BC}$ (see figure 7.3.)

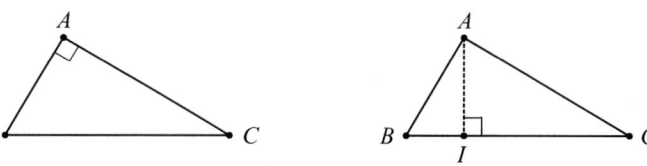

Figure 7.3. A right triangle

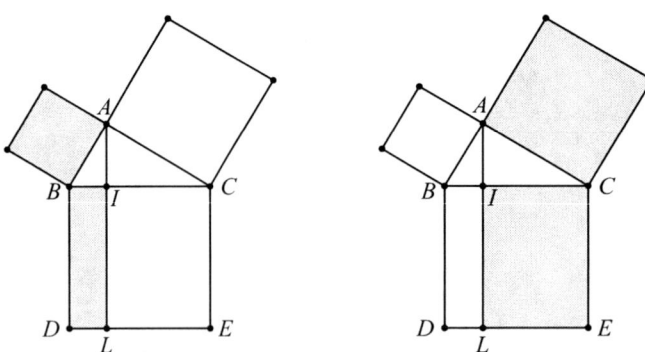

Figure 7.4. Euclid's observation

We now build a square on each side of the triangle. Let $\square BCED$ be the square on the hypotenuse and extend segment $\overline{AI}$ to $\overline{DE}$. Denote the intersection of the two by L. Euclid's brilliant observation at this point was that rectangle $BILD$ has the same area as the square on side $\overline{AB}$ (See Figure 7.4-left). Similarly, rectangle $ICEL$ has the same area as the square on side $\overline{AC}$ (See Figure 7.4-right). Since $\square BCED$ is obtained by putting rectangles $BILD$ and $ICEL$ together, the desired result follows.

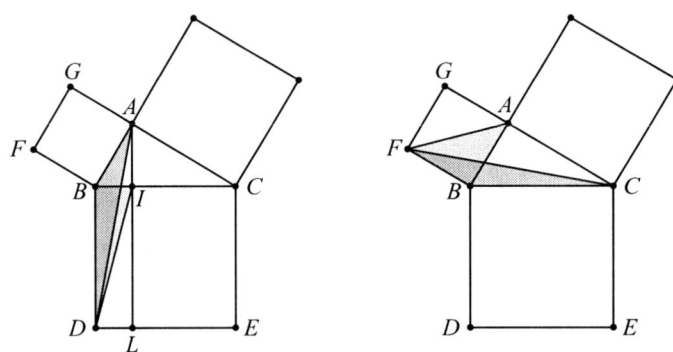

Figure 7.5. Euclid's "sheared" triangles

How did Euclid prove that rectangle $BILD$ does indeed have the same area as the square on side $\overline{AB}$? Let's denote this square by $\square ABFG$ and play the shearing game (See Figure 7.5). First, notice that the triangle with base $\overline{BD}$ and vertex I has half the area of rectangle $BILD$[4]. Since $\overline{BD}$ is parallel to $\overline{AL}$ we can shear this triangle along $\overline{AL}$ and the area will not change (See Figure 7.5-left). Hence $\triangle ABD$ will have the same area as $\triangle IBD$, namely half the area of rectangle $BILD$.

Similarly, the triangle with base $\overline{FB}$ and vertex A has half the area of $\square ABFG$. Since $\overline{GA}$ is parallel to $\overline{FB}$ and the points G, A, and C are collinear, we have $\overline{GC}$ parallel to $\overline{FB}$. Thus we can shear $\triangle FBA$ along $\overline{GC}$ (See Figure 7.5-right). It follows that $\triangle FBC$ has the same area as $\triangle FBA$, namely half the area of $\square ABFG$. Euclid then invoked a standard side-angle-side argument (Proposition I-4) to establish the congruence of $\triangle ABD$ and $\triangle FBC$, from which it follows that rectangle $BILD$ and $\square ABFG$ have the same area. The same argument will apply to rectangle $CELI$ and the square off the side $\overline{AC}$, thereby completing the proof.

7.2.3 The Volume of a Pyramid á la Euclid

We now consider shearing in the third dimension by investigating the volume of a pyramid with a polygonal base. Prior to its derivation in integral calculus, the formula for the volume of a pyramid is often introduced with little or no motivation. Nevertheless, it is a formula often used in foundational algebra and geometry courses. Certainly it is not difficult to use a tool such as unit cubes to motivate the fact that the volume of a right rectangular prism should be the product of its base area with its height, but it is not obvious why the volume of a pyramid with the same base and height should be one-third of this.

---

[4]Euclid proves this result in Proposition I-41.

7.2. Historical Background

A useful introduction to the relationship between a square-based pyramid and a cube with the same base and height is a demonstration with one of those nice geometry sets that include hollow versions of each. We fill the pyramid with a substance such as rice or water and show that if we pour the contents into the cube, and then repeat another two times, the cube appears to be filled. Although this demonstration may help to reinforce the formula for the volume of a prism, it does little to quench our curiosity about why such a relationship should hold, nor does it answer the question of whether the relationship will hold in more general cases.

How fortunate we are to have Euclid to provide enlightenment! By applying the concept of shearing – this time in three dimensions – to Euclid's derivation of the relationship between a prism and a pyramid, we gain insight and intuition using purely geometric concepts. We start by discussing triangle-based pyramids. For ease of notation, we will denote, for example, the pyramid with base $\triangle ABC$ and vertex D by $D\text{-}ABC$.[5]

In Proposition 5 of Book XII, Euclid tells us:

Proposition XII-5. *Pyramids which are of the same height and have triangular bases are to one another as the bases.*[6]

One sees a glimmer of the shearing concept in this statement, for if two pyramids happen to have congruent bases we obtain a three-dimensional analog of Proposition I-37, which is: If two pyramids of equal height have congruent triangular bases then their volumes will be the same. In dynamic terms, this is equivalent to the statement that shearing a pyramid leaves its volume unchanged.

How does Proposition XII-5 help us determine the volume of a triangular prism? Euclid answers this question in his proof of Proposition XII-7 by showing us how to dissect a triangular prism into three equal-volumed pyramids.

Proposition XII-7. *Any prism which has a triangular base is divided into three pyramids equal to one another which have triangular bases.*

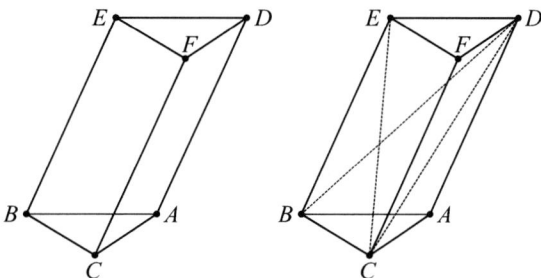

Figure 7.6. Dissecting a triangular prism into three pyramids

Let's look at Euclid's argument. Consider the prism $ABCDEF$ with triangular bases $\triangle ABC$ and $\triangle DEF$. We can dissect this prism into three tetrahedra, $ABCD$, $BCDE$, and $CDEF$ as illustrated in Figure 7.6. Euclid now shows us in two steps how to view these three tetrahedra as pyramids with equal volume.

1. Consider the pyramids $C\text{-}ABD$ and $C\text{-}EDB$. Their respective bases $\triangle ABD$ and $\triangle EDB$ are opposite halves of the parallelogram $ADEB$. Therefore these bases are congruent and lie in the same plane. Since the pyramids share vertex C and have bases in the same plane, they must have the same altitude. It follows from Proposition XII-5 that these pyramids have the same volume.

2. Consider the pyramids $D\text{-}EBC$ and $D\text{-}CFE$. Their respective bases $\triangle EBC$ and $\triangle CFE$ are opposite halves of the parallelogram $BCFE$. Therefore these bases are congruent and lie in the same plane. Since the pyramids share vertex D and have bases in the same plane, they must have the same altitude. It follows from Proposition XII-5 that these pyramids have the same volume.

In other words, the first step tells us that the tetrahedra $ABCD$ and $BCED$ have the same volume, while the second step tells us that the tetrahedra $BCED$ and $CDEF$ have the same volume. Since these three equal-volumed tetrahedra

---

[5]This notation is used in Heath's commentary on Proposition XII-7 [7, p. 395].
[6]Euclid proves this proposition by contradiction, considering what would happen if the volumes of two prisms of the same height were not in the same ratio as the areas of their respective bases.

together form the prism, each must have one-third the volume of the prism and the conclusion of Proposition XII-7 follows.

From here it is a straightforward matter to derive the formula for the volume of a triangle-based pyramid. If we start with the pyramid D-ABC, we can construct a triangular prism $ABCDEF$ with the same base and altitude. By Proposition XII-7, the volume of pyramid D-ABC is one-third the volume of the prism, and the standard volume formula for the pyramid follows by Proposition XII-5.

Moreover, any pyramid with an n-sided polygonal base can be dissected into $n - 2$ pyramids with triangular bases. In fact, this is the idea behind the proof of Euclid's more general sixth proposition of Book XII.

Proposition XII-6. *Pyramids which are of the same height and have polygonal bases are to one another as the bases.*

For example, the pyramid with quadrilateral base E-$ABCD$ can be dissected into the two pyramids E-ABC and E-ADC as shown in Figure 7.7. For a more general polygonal base, triangulate the base and construct pyramids with these triangular bases and a common vertex. Since each triangle-based pyramid has volume equal to one-third that of its corresponding prism, so does the pyramid with polygonal base. From this dissection we conclude that if we can determine the volume of a triangle-based pyramid, then we can extend this idea to determine the volume of any pyramid with a polygonal base. Further, by considering the circle as a limit of regular polygons, we see why the volume of a cone is one-third of the volume of the cylinder with the same base and height.

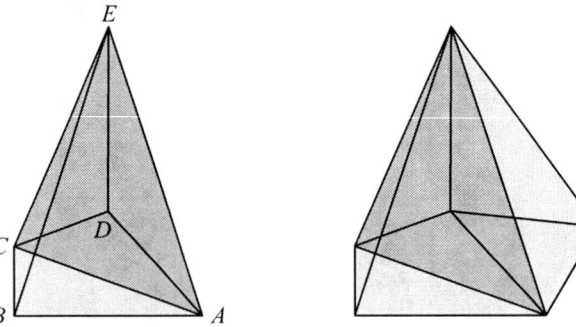

Figure 7.7. Dissecting a pyramid into tetrahedra

7.3 In the Classroom

The core activities described in this section can be carried out in an hour of classroom time. However, shearing is a new concept, and may need some digesting before being applied to three-dimensional shapes.

7.3.1 Shearing in Two Dimensions

There are a number of ways to enhance a lesson on Euclid's proof of the Pythagorean Theorem with classroom demonstrations. For example, here are two options for demonstration and activities on these topics:

1. **Video demonstrations.** The classic educational video *The Theorem of Pythagoras* produced by Tom Apostol for *Project MATHEMATICS!* includes a lovely dynamic illustration of Euclid's proof [1].

2. **Dynamic Geometry Software.** Packages such as *Geometer's Sketchpad* (GSP) or *GeoGebra* may be used to create teacher demonstrations as well as hands-on opportunities for students to shear triangles, rectangles, and other polygons.[7] The appendix includes instructions for creating such a demonstration for the area of a triangle.

7.3.2 Shearing in Three Dimensions

Examining Euclid's proof of Proposition XII-7, we see that with a proper dissection of the pyramid, the proposition follows directly from Proposition XII-5. This proof utilizes an inherently static view of geometry. However, our experience in teaching these concepts has been that adding a dynamic viewpoint often helps the "lights go on" for many

---

[7] A word of warning regarding geometry software: students might confuse the roles of demonstration and proof in geometry. When using geometry software in the classroom, it is a good idea to emphasize that a demonstration doth not a proof make.

of our students. Even better, a dynamic view appeals to our students as a possible answer to the perennial, and always intriguing, question: "Where did he get that idea?" – in this case, in regard to the dissection of the prism. Thus we suggest the following shearing activities to enhance students' understanding and intuition about the material, and possibly about the mathematical process.

Students who have been introduced to and gained an understanding of shearing in two dimensions, often find the concept quite plausible in three dimensions as well. In particular, students who have acquired sufficient practice with two-dimensional shearing and have experienced the shearing of a prism, are often ready to accept the notion that three-dimensional shearing preserves volume in general.

Having arrived at the point where students accept that pyramids with congruent bases and equal heights have equal volumes, we are ready to use these concepts for a hands-on dissection of a prism á la Euclid. Please see Figure 7.11 in the Appendix for nets that can be used to construct a three-dimensional version of Euclid's dissection for a triangular prism, and for suggested questions to facilitate discussion and discovery. Cut and constructed accurately, they will allow students to assemble three pyramids that will combine to create a triangular prism. Each pair of these pyramids has a pair of congruent bases and equal heights, although the congruent bases may not be shared by different pairs.

Using the nets in Figure 7.11, we translate Euclid's dissection to the language of shearing as follows: Euclid's first step was to show that the tetrahedra $ABCD$ and $CDEB$ have the same volume. To view this through the lens of shearing, observe that $\overline{DE}$ lies on a plane parallel to the plane defined by $\triangle ABC$, so the pyramid E-ABC may be obtained from D-ABC by a shear along $\overline{DE}$. On the other hand, $\overline{AD}$ lies parallel to the plane defined by $\triangle BCE$, so the pyramid A-BCE may be obtained from D-BCE by a shear along $\overline{AD}$. However, as illustrated in Figure 7.8, the pyramids E-ABC and A-BCE are the same tetrahedron. Thus, since shearing preserves volume, the pyramids D-ABC and D-BCE have the same volume.

Now is a good time to challenge students to derive a similar comparison for tetrahedra $CDEF$ and $CDEB$. One such comparison is illustrated in Figure 7.9. Of course, other comparisons are possible.

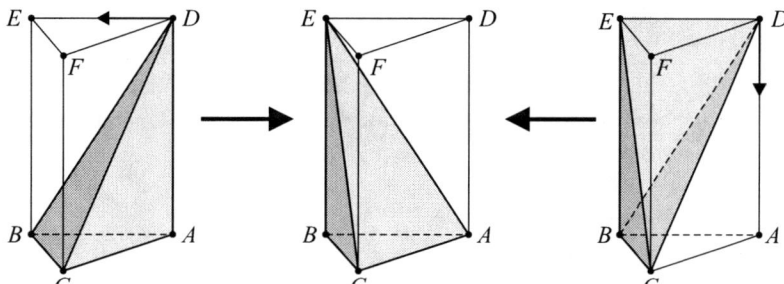

Figure 7.8. Shearing D-ABC and D-BCE to obtain $ABCE$

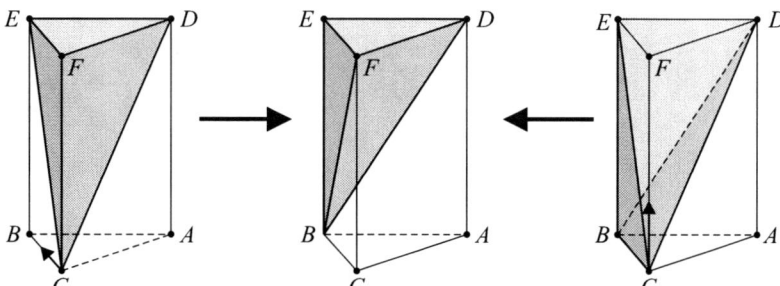

Figure 7.9. Shearing C-DEF and C-BDE to obtain $BDEF$

7.4 Conclusion

Many students have very little formal or informal experience in either a dynamic view of geometry or in three-dimensional geometry. The hands-on shearing demonstrations and dissections described here provide an opportunity to enhance learning in a geometry or algebra classroom. Using them we may broaden students' mathematical understanding to include the dynamic viewpoint as well as some three-dimensional concepts, while helping to add intuition to formulae that may not seem intuitive.

Through shearing we are able to link traditional and time-honored ideas to more contemporary approaches to the subject. We see how to extend two-dimensional ideas into three-dimensions. And we make a connection between two seemingly unrelated and ubiquitous geometry topics: the Pythagorean Theorem and the volume formula for a pyramid.

For more inspiration one may look to the many wonderful resources for reading Euclid's *Elements*. A good place to start might be David Joyce's comprehensive and user-friendly website [8], which includes the *Elements* along with commentary and dynamic diagrams to demonstrate each of the propositions. Another wonderful pedagogical tool is Byrne's beautiful edition of the first six of Euclid's books [2].

Appendix

Algebraic Notation for Shearing

The concept of shearing can be translated to algebraic notation for a two-dimensional shear transformation as follows: Suppose we orient a rectangle in the xy-plane so that its vertices are at $(0, 0)$, $(x_o, 0)$, $(0, y_o)$, and (x_o, y_o). Now transform this rectangle into a parallelogram with the same base vertices, $(0, 0)$ and $(x_o, 0)$, and new top vertices, (a, y_o) and $(x_o + a, y_o)$, by using the following transformation:

$$x_{\text{new}} = x_o + ay_o \quad \text{or} \quad \begin{bmatrix} x_{\text{new}} \\ y_{\text{new}} \end{bmatrix} = \begin{bmatrix} 1 & a \\ 0 & 1 \end{bmatrix} \begin{bmatrix} x_o \\ y_o \end{bmatrix}.$$
$$y_{\text{new}} = y_o$$

Similarly, a three-dimensional shear with a fixed base on the xy-plane would be described by

$$x_{\text{new}} = x_o + az_o$$
$$y_{\text{new}} = y_o + bz_o \quad \text{or} \quad \begin{bmatrix} x_{\text{new}} \\ y_{\text{new}} \\ z_{\text{new}} \end{bmatrix} = \begin{bmatrix} 1 & 0 & a \\ 0 & 1 & b \\ 0 & 0 & 1 \end{bmatrix} \begin{bmatrix} x_o \\ y_o \\ z_o \end{bmatrix}.$$
$$z_{\text{new}} = z_o$$

Note that neither the base nor the height changes under either transformation. Visualizing these transformations dynamically as one shape morphing into the other provides insight for many students as to why area and volume are preserved under shear transformations.

GSP Demonstration of Proposition I-37

1. Create a line segment $\overline{AB}$ in GSP. This will be the (fixed) base of the triangle.
2. Construct a point P not on $\overline{AB}$.
3. Construct a line parallel to $\overline{AB}$ through P. Label this line m. Hide point P. This creates a parallel to the base of the triangle.
4. Construct a point C on m. Construct segments $\overline{AC}$ and $\overline{BC}$. This creates the triangle $\triangle ABC$.
5. We now wish to measure the area of the triangle $\triangle ABC$:
   (a) Highlight the vertices A, B, C.
   (b) Construct the interior of the triangle (and leave it highlighted).
   (c) Measure the interior of the triangle.

The demonstration is now set up. Just slide the point C along the line m, noting that the area of the triangle does not change. An example of this demonstration is shown in Figure 7.10.

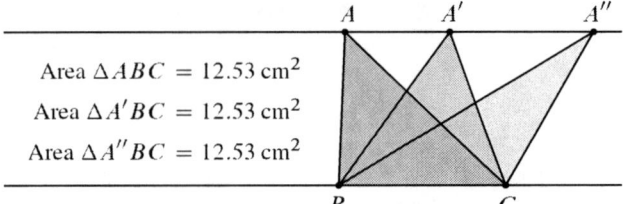

Figure 7.10. Shearing a triangle with geometry software

7.4. Conclusion

Nets for Dissecting a Pyramid

The nets in Figure 7.11 can be used to construct a three-dimensional version of Euclid's dissection for a triangular prism. The nets can be enlarged and printed out on different colored cardstock, then cut, folded and taped into the pyramids D-ABC, C-DEF, and D-BCE.

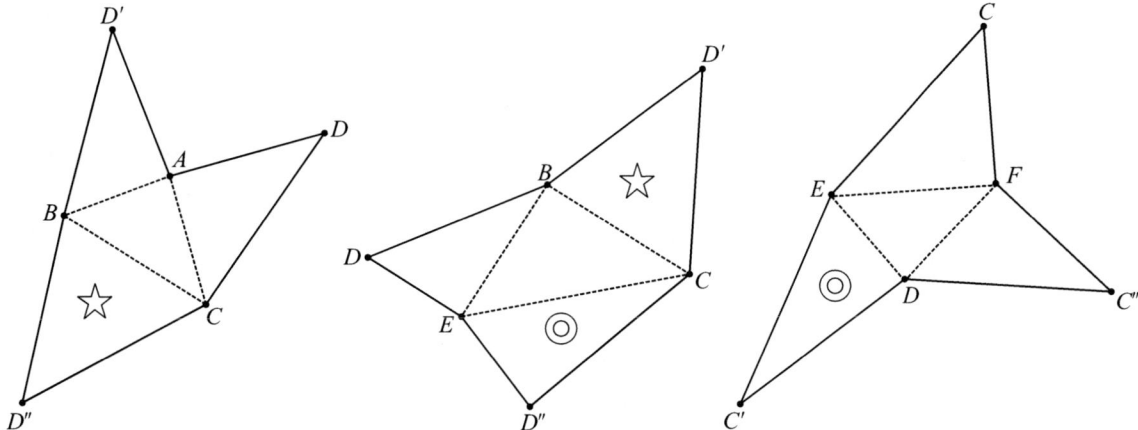

Figure 7.11. Nets for constructing prism $ABCDEF$

If constructed carefully, the three pyramids can be assembled into a right triangular prism. (Pyramid faces with the same mark will align on the interior of the prism.) Once students are convinced that the three pyramids can indeed be used to create a prism, we can explore their volumes. Choose two of the pyramids and identify congruent bases. Set each on its base and compare heights. Have students use the pyramids to follow along with Euclid's proof of Proposition XII-7.

The pyramids can also be used motivate some geometry review. Here is a sample of some of the types of problems that may be presented to students.

1. Using a ruler for measurements, determine the volume of the prism and each pyramid.

2. Given the lengths of the sides of $\triangle ABC$ and the length of $\overline{AD}$, determine the length of each pyramid edge.

3. Are any of the pyramids in the dissection congruent? If not, under what conditions (if any) could we have two congruent pyramids? Three?

4. Which pyramid has the greatest surface area? The least?

5. Construct nets for pyramids that can be assembled to create a prism with an equilateral triangle for its base.

6. Use the prism constructed in problem 5 above to help determine the volume of a regular tetrahedron.

Bibliography

[1] Tom Apostol (prod.), "The Theorem of Pythagoras (video)," *Project MATHEMATICS!*, California Institute of Technology, Pasadena, 1988. <www.projectmathematics.com/pythag.htm>

[2] Oliver Byrne, *The Elements of Euclid in which Coloured Diagrams and Symbols are used instead of Letters for the Greater Ease of Learners*, Chiswick Press, London, 1847. Available online: University of British Columbia `sunsite.ubc.ca/DigitalMathArchive/Euclid/byrne.html`, Copyright 1997.

[3] William P. Berlinghoff and Fernando Q. Gouvêa, *Math Through the Ages: A Gentle History for Teachers and Others*, 2nd edition, Oxton House, Farmington, ME, 2002.

[4] Charles Dodgson, *Curiosa Mathematica, Part I: A New Theory of Parallels*, 3rd Ed., Macmillan and Co., London, 1890.

[5] William Dunham, *Journey Through Genius: The Great Theorems of Mathematics*, Penguin, New York, 1991.

[6] Howard Eves, "Two Surprising Theorems on Cavalieri Congruence," *The College Mathematics Journal*, Vol.22, No.2 (1991) 118–124.

[7] Thomas L. Heath (ed. and trans.), *The Thirteen Books of Euclid's Elements, Volume I*, 2nd edition, Cambridge University Press, 1926.

[8] David Joyce, *Euclid's Elements*, Clark University, `aleph0.clarku.edu/~djoyce/java/elements/elements.html`, Copyright 1998.

[9] Victor J. Katz, *A History of Mathematics*, Addison-Wesley Educational Publishers, Inc., 1998.

[10] Elisha S. Loomis, *The Pythagorean proposition; its proofs analyzed and classified and bibliography of sources for data of the four kinds of proofs*, NCTM, 1968.

[11] Glenn R. Morrow (trans.), *Proclus: A Commentary on the First Book of Euclid's Elements*, Princeton University Press, 1992.

[12] Eric W. Weisstein, "Shear," *MathWorld* (A Wolfram Web Resource), `mathworld.wolfram.com/Shear.html`. Accessed May 7, 2007.

8

The Mathematics of Measuring Time: Astronomical Timekeeping and the Sinking-Bowl Water-Clock in India

Kim Plofker
Union College

8.1 Introduction

In today's world of electronic clocks and universal calendars, it's easy to forget how important mathematics used to be just for the fundamental task of figuring out what time it was. The standard rigorous approach to the problem involved applying trigonometry to observed positions of the sun or the stars, as described below ("In the Classroom"). But several simpler methods were also developed for use when observations were unavailable or calculation was unappealing. One such practical device was the sinking-bowl water-clock, used for many centuries in India. Students (and teachers) will be impressed by how easy such a clock is to construct and adjust, and how much mathematical labor it can save.

This activity and discussion can be used as part of a module on trigonometry. A more advanced class in calculus may be interested in the theoretical modeling of water-clock construction, and especially in comparing the real mathematics of water-clock design with the artificial assumptions made in typical "related rates" problems about filling and draining water tanks. The construction and testing of the sinking-bowl model can take as little as ten or fifteen minutes (depending on the length of its period): exploring the trigonometry of time-telling may involve fifteen or twenty minutes more.

8.2 Historical Background

Any water-clock (or "clepsydra", Greek for "stealing water") works on more or less the same principle as an hourglass: it measures a fixed period or interval of time by means of a substance flowing through a hole in a container, and at the end of that interval it must be reset manually to measure another period of the same length. But instead of containing sand flowing through a tube between two glass bulbs, the sinking-bowl water-clock is an open metal bowl with a small puncture at the bottom, which is set afloat in a larger vessel full of water. The water seeps into the bowl through the hole, and after a certain interval the bowl becomes filled with water and sinks to the bottom of the larger container. The clatter of the bowl on the bottom of the container acts like a timer alarm to announce that the interval is up. At that point, whoever has the job of "clock-watcher" has to fish the bowl out of the water, empty it, and set it on the surface again to measure another interval.[1]

Indian use of water-clocks of the sinking-bowl variety is first definitely attested in texts from the early first millennium of this era.[2] The Indian clepsydras conventionally measured an interval equal to the time-unit called a *ghaṭikā* (from *ghaṭa*, "water-jar"), namely one-sixtieth of a day or 24 minutes. Since the bowl wasn't graduated with markings to measure smaller sub-intervals, the technical requirements for its construction were pretty simple: make an appropriately-sized hole in the bottom of an appropriately-sized bowl so that it takes just 24 minutes to fill with water and sink.

The construction of the sinking bowl would probably always have involved some trial and error to get the size of the hole right. But some Indian treatises attempted to provide more exact specifications about the bowl's size and weight and the size of its hole. Usually, the desired diameter of the hole was stated as equal to the thickness of a pure gold wire of a given length made from a given weight of gold, as in the following description from an eighth-century text:

> The bowl, which resembles half a pot..., which is made of ten *palas* of copper, which is half a cubit... in diameter at the mouth and half... as high, which is evenly circular, and which is bored by a uniformly circular needle, made of three and one-third *māṣas* of gold and of four [finger-breadths] in length, sinks into clear water in one *ghaṭikā*. [15, p. 305]

To the best of my knowledge, such descriptions may represent some of the earliest surviving attempts in mathematical treatises to provide technical specifications for timekeeping mechanisms. Unfortunately, the extant examples vary so widely in the measurements they state that it is hard to believe that they served any real practical purpose. The most famous of medieval Indian mathematicians, the renowned twelfth-century scientist Bhāskara, didn't bother with quantitative specifications at all, remarking merely, "The duration of a day and night divided by the number of immersions [of this bowl] gives the measure of the water clock" [15, pp. 303–307].

The Indian sinking-bowl water-clock is best known in the realm of general history of mathematics not for its design or its historical development, but for a legend that associates it with a famous Indian mathematical textbook.[3] This textbook, composed in Sanskrit by the above-mentioned mathematician-astronomer Bhāskara and named by him the *Līlāvatī* (a feminine adjective meaning "beautiful" or "playful"), is widely thought to have been named for a girl or woman whom Bhāskara addresses at some points in the text. In 1587, a courtier of the Emperor Akbar translated the *Līlāvatī* into Persian, adding a Persian preface that included the following story (which does not appear anywhere in the original text or known commentaries of the *Līlāvatī*):

> Lilawati was the name of the author's (Bhascara's) daughter, concerning whom it appeared, from the qualities of the Ascendant at her birth, that she was destined to pass her life unmarried, and to remain without children. The father ascertained a lucky hour for contracting her in marriage, that she might be firmly connected, and have children. It is said that when that hour approached, he brought his daughter and his intended son near him. He left the hour cup on the vessel of water, and kept in attendance a time-knowing astrologer, in order that when the cup should subside in the water, those two precious jewels should be united. But, as the intended arrangement was not according to destiny, it happened that the girl, from a curiosity natural to children, looked into the cup, to observe the water coming in at the hole; when by chance a pearl separated from her bridal dress, fell into the cup, and, rolling down to the hole, stopped the influx of the water. So the astrologer waited in expectation of the promised hour. When the operation of the cup had thus been delayed beyond all moderate time, the father was in consternation, and examining, he found that a small pearl had

---

[1] The sinking-bowl design is a variant of the so-called "inflow" type of water-clock, as distinct from the "outflow" variety which consists of a large graduated container with a hole at the bottom from which water flows out. In an outflow clepsydra, the passage of time is measured by the amount of water remaining in the container.

[2] The outflow water-clock, on the other hand, had been known in India much earlier, at least from the middle of the first millennium BCE. It may have been adopted there from Babylonian examples via transmission within the Achaemenid Empire [12]; outflow water-clocks had become common in both Mesopotamia and Egypt by the early second millennium BCE [13, p. 148], [10, pp. 38–39]. It is not clear exactly when and where the sinking-bowl inflow clock originated; there are some inconclusive indications that it was used in Mesopotamia in the Neo-Assyrian period (late tenth to late seventh centuries BCE) [3, pp. 119–120], or it may have been an Indian invention. In any case, it became the standard timekeeping device of medieval and early modern India [14, pp. 241–243], [15, p. 302].

[3] I am indebted for much of the material in the remainder of this section to a paper written in 2002 by E. Allyn Smith, then an undergraduate student in a course I taught at Brown University, who co-authored with me the presentation "The Mathematics Textbook and the Disappointed Daughter: History of a Mathematical Urban Legend", given at the AMS/MAA Joint Mathematics Meetings on 17 January 2003 in Baltimore.

stopped the course of the water, and that the long-expected hour was passed. In short, the father, thus disappointed, said to his unfortunate daughter, I will write a book of your name, which shall remain to the latest times... [8, vol. 2, pp. 177–178]

The eagerness with which later historians adopted and modified this story, on the basis of this one dubious account occurring in a translation some four centuries later than the original, should be a warning to all mathematics teachers looking for historical tidbits to brighten up a lesson plan! The following quotes contain versions found in a few general histories of math published in the twentieth century:

> [T]he name in the title is that of Bhaskara's daughter who, according to legend, lost the opportunity to marry because of her father's confidence in his astrological predictions. Bhaskara had calculated that his daughter might propitiously marry only at one particular hour on a given day. On what was to have been her wedding day the eager girl was bending over the water clock, as the hour for the marriage approached, when a pearl from her headdress fell, quite unnoticed, and stopped the outflow of water. Before the mishap was noted, the propitious hour had passed. [2, p. 244]

> According to the tale, the stars foretold dire misfortune if Bhaskara's only daughter Lilavati should marry other than at a certain hour on a certain propitious day. On that day, as the anxious bride was watching the sinking water level of the hour cup, a pearl fell unknowingly from her headdress and, stopping the hole in the cup, arrested the outflow of water, and so the lucky moment passed unnoticed. [5, p. 185]

> Bhaskara was celebrated as an astrologer no less than as a mathematician. He learnt by this art that the event of his daughter Lilavati marrying would be fatal to himself. [1, p. 158]

Note how the nature of the prophecy changes in the different retellings, and also how the clearly described sinking-bowl water-clock of the original Indo-Persian narrative has been transformed into the more familiar outflow clock in the modern versions. (Some forms of the story recently appearing on the Internet carry the transformation even further by making the clock into a sand-filled hourglass!)

In fact, it is quite plausible that this sad tale of a malfunctioning water-clock and a doomed marriage was simply a current urban legend of the time that became attached to the name of the famous Bhāskara. A variant version appearing in an Indian story composed in 1600, nearly contemporary with the Persian translation of the *Līlāvatī*, tells of a Hindu priest setting up a water-clock for a marriage ceremony at which he was to officiate, with his forehead ritually ornamented with a decoration of uncooked rice grains and saffron paste. In this version, it was not one of the bride's pearls but one of the rice grains on the priest's forehead that fell unnoticed and blocked up the hole in the sinking bowl [15, pp. 314–315]. It would be interesting to see how far back this motif extends in medieval or early modern Indian literature, but at present the evidence for accepting it as a reliably reported event in the life of Bhāskara seems flimsy.

8.3 In the Classroom

In India as elsewhere, before the modern period, the gold standard of time measurement was astronomical observation and calculation. The periodic apparent motions of the sun, despite their slight irregularities, were far more constant and reliable than even the most carefully crafted hourglass or water-clock. Ultimately, the ability to test and calibrate any non-astronomical timekeeping device depended on understanding the relationship between the passage of time and the position of the sun, which required some rather elaborate mathematics.

The everyday concept of telling time imagines the sun going around the earth. When the sun appears to be rising in the east, we say it's early morning, and when we see the sun sinking low in the west, we know the end of the day is near. A complete day is considered to be the time interval between two successive occurrences of the same position of the sun with respect to the observer: e.g., the interval from one noon to the next (when the sun is visible on the meridian arc extending between the north and south points of the horizon), or as we now demarcate days, from one midnight to the next (when the sun is on the meridian circle but below the horizon).[4]

---

[4] This explanation ignores the difference between so-called "true solar" time measured by observed positions of the sun, in which some days are slightly longer than others due to small periodic variations in the earth's motion, and "mean solar" time in which the day has a constant length equal to the average value of the true solar day. All our modern standard time units are based on mean solar time.

The constantly cycling solar positions that we see, of course, are produced by the daily rotation of the earth. As our planet spins, what we see as the sphere of the heavens all around us appears to spin about an axis passing through the north and south celestial poles. Both the sphere and the axis are obliquely bisected by the circular plane of our horizon, shown in Figure 8.1 as circle $NESW$ passing through the four cardinal direction points, and centered on the observer's location O. If we imagine a great circle on this sphere ninety degrees away from the poles—the so-called "celestial equator"—we can think of it as marked with a scale of 360 degrees, called "time-degrees", whose continual rising and setting measures the flow of time. (The visible half of the celestial equator is shown as semicircle $\overparen{EMW}$ in the figure; the north celestial pole is the point P on the celestial sphere and the zenith directly overhead is point Z.) Thus if a 360° rotation of the celestial equator constitutes one day, the rising or setting of one degree along the equator takes 1/360 of a day, or four minutes.[5]

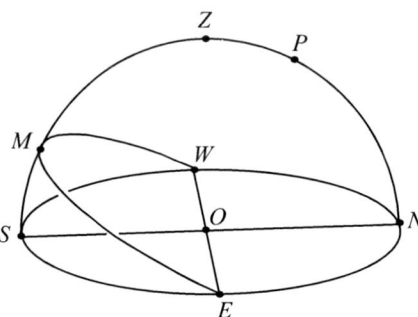

Figure 8.1. The celestial hemisphere above the observer's horizon for a location north of the equator.

The simplest form of the problem of telling time occurs on the day of an equinox, when the apparent daily path of the sun coincides (more or less) with the celestial equator. On that day, the sun will appear on the horizon at the east point E, and approximately follow the half-circle of the equator $\overparen{EMW}$ till it sets in the west at W. In this case, knowing the length of the time interval between two given moments during the day requires finding the extent of the equatorial arc that the sun has traversed in the sky in the meantime. The size of the arc tells us directly how many time-degrees (each equivalent to four minutes of time) have passed in that interval. But figuring out even this straightforward task requires the timekeeper (who was usually a practicing astronomer) to be competent in basic geometry and trigonometry, as the following demonstration shows.

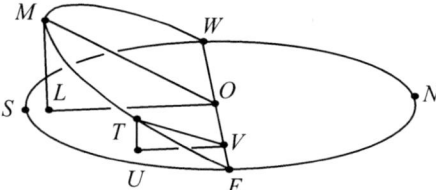

Figure 8.2. Similar triangles relating solar altitude to time.

Consider the example pictured in Figure 8.2, where the altitude of the sun above the horizon is observed from O sometime in the morning at point T and later at noon at point M. What was the time of the initial observation at T? In other words, how much time had elapsed between sunrise and the moment of the first observation? To find out, we have to determine the size of the time-arc $\overparen{ET}$ between the sun's first appearance on the horizon at E and its arrival at T.

We find the answer by first computing the sines of the observed altitudes of the sun at the positions T and M. We can then imagine dropping perpendicular lines TU and ML from the equator to the plane of the horizon at those

---

[5] We are making another simplification here by ignoring the difference between solar time—measured by the apparent motion of the sun—and sidereal time, which is measured by the apparent motion of the stars which seem to be fixed on the celestial sphere. Actually, the sun appears not only to spin westwards every day along with the heavenly sphere but also to shift its position slightly eastwards against the background of fixed stars (at the rate of about one degree per day). So during a full revolution of the sun through 360°, the stars have revolved only about 359°, meaning that a solar day is about four minutes shorter than a sidereal day. For the purposes of this discussion, though, we will consider the sun to be fixed with respect to the stars, so that one revolution of the sun and one revolution of the celestial equator mean the same thing.

8.3. In the Classroom

positions. Those perpendicular lines will be proportional to the sines of the corresponding altitudes:

$$\frac{\sin(\text{altitude at } T)}{TU} = \frac{\sin(\text{altitude at } M)}{ML} \quad \text{or} \quad \frac{\sin(\text{altitude at } T)}{\sin(\text{altitude at } M)} = \frac{TU}{ML} \tag{8.1}$$

Now let us picture TU and ML as the vertical legs of two similar right triangles TVU and MOL, whose bases in the plane of the horizon are perpendicular to the east-west line $EVOW$. (The right triangles are similar because the acute angle in each of them is the same: namely, the angle between the plane of the horizon and the plane of the celestial equator.)

The triangle hypotenuses TV and MO lie in the plane of the equator. They are proportional, respectively, to the sine of the desired time-arc $\widehat{ET}$ and the sine of the quarter-circle $\widehat{EM}$ or 90°, allowing us to express $\widehat{ET}$ solely in terms of TV and MO, as follows:

$$\frac{\sin(\widehat{ET})}{TV} = \frac{\sin(\widehat{EM})}{MO} = \frac{\sin(90°)}{MO} = \frac{1}{MO}, \quad \text{so} \quad \sin(\widehat{ET}) = \frac{TV}{MO} \tag{8.2}$$

We don't know the lengths of the hypotenuses TV and MO. But we can use the similarity of the two right triangles to express the ratio TV/MO in terms of the ratio of the corresponding vertical legs TU and ML, which we found previously:

$$\sin(\widehat{ET}) = \frac{TV}{MO} = \frac{TU}{ML} = \frac{\sin(\text{altitude at } T)}{\sin(\text{altitude at } M)} \tag{8.3}$$

In other words, the timekeeper wishing to know the time when the sun was at T just needs to find the ratio of the sines of the arcs from the two altitude observations, take the arcsine (in degrees) of the result, and convert it to minutes of time by multiplying by 4, and the job is done.

A closer look at Figure 8.1 will reveal that in fact, the equinoctial noon altitude $\widehat{MS}$ is just the complement of the observer's terrestrial latitude $\widehat{PN}$ (because $\widehat{PM} = 90°$ and $\widehat{ZS} = 90°$, so $\widehat{PZ}$, the complement of the latitude, equals $\widehat{PM} - \widehat{ZM} = \widehat{ZS} - \widehat{ZM} = \widehat{MS}$). So a timekeeper who knows the value of the local latitude could just plug in the latitude complement in place of the solar altitude at M in equation 8.3, without bothering to wait for a noon observation. For example, imagine that the terrestrial latitude is 40° (approximately correct for, say, Philadelphia), and the sun is observed on an equinox morning at 30° altitude above the horizon. Then the time of the observation is easily found as

$$\widehat{ET} = \arcsin\left(\frac{\sin(30°)}{\sin(50°)}\right) \approx 40.75° \approx 163 \quad \text{minutes}, \tag{8.4}$$

or 2 hours and 43 minutes after sunrise.

On a non-equinoctial day, though, the task is harder: the sun's apparent westward path in the sky is not the celestial equator itself but some small circle parallel to the equator, so some more laborious spherical geometry and trigonometry are needed. And at night, of course, the sun is not available for observational purposes at all, so the astronomer would have to use measurements of a star with known celestial coordinates instead. (See a text on the history of mathematical astronomy such as [4] for detailed explanations of how time is computed in these more complicated situations.)

Ancient astronomers simplified their timekeeping calculations by designing instruments based on geometric projections of the celestial sphere, which allowed time measurements to be read directly off a graduated scale. These instruments are what we now call sundials, which have a shadow-casting gnomon attached to a calibrated plate. (Timekeeping instruments constructed for nighttime use, which also depend on the mathematics of spherical astronomy, are known as nocturnals or moondials; the ancient instrument called the astrolabe is a more elaborate general-purpose mechanism, usable during both night and day, that works on the same trigonometric principles.) But what is the poor astronomer supposed to do when it's cloudy or raining? Because the celestial bodies aren't always visible, it's necessary to have some kind of timekeeping device that doesn't depend on astronomical observation. Hence, the persistent importance of the water-clock.

As noted in the previous section, the sinking-bowl water-clock measuring one fixed-length time interval is a fairly simple device, and can be constructed and calibrated quite easily by trial and error. Take a cylindrical metal can which

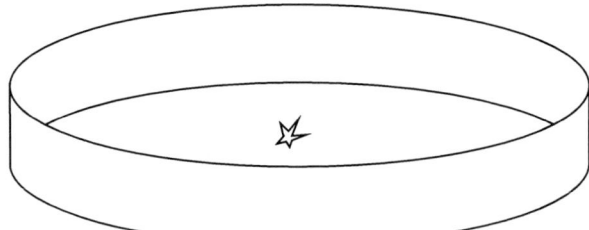

Figure 8.3. A can with a hole punched in the bottom makes an adequate sinking-bowl water-clock.

will float on its bottom without tipping over (such as a tuna-fish or cat-food can, or a sawed-off juice or soup can), and punch a hole in the bottom of it as illustrated in Figure 8.3. Set it on a vessel of water and compare the time it takes to sink with the desired time interval you want your clepsydra to measure. If the can fills up too slowly, enlarge the hole in the bottom; if it fills too quickly, partially cover the hole with waterproof adhesive tape.

Alert students may wonder how the instrument could have been calibrated if there were no accurate clocks to measure its sinking rate against. This is where the astronomer comes in again, using a sundial gnomon to mark the passage of the desired time interval. As a seventh-century Indian text explains,

> The more accurate method is to measure the *ghaṭikā* by marking the shadow of one *ghaṭikā*, cast by a gnomon of specified shape that has been set up on level ground. The perforation in [the bowl of] the water-clock should be made skilfully according to the period measured by the shadow. [15, p. 309]

So ultimately, the mechanism of the water-clock still depends on the astronomical determination of time-units, although once the clock is properly calibrated it can operate independently of the celestial cycles.

Students might be asked to compare the work involved in the time calculation shown in equation 8.4 with the effort required to measure the same interval with their sinking-bowl water-clock. How many times would they have to re-set the bowl between sunrise and the desired time 2 hours and 21 minutes later? Considering the delay caused by fishing the bowl out of the water to re-set it each time, would the water-clock be likely to give an accurate measurement for the total interval? (Note that a sinking bowl with a longer period would be more accurate and less fussy to use, but would take more time to calibrate to a desired time-interval.)

8.4 Taking It Further

Filling or emptying a reservoir of water is a common "real-world" application of calculus methods. Many related-rates problems in differential calculus blithely assume that water is flowing into or out of a tank at some constant rate. A few chapters on, problems in integral calculus demand the computation of the amount of work required to pump a given volume of water out of a container with a given shape. But the operation of a water-clock depends on the mathematics of how water *really* behaves when flowing through a container.

An ideal water-clock would be what is called a "linear clepsydra", measuring time at a constant rate via gradations showing linear differences in the height of the water within it. Unfortunately, getting water to drain from a reservoir at a constant rate is by no means so simple as the related-rates problems pretend. Real-world water flows through an opening at a rate determined partly by the depth of the water above the opening, so an outflow water-clock drains more slowly as its water level decreases.

Ancient timekeepers didn't try to model this phenomenon theoretically, focusing instead on empirical calibration of individual instruments. But as new mathematical methods developed in the early modern period, scientists began to apply them to a variety of physical problems, including some in the area we now call hydrodynamics, or the behavior of moving liquids. Following Galileo's pioneering work on the speeds of falling bodies under gravity, Evangelista Torricelli in the mid-seventeenth century came up with a relation between the velocity v of water flowing through a pierced container and the height h of the water level above the hole that became the modern "Torricelli's theorem" [10, pp. 42–43]:

$$v^2 = 2gh \tag{8.5}$$

where g is, as usual, the acceleration due to gravity.

This formula neglects issues of turbulence, viscosity, compression, temperature, and so on which make a complete analytical solution very problematic. But for the purposes of this discussion we can take it to be more or less the case that the water's velocity v is proportional to the square root of h. So for any two given water-levels h_1 and h_2, the relationship between the corresponding velocities v_1 and v_2 will be given by

$$\frac{v_1}{v_2} = \frac{\sqrt{h_1}}{\sqrt{h_2}} = c_1, \tag{8.6}$$

where c_1 is some constant. Then if we want the velocity and consequently the rate of flow—i.e., the change in volume with respect to time, $\frac{dV}{dt}$—to be linear, the above relation gives

$$\sqrt{h} = c_1 v = c_1 c_2 \frac{dV}{dt} = c_1 c_2 \frac{dV}{dh} \cdot \frac{dh}{dt}, \tag{8.7}$$

where c_2 is some other constant relating the linear velocity and flow rate. But if the clepsydra is linear, as we specified for our ideal clock, then the change $\frac{dh}{dt}$ of the water level with respect to time is also a constant. And if its shape is some solid of revolution, then its rate of change of volume $\frac{dV}{dh}$ is a circular slice of area πr^2 where r is the radius of the container at the height h. Consequently, the square root of the height is linearly proportional to the square of the corresponding value of the container radius, and the height itself likewise is proportional to the fourth power of the radius:

$$\sqrt{h} = c_1 c_2 \frac{dh}{dt} \pi r^2 \quad \text{or} \quad h = \left(c_1 c_2 \frac{dh}{dt} \pi\right)^2 r^4. \tag{8.8}$$

So a linear outflow tank should really have the shape of a "higher-order paraboloid", specifically a quartic curve [10, pp. 39–47], rather than the cones and cylinders that show up in calculus textbooks.

An alternative is to keep the rate of outflow constant by keeping the water level in the pierced container constant (e.g., if it is constantly refilled by a stream or fountain) and using this constant flow as the linear *inflow* of a second container, with no hole. Then the water level in the second container will rise linearly with time. A cylindrical sinking bowl in a larger vessel with a constant water level, like the pierced can described in the previous section, will also fill with water at a constant rate [10, p. 40].

8.5 Conclusion

Measuring time astronomically takes a lot of complicated trigonometric calculation. Measuring time mechanically takes a lot of careful calibration and testing. In either case, the importance of mathematics in timekeeping—and the extent to which the basic mathematics required is accessible to students of calculus and precalculus—may surprise and intrigue today's students, for whom telling time usually means nothing more than reading numbers off a clock face.

Constructing a simple sinking-bowl water-clock and comparing it with trigonometrically accurate methods of measuring time will bring home to them how vital a knowledge of mathematics is in truly understanding this formerly routine experience. More advanced students dipping a cautious toe into the calculus of fluid dynamics will gain respect for the complexity of the mathematics operating at a slightly deeper level in the functioning of water-clocks. And the often mysterious narrative of the history of this simple instrument will teach students (and their instructors) respect for the passage of time in a different sense, as they realize how much we still don't know about the history of mathematics, where many of the "facts" in our standard textbooks are merely oft-repeated ignorance.

Bibliography

[1] W. W. Rouse Ball, *History of Mathematics*, Macmillan and Co., London, 1901.

[2] Carl Boyer, *A History of Mathematics*, Wiley & Co., New York, 1968.

[3] David Brown, "The Cuneiform Conception of Celestial Space and Time", *Cambridge Archaeological Journal* 10, 2000, 103–122.

[4] James Evans, *The History and Practice of Ancient Astronomy*, Oxford University Press, New York, 1998.

[5] Howard Eves, *Introduction to the History of Mathematics*, Holt, Rinehart and Winston, New York, 1969.

[6] J. Fermor, and J. M. Steele, "The Design of Babylonian Waterclocks: Astronomical and Experimental Evidence", *Centaurus* 42, 2000, 210–222.

[7] Jens Høyrup, "A Note on Water-Clocks and on the Authority of Texts", *Archiv für Orientforschung* 44–45, 1997–98, 192–194.

[8] Charles Hutton, *Tracts on Mathematical and Philosophical Subjects*, 3 vols., Rivington, London, 1812.

[9] C. Michel-Nozières, "Second Millennium Babylonian Water Clocks: a Physical Study", *Centaurus* 42, 2000, 180–209.

[10] A. A. Mills, "Newton's Water Clocks and the Fluid Mechanics of Clepsydrae", *Notes and Records of the Royal Society of London* 37, 1982, 35–61.

[11] Otto Neugebauer, "Ancient Astronomy, VIII. The Water Clock in Babylonian Astronomy", *Isis* 37, 1947, 37–43.

[12] David Pingree, "The Mesopotamian Origin of Early Indian Mathematical Astronomy", *Journal for the History of Astronomy* 4, 1973, 1–12.

[13] Colin Ronan, and Joseph Needham, *The Shorter Science and Civilisation in China: An Abridgement of Joseph Needham's Original Text*, Cambridge University Press, Cambridge, 1994.

[14] Sreeramula Rajeswara Sarma, "Astronomical Instruments in Mughal Miniatures", *Studien zur Indologie und Iranistik* 16, 1992, 235–276.

[15] ——, "Setting up the Water Clock for Telling the Time of Marriage", in C. Burnett et al. (eds.), *Studies in the History of the Exact Sciences in Honour of David Pingree*, Brill, Leiden, 2004, pp. 302–330.

9

Clear Sailing with Trigonometry: Navigating the Seas in 14th-Century Venice

Glen Van Brummelen
Quest University

9.1 Introduction

Does anyone care about trigonometry? Certainly many of our students don't, aside from the exigency of getting through their exams. As mathematics teachers, we have passion for our subject for its own sake — but we often justify ourselves to our students in terms of what the mathematics can accomplish elsewhere. For trigonometry as for many other topics, this takes the form of the widespread "word problems": how high is that pine tree across the street? How far did that motorboat travel when it went across the lake? And here we reach a crucial pedagogical problem: few of us really care precisely how tall the tree is, or how far the boat went. We find ourselves forced into producing "baby" problems like these with little real relevance, assuring our students (with fingers crossed behind our backs) that the genuine applications — too complex for their immature minds — hopefully work kind of like these ones do.

Meaningful contexts are surprisingly hard to find. Some pedagogical efforts are searching for realistic classroom-friendly projects, and are having some success. However, one source that might easily be overlooked is the history of the subject. Two thousand or more years of human experience is a powerful resource on which to draw. Mathematical subjects arise for good reasons, and bringing these reasons to light can motivate more honestly what otherwise might appear dull, even deceptive in its fake "applications". Trigonometry, for instance, came into being over 2100 years ago not to determine the heights of trees or distances across lakes, but rather to quantify the motions of the Sun, planets, and other heavenly bodies. This was immediately useful for a number of reasons, among them the need for farmers to have reliable calendars to schedule their crops, and the crucial business of predicting the future through astrology so that rulers knew when to go to war.

In fact, trigonometry was a branch of astronomy for over 1,000 years. It found some use in mathematical geography especially in the medieval period, but it was not until the Renaissance and later that there arose good new reasons to fill our children's education with triangles. As the Age of Exploration took hold, the need to find one's way across a featureless ocean became critical — not just for discovery, but also for reasons of trade. The importing/exporting community was growing dramatically, and safe passage meant safe profits. But oddly enough, the first known use of trigonometry in navigation (or, indeed, anywhere outside of astronomy and geography) happened well before the Age of Exploration, in an isolated circumstance centuries earlier. This episode provides a meaningful context for trigonometry that could both deepen and enliven the experience of students in trigonometry or precalculus courses, at the high school or early undergraduate level.

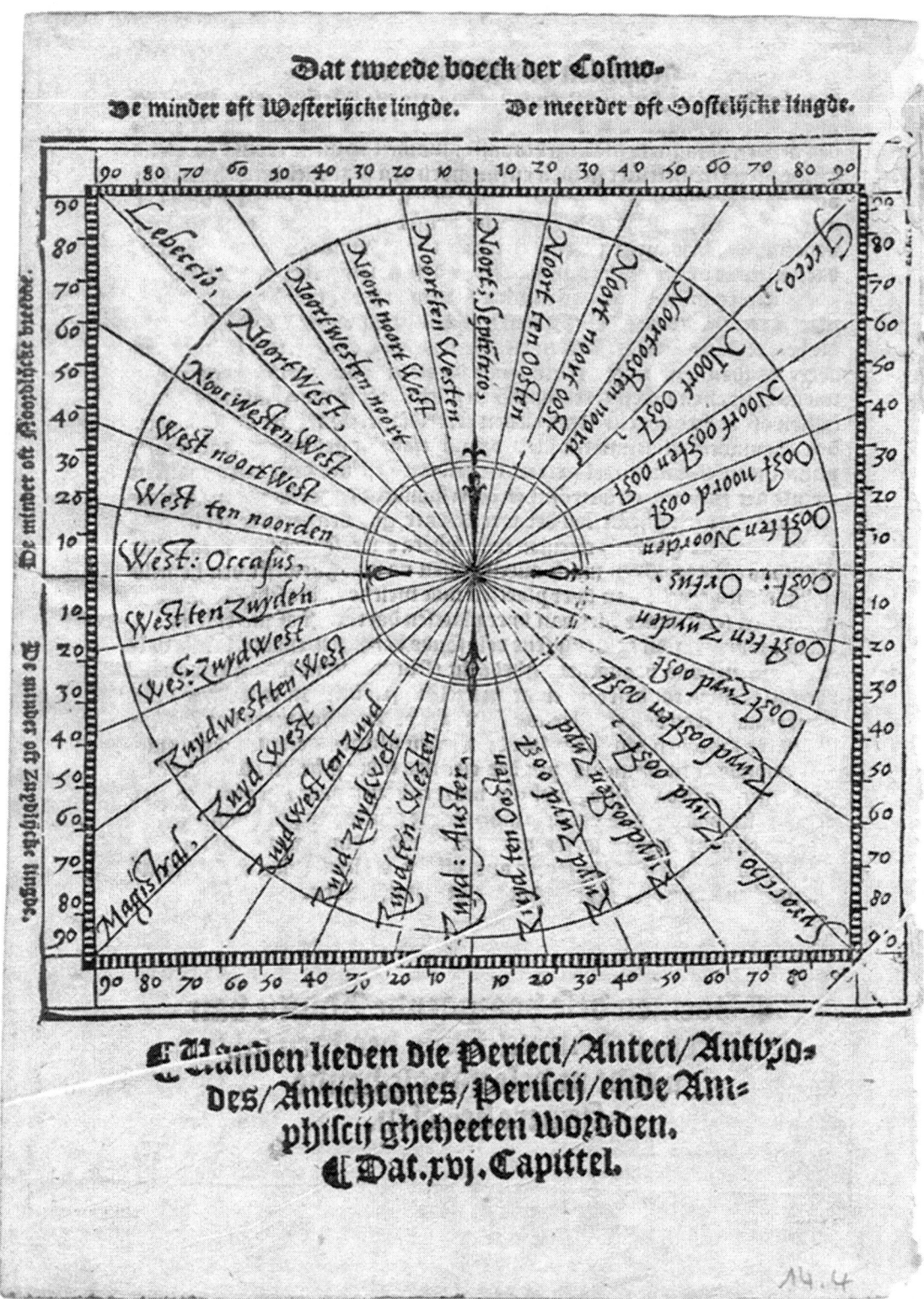

Figure 9.1. The windrose, depicted in Peter Apian's *Cosmographia* (1574). (Image reproduced with permission of Leen Helmink Antique Maps & Atlases, www.helmink.com)

9.2 Historical Background

The Republic of Venice, at the northern end of the Adriatic Sea in northeast Italy, was a potent political force through most of the medieval period. Much of its power derived from its position at the center of a large network of seafaring mercantile trade. With over 3,300 ships at its peak, the Venetian enterprise plied shipping routes that extended across not just the Mediterranean, but the Black Sea as well. Clearly any technological edge in nautical affairs would be of great commercial value. These included the marine chart, which was revolutionary in its accurate portrayal of coastlines; and the portolan, a written description of distances, directions, and local landmarks [10, pp. 5–11]. Venetian navigators kept track of their knowledge of the technical side of seafaring in personal notebooks, several of which survive today. In some of these notebooks we find an even more remarkable navigational method, called the **marteloio**,[1] which turns out to be essentially a primer of basic trigonometry and an accompanying set of trigonometric tables. This gives teachers a unique pedagogical opportunity: the marteloio provides a fascinating historical motivation to study trigonometry, one that is completely genuine, and yet is accessible to beginning students.

Where the *marteloio* came from is anybody's guess.[2] Seafarers had the run of the coasts of the civilized world, which admits several possibilities. Scholars in Spain, the Middle East, and even in Italy itself were studying trigonometry for astronomical purposes. It has been speculated that the *marteloio* was a product of Leonardo of Pisa's (Fibonacci's) school ([7, LI–LII]), although there is no evidence to support this claim. It doesn't seem likely that scholars and seamen had much to say to each other generally. But *someone* must have put significant work into designing the *marteloio*, since as we shall see its form does not match any known trigonometry. The notebooks themselves don't help much; they were copied from each other without concern for academic credit, so their provenance is hard to determine.

9.3 Navigating with Trigonometry

The *marteloio*'s difference from other trigonometric practices starts with the measurement of direction; there is no hint of the use of degrees. Points of the compass were measured using the **windrose** (often called a "compass rose"), which divides every right angle into eight parts, called "quarters" (Figure 9.1), corresponding to $11\frac{1}{4}°$ each. Early windroses from Roman times had an apparently more natural division of the right angle into three parts of 30°, but it is actually easier to work out trigonometric values by dividing 90° into a power of 2. So, in medieval times the windrose had either 4 or 8 divisions per right angle.

The directions and the distances that boats traveled along them needed to be recorded carefully. A boat could not sail directly into the wind, but rather had to start its approach to a windward destination by traveling in a direction almost 90° removed from its desired direction (the *alargar*; see Figure 9.2), usually advancing only slightly toward its destination (the *avanzar*). The ship would then need to change its direction, again and again, eventually crawling in a zig-zag fashion almost crab-like toward its goal.

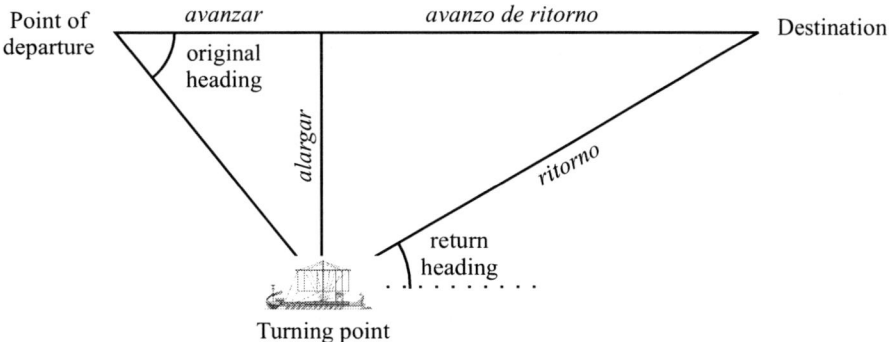

Figure 9.2. Setup for the *marteloio*

---

[1] The origin of the word "marteloio" is uncertain. It might mean "hammering", an idiom referring to the change of watch on a ship ([3, 40, 48–50]).

[2] It has been suggested that the *marteloio* originated in the 13th century, based on a passage in the Spanish writer Ramó Lull's *Arbre de Sciencia* (see [8, 271]; for more information on the passage see [9, 117–119], [2, vol. 1, 206–207], and [1, vol. 1, 441–442]). But Kelley disagrees, arguing that Venice was the "center of use" of the *marteloio* ([3, 144–146]).

Quarter	*Alargar* (distance off course)	*Avanzar* (advance)	*Ritorno* (return)	*Avanzar de ritorno* (advance on return)
1	20	98	51	50
2	38	92	26	24
3	55	83	18	15
4	71	71	14	10
5	83	55	12	$6\frac{1}{2}$
6	92	38	11	4
7	98	20	$10\frac{1}{5}$	$2\frac{1}{5}$
8	100	0	10	0
	For every 100 miles		For every 10 miles *alargar*	

Figure 9.3 A typical *toleta de marteloio*.
brunelleschi.imss.fi.it/michaelofrhodes/navigate_toolkit_basics.html

To keep track of the ship's position and direction, a small set of tables known as the "*toleta de marteloio*" was used.[3] These tables, which varied little from one to the other (see Figure 9.3),[4] gave four quantities to help the navigator know where he was and where he should go. For instance, suppose a ship travels 100 miles from its point of departure in a heading 6 points south of east (Figure 9.2). The *alargar* entry tells us that the ship is 92 miles off course, and the *avanzar* shows that we have progressed 38 miles toward our goal. In modern terms, the *alargar* and *avanzar* are simply 100 times the sine and cosine of the heading.

Suppose now that the ship is able to turn and head directly for its destination, either because the wind's direction has become favorable or because some obstacle has been cleared. Given the direction of the destination, the *ritorno* tells us how far the ship must travel to reach its destination for every 10 miles of *alargar*. Similarly, the *avanzo de ritorno* gives the advance corresponding to the *ritorno* leg, along the originally intended path. So, the *ritorno* and *avanzo de ritorno* are just 10 times the cosecant and cotangent functions.

Now, while the *avanzar* and *alargar* are eminently practical for a navigator to know, at first glance the *ritorno* and *avanzo de ritorno* seem less useful. Wouldn't it be more important to find the return heading than how far you need to travel? After all, if you know which direction to face, if you head that way you are bound to get there eventually. And in fact, the techniques of the *marteloio* are more sophisticated; they contain various combinations of uses of the *toleta* to solve real problems. We shall work through an example from the notebook of Michael of Rhodes, one of the best known of the surviving Venetian manuscripts.[5]

Suppose that we wish to travel from A to B, a distance of 100 miles eastward (Figure 9.4), but we are forced initially to travel in a direction $\alpha = 2$ quarters south of east. We decide that we will turn the ship when the direction to the destination is $\beta = 7$ quarters north of east. How far should we permit ourselves to travel along the original course (AC) before turning?

To answer the question, Michael instructs us to apply the following pattern of calculation:

$$AC = \frac{alargar(\beta)}{10} \cdot ritorno(\alpha + \beta). \tag{9.1}$$

He does not tell us precisely what the terms in (9.1) mean, but it is clear that the *alargar* and *ritorno* are being used more abstractly than Figure 9.2 might have indicated. For instance, our *alargar* is defined here with respect to β rather than the original heading α, so it is AD. In our case the calculation comes out to

$$AC = \frac{alargar(7)}{10} \cdot ritorno(2+7) = \frac{98}{100} \cdot 10\frac{1}{5} = 99\frac{48}{50} \text{ miles.} \tag{9.2}$$

---

[3] The actual table used by Michael of Rhodes may be viewed in the original manuscript; it is on page 48b at the site brunelleschi.imss.fi.it/michaelofrhodes/manuscript.html.

[4] See [6, p. 403], which reproduces five marteloio tables differing in only a few entries.

[5] Michael of Rhodes's notebook has been published recently in [5]. See also the associated web site, brunelleschi.imss.fi.it/michaelofrhodes/.

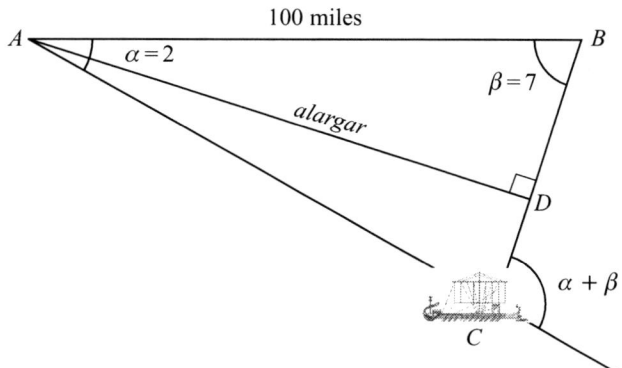

Figure 9.4. A typical *marteloio* problem

If we convert the *alargar* and the *ritorno* back to their modern equivalents, Michael's formula becomes

$$AC = 100 \sin \beta \cdot \csc(\alpha + \beta) = \frac{100 \sin \beta}{\sin(\alpha + \beta)}. \tag{9.3}$$

Now, in $\triangle ABC$, the sine of the external angle at C (labeled $\alpha + \beta$ in Figure 9.4) is the same as the sine of the internal angle. So, Michael's formula for finding AC is simply an application of the Law of Sines to angles B and C in $\triangle ABC$.

Or is it? We reach here a delicate point in interpreting historical mathematical texts. Just because a text gives a calculation mathematically equivalent to a particular result doesn't mean that the author is aware of it. In cases similar to our example, the *marteloio* texts take two steps to arrive at their result. Here, Michael has us find $alargar(\beta)$ first, before applying the rest of (9.1). Why would he do this? The reason may be as follows: drop a perpendicular AD onto BC. Then, since the hypotenuse AD of $\triangle ABD$ is equal to 100 miles, AB is equal to the *alargar* of β. Next, look at $\triangle ACD$. If we assume that AD is 10 miles, then $AC = ritorno(\alpha + \beta)$. But of course AD is not 10 miles, so we need to do a bit of rescaling:

$$\frac{alargar}{AC} = \frac{10}{ritorno}. \tag{9.4}$$

This unit conversion, known as the Rule of Three,[6] was popular both in mathematics and in trade. If we solve (9.4) for AC, we arrive immediately at (9.1).

The unknown inventor of the *marteloio*, faced with a situation where the Law of Sines might be used, always goes through the process of breaking the triangle into two pieces and solving them separately. If he had codified the result and applied it in one fell swoop, we might have reason to credit him with the Law of Sines. All we can say here is that he knew what to do in these circumstances using elementary trigonometry, but *not* explicitly the Law of Sines.[7]

The question remains whether seamen of Michael's time were up to the mathematics required to implement the *marteloio*. The principles of decimal arithmetic had been introduced to Europe in the early 13th century by Leonardo of Pisa (Fibonacci) in the *Liber abaci* [4], so the calculations might have been accessible. A related manuscript by Pietro de Versi (now known to be written by Michael of Rhodes) says: "This is the *raxion* called *del marteloio* for navigating mentally... This is done by an artful method for the man who may have the intellect to learn and who might take pleasure in learning the theory." This might imply that the *marteloio* was more for mathematical pleasure than for use at sea ([3, 150]). But the existence of a number of *marteloio* tables in the manuscripts seems to give reason to award the Venetian navigators a little more credit than that.

9.4 In the Classroom

Instructors in elementary trigonometry might find in the *marteloio* a fascinating narrative to follow when developing both the sine/cosine and the more advanced functions — with the crucial advantage that this technique probably was

---

[6] The Rule of Three is essentially the statement that if $a/b = c/x$ (where only x is unknown), then $x = bc/a$.

[7] I am grateful for conversations with Joel Silverberg that led me to this view. Kelley ([3, 43–45]) assumes that the Law of Sines is present, and that the use of the product of *alargar* by *ritorno* (rather than the division of one *alargar* by another) is implemented in order to avoid division. Obviously, we disagree.

really used by navigators. In addition, this episode provides a surprisingly simple lead-in to the Law of Sines and its proof, with the added interest of a meaningful question of historical interpretation. Some possible specific uses:

- The left side of Figure 9.2 provides an immediate motivation, perhaps even a method of definition, for the sine and cosine functions as the *alargar* and *avanzar*. The sine and cosine are expressed as lengths rather than the more common ratios, but lengths are in any case more intuitive at first exposure. The instructor may choose to set the ship's initial journey to one mile in length rather than 100 so that the *alargar* and *avanzar* coincide with the modern functions; alternately students might be asked to work out the conversion. This leads directly to the Rule of Three, which in any case is important even for modern students to learn.

- It is easy to come up with new *marteloio* problems to help students practice their skills. With a bit of creativity, the *marteloio* tables can deal with almost every possible situation where solving a triangle is required.

- The more advanced problems involving the *ritorno* and *avanzo de ritorno* are obviously important to navigation, and show immediately why the usually obscure cosecant and cotangent are worth thinking about.

- Michael of Rhodes's solution of the problem given above may be seen as an implicit demonstration of the Law of Sines. Students might fruitfully debate (either in class or in groups) the point we raised earlier: whether or not the *marteloio* practitioners may be credited as being among the first people to have proved and used this important theorem.

Included in an appendix at the end of this article is a guided classroom activity that takes students through Michael's solution of the *marteloio* problem of Figure 9.4. It assumes that the teacher has introduced the basic definitions of Figure 9.2 and the table of Figure 9.3. The second page leads students to compare with the Law of Sines solution of the same problem, and asks them to reach their own conclusions. Instructors may take or leave this last step of the exercise as they choose. Here are typical responses from my test run with a class of math-phobes:

- "Although he pretty much did the same thing in terms of procedure, as the sine law would have him do, the very fact that he approached the question with a certain mind set and would not have been able to even comprehend the Sine law shows that he was using his own theorem."

- "A theorem is a mathematical truth that under the given conditions will always be true. To know a theorem does not mean that you need to quote its name; it simply means that you need to imply the ideas that it incorporates in order to solve a problem. Using this logic, we can determine that Michael of Rhodes did in fact use the Law of Sines. Michael may not have had the knowledge of the Law of Sines, but still implied its ideas and a similar process. His method for solving the triangle came to the same conclusions that the Law of Sines did."

- "Our answer to this question is predicated upon the definition of the term theorem. One viewpoint holds that as long as two theorems always agree, with 100% accuracy, on the solution to the same problem, than the two theories are held to be the same and the intervening steps are largely irrelevant. The other argument takes the opposite stance, emphasizing the importance of the qualitative nature of the intervening steps and the preceding knowledge upon which the operation is based. Following the first line of reasoning, Michael of Rhodes would be credited as the discoverer of the Law of Sines, while according to the second argument his method would not be considered a full explication of the Law of Sines."

Instructors who use the *marteloio* in the classroom may choose to measure angles in degrees, or with the windrose. The windrose has the advantage of authenticity, but it introduces the complication of another angular measure.

9.5 Conclusion

Our students will not be navigating ocean trade routes any time soon using the *marteloio*. But they can at least reproduce a technique that was actually used to traverse the seas in the 14th century. This unit allows instructors to replace the simplified, obviously artificial settings in which trigonometry is usually cast with an application demonstrating immediate practicality, albeit 600 years out of date. It illustrates that certain historical episodes can bring mathematical topics to the classroom in a lively way, maintaining an authentic feel while remaining accessible to our students.

Bibliography

[1] Tony Campbell, "Portolan charts from the late thirteenth century to 1500", in *The History of Cartography*, J. B. Harley and D. Woodward, eds., University of Chicago Press, Chicago, 1987.

[2] Armando Cortesão, *History of Portuguese Cartography*, Junta de Investiações do Ultramar-Lisboa, Coimbra, 1969/1971.

[3] James E. Kelley Jr., *Analog and Digital Navigation in the Late Middle Ages*, Sometime Publishers, Melrose Park, PA, 2000.

[4] Leonardo of Pisa, *Fibonacci's Liber Abaci*, transl. L. E. Sigler, Springer-Verlag, New York, 2002.

[5] Pamela O. Long, David McGee, and Alan M. Stahl, eds., *The Book of Michael of Rhodes, a Fifteenth Century Mariner in Service to Venice*, MIT Press, Cambridge, MA, 2009.

[6] Franco Masiero, "Le Raxon de Marteloio", *Studi Veneziana*, (2) 8 (1984) 393–412.

[7] Bacchisio R. Motzo, "Il Compasso da Navigare, Opera Italiana della Metà del Secolo XIII", *Annali della Facoltà di Lettera e Filosofia della Università di Cagliari*, 8 (1947) I-CXXXII, 1–137.

[8] Michael Richey, "The navigational background to 1492", *Journal of Navigation*, 45 (1992) 266–284.

[9] Eva G. R. Taylor, *The Haven-Finding Art*, Abelard-Schuman, New York, 1957.

[10] ——, "Mathematics and the Navigator in the Thirteenth Century", *Journal of Navigation*, 13 (1960) 3–14.

[11] Glen Van Brummelen, *The Mathematics of the Heavens and the Earth: The Early History of Trigonometry*, Princeton University Press, Princeton, NJ, 2009.

Appendix
Michael of Rhodes: Did He Know the Law of Sines?

REQUIRED:

- Figure 9.2, the definitions of the *marteloio*
- Figure 9.3, the *marteloio* table
- Figure 9.4, Michael of Rhodes's new problem
- For the last two questions, knowledge of the Law of Sines

Solving Michael's Navigation Problem

Michael now needs to sail 100 miles from A to B, but he can only sail into B's harbor at heading $\beta = 7$ quarters North of East. He decides to sail first from A to C at heading $\alpha = 2$ quarters South of East. How far does Michael need to sail before turning; i.e., how long is AC?

Unfortunately our new problem doesn't fit the diagram of the *marteloio* right away … but, if you turn your head sideways so that BC is at the top, you can make it look just like the original! The ship may no longer be traveling along the same side of the triangle as before, but that's OK — the *marteloio* table still works.

Q1: Use the *marteloio* table to figure out the length of line segment AD, in miles.

Now, if you used the *alargar* table, you were right. The length we're really after, AC, looks like it corresponds to the *ritorno* on the original diagram. But before we use the *ritorno* table, we're going to need to know what heading to look up.

Q2: Can you work out all the angles in the diagram — in particular, $\angle DCA$?

(*Hints*: The angles of a triangle add up to 16 quarters. A right angle is equal to 8 quarters.)

Now compare your new, tilted diagram with the original *marteloio* definition diagram. It should be clear that the return heading you need the same as $\angle DCA$.

Q3: According to the *marteloio* table for our heading, for every 10 miles of *alargar*, what's the *ritorno*?

But we don't have 10 miles of *alargar*; we have 98. So …

Q4: How long is **our** *ritorno*? This is the answer we seek — AC, the distance we must sail before changing the boat's heading.

Now let's bring the whole process together by doing a similar problem with different numbers:

Q5: Now suppose that the final heading is $\beta = 5$ quarters North of East, and the original heading is $\alpha = 4$ quarters South of East. Now how far does Michael have to sail before changing course?

Comparing with a Modern Solution

How might we solve the problem today? You have already learned the Law of Sines … and this problem seems to fit it perfectly.

9.5. Conclusion

Q6: Apply the Law of Sines to Michael's navigation problem. Do you get the same result for AC?

OK; if you did the calculations correctly, you got the same answer. But is it the same **method**? Here's an argument that's been made that claims that it is:

> The *alargar* is a sort of a sine, since in Figure 9.2 it forms a kind of ratio of the opposite side of a triangle to its hypotenuse. The *ritorno* is sort of a reciprocal of the sine, since it is kind of a ratio of hypotenuse to opposite side. Therefore, Michael is effectively multiplying and dividing the sines in the same way that we did in Question 6.

Is this a fair conclusion?

Q7: Compare your *alargar/ritorno* calculations of AC with your Law of Sines calculations. In what way are the calculations the same? In what way are they different?

Given your answers so far, do you think that it's fair to say that Michael of Rhodes knew the Law of Sines? Why or why not?

10

Copernican Trigonometry

Victor J. Katz
University of the District of Columbia

10.1 Introduction

In most trigonometry courses, the instructor begins by defining the sine, cosine, and tangent of an angle as ratios of certain sides in an appropriate right triangle. She then proceeds to calculate, using elementary geometry, the sine, cosine and tangent of angles of 30°, 45°, and 60°. But once students need to calculate the sine of 27°, they are told to punch some buttons on their calculators. What do students think happens when they do that? Do they imagine that somewhere inside the calculator, someone draws a miniature right triangle with one base angle 27°, then measures the sides and divides? Where do these numbers come from that so miraculously appear on the calculator screen in half a second?

Fifty years ago, no one had calculators. Then, the trigonometry texts simply told the students to consult the table at the back of the book to find the sine of 27°. That took a bit longer, but still, there was little in the text to show students where those numbers came from. They just "were". Whether one uses tables or uses calculators, it still seems that there is a mystery in these numbers that should not exist. Most teachers certainly want their students to be fluent in calculator use – and these are generally easier to use than tables. But still, we do not want students thinking that calculators are magic. Someone at sometime figured out how to calculate these values and, later, someone else figured out how to get the calculator to spit out the numbers when appropriate buttons are pushed.

Thus, I firmly believe that an introduction to trigonometry should show students where these values, in fact, come from. Such an introduction has the additional benefit of giving some real motivation for learning some basic trigonometric identities as well as some of the basic properties of the sine function. The following pages provide details of a few class lessons for the beginning of a trigonometry course that will enable students to understand how the values of trigonometric functions were calculated initially. (These lessons do not, unfortunately, show how the calculator calculates the numbers; that must be the goal of another set of lessons after students understand some calculus.)

10.2 Historical Background

The earliest extant source of detailed calculations of a trigonometric function is the most famous ancient text on astronomy, the *Almagest* of Claudius Ptolemy, written in the middle of the second century C.E. in Alexandria, in Egypt. Ptolemy was able to calculate values to approximately five-decimal-place accuracy by using geometry and an approximation technique. As knowledge of trigonometry spread to India and later to the lands of Islam, other mathematicians modified and extended Ptolemy's calculations. In particular, Indian mathematicians used various sophisticated interpolation methods and later methods based on power series that could calculate values of the sine and cosine functions to as high a degree of accuracy as desired. In Islam, Abū al-Wafā (940–997) improved Ptolemy's method to calculate sine values accurate to approximately eight decimal places, while Ghiyāth al-Dīn al-Kāshī (d. 1429) developed an approximation technique that could give accuracy to more than fifteen decimal places. Rather than study any of these works, however, it will be easier to look at the trigonometric calculations of Nicolaus Copernicus (1473–1543) that appeared in his 1543 work, *On the Revolutions of the Heavenly Spheres*[1], the work that introduced heliocentric astronomy to Europe. Copernicus was probably not aware of any of the trigonometric work accomplished in India or Islam, so he modeled his work very closely on that of Ptolemy. Thus, in what follows, we will study in detail section 12 of the opening chapter of *De Revolutionibus* to see how Copernicus developed his table of sines by a method not very different from Ptolemy's own method of fourteen hundred years earlier. Whenever necessary, we will translate Copernicus's sixteenth century language into more modern language. And we provide activities and exercises for students to attempt in the process of working through this material.

10.3 In the Classroom

Copernicus began by setting out his initial goal, "by means of the arc to determine the straight line, or chord, which subtends the angle."[1, p. 532] In other words, in a circle of given radius, he wanted to be able to determine the length of the chord subtending any given arc. It will, however, be somewhat easier to modify the goal to be able to calculate sines associated to any arc, rather than chords. To see the relationship between Copernicus's chords and modern sines, recall that the sine of an acute angle θ in a right triangle is defined to be the ratio of the leg opposite the angle to the hypotenuse of the triangle. In Figure 10.1, this means that $\sin \theta = \frac{a}{c}$. We know that this ratio is the same no matter what size the triangle, for any two right triangles with the same acute angle will have the same ratio of the opposite side to the hypotenuse. On the other hand, the chord of θ, where θ is a central angle in a circle, is the line in the circle subtending the angle θ. To connect the chord with the sine we draw a circle of radius 1, with θ a central angle, and draw the right triangle AOB as in Figure 10.2. In triangle AOB, we have $\sin \theta = \frac{AB}{OA} = \frac{AB}{1} = AB$. Then the chord AE subtending the angle 2θ is twice the length of the line equal to the sine of θ. This fact provides the basic relationship between modern trigonometry and the work of Copernicus:

$$\sin \theta = \frac{1}{2} \operatorname{crd} 2\theta \qquad \text{or} \qquad \operatorname{crd} \theta = 2 \sin \frac{\theta}{2}.$$

Furthermore, it will prove convenient to define two other functions of angle θ, the cosine and the tangent. The **cosine** of θ, written $\cos \theta$, is the ratio of the leg adjoining the angle to the hypotenuse. Also, we define the **tangent** of θ, written $\tan \theta$, to be the ratio of the leg opposite the angle to the leg adjoining the angle. In Figure 10.1, we have

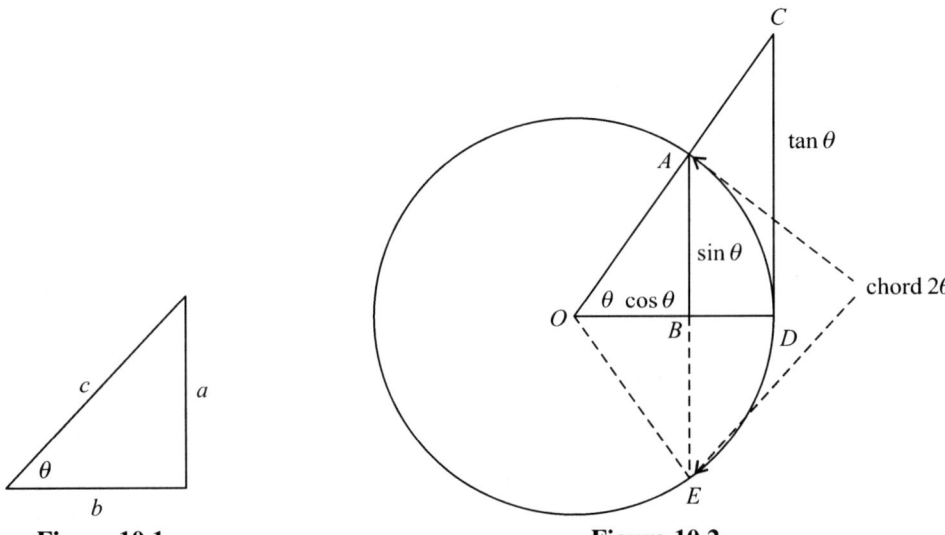

Figure 10.1. **Figure 10.2.**

10.3. In the Classroom

$\cos\theta = \frac{b}{c}$ and $\tan\theta = \frac{a}{b}$. Both of these functions may also be represented in a circle of radius 1. Namely, in Figure 10.2, since $\cos\theta = \frac{OB}{OA} = \frac{OB}{1} = OB$, the length OB is equal to the cosine of the angle θ. Also, in right triangle OCD, we have $\tan\theta = \frac{CD}{OD} = \frac{CD}{1} = CD$, and so the tangent of θ is given by the length of the line CD. Thus, each of the three basic functions of the angle θ are represented as lines connected to a circle of radius 1.

There are a few basic relationships among the sine, cosine, and tangent that we can see from the definition. First, we know from the Pythagorean Theorem that, in Figure 10.1, $a^2 + b^2 = c^2$. If we divide this equation by c^2, we get $(\frac{a}{c})^2 + (\frac{b}{c})^2 = (\frac{c}{c})^2$, an equation which can be rewritten in the form

$$\sin^2\theta + \cos^2\theta = 1. \tag{1}$$

Second, we can rewrite the definition of the tangent as follows: $\tan\theta = \frac{a}{b} = \frac{a/c}{b/c}$, or

$$\tan\theta = \frac{\sin\theta}{\cos\theta}. \tag{2}$$

Third, if θ is an acute angle in a right triangle, then the other acute angle is $90 - \theta$. Again using figure 1, we see that $\sin(90 - \theta)$ equals the ratio of the leg opposite $90 - \theta$ to the hypotenuse. This means that $\sin(90 - \theta) = \frac{b}{c}$. But this ratio is equal to the cosine of θ. Therefore

$$\sin(90 - \theta) = \cos\theta. \tag{3}$$

This equation explains the name "cosine", for the cosine of an angle is equal to the sine of the complement of that angle. Interchanging sine and cosine in this result gives us one further result:

$$\cos(90 - \theta) = \sin\theta. \tag{4}$$

Finally, since the tangent of $90 - \theta$ equals the ratio of the leg opposite that angle to the leg adjoining that angle, we have $\tan(90 - \theta) = \frac{b}{a}$. But $\frac{b}{a}$ is the reciprocal of $\frac{a}{b}$, the tangent of θ. It follows that

$$\tan(90 - \theta) = \frac{1}{\tan\theta}. \tag{5}$$

It is now time to calculate the sine, cosine, and tangent of various angles, essentially following Copernicus's methods. We will do so by the use of geometry, the four basic arithmetic operations, and the square root. But to perform the basic arithmetic operations and the square root, we will use a calculator, because we know it is possible, if we had the time, to do these calculations by hand.

We quote Copernicus's first theorem:

The diameter of a circle being given, the sides of the triangle, square, hexagon, pentagon, and decagon, which the same circle circumscribes, are also given.[1, p. 533]

What Copernicus meant here is that if we know the diameter of a circle, we can calculate the sides of these five regular polygons inscribed in the circle. For our purposes, we will assume that the diameter of the circle is 2, so the radius is 1. A side of a regular polygon of n sides is a chord subtending an angle of $\frac{360}{n}$ degrees at the center of the circle or, alternatively, twice the sine of half of that angle, or $\frac{360}{2n}$ degrees. In particular, the sides referred to subtend angles of $120°$, $90°$, $60°$, $72°$, and $36°$ respectively. It follows that we can calculate the sines of angles equal to half of these angles, namely, $60°$, $45°$, $30°$, $36°$, and $18°$. We will therefore, following Copernicus, determine the relevant sides and thus calculate these sines. Of course, we can then use formulas 1 and 2 to calculate cosines and tangents as well.

We begin with the hexagon. (See Figure 10.3.) Since a hexagon has six sides, each side subtends an angle of $60°$ at the center of the circle. Furthermore, since a triangle formed by any side of the hexagon and the two radii to the ends of that side is equilateral, we know that the side of the hexagon is equal to the radius, namely 1. It follows that the sine of $30°$ is equal to half of the side of the hexagon, namely, 0.5. We write this as

$$\sin 30° = 0.5.$$

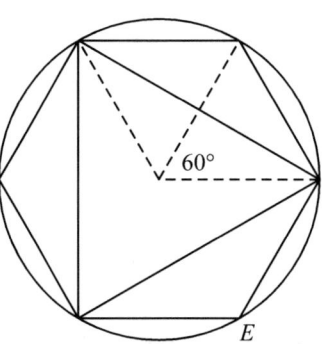

Figure 10.3.

We can now calculate the cosine of 30° using formula 1. We have $\sin^2 30° + \cos^2 30° = 1$. Therefore, $(0.5)^2 + \cos^2 30° = 1$ and $\cos^2 30° = 1 - (0.5)^2 = 1 - 0.25 = 0.75$. It follows that

$$\cos 30° = \sqrt{0.75} = 0.866025404.$$

This value comes from my calculator by using the square root button. Your calculator may give you a value with fewer or with more decimal places. In any case, for now it will be worthwhile to keep all of the decimal places.

To calculate the tangent of 30°, we use formula 2. We just divide sin 30° by cos 30°. We get

$$\tan 30° = 0.5/0.866025404 = 0.577350269.$$

For future reference, you should keep a table of sines, cosines, and tangents of various angles, all carried out to the maximum number of decimal places available on your calculator.

We move to the equilateral triangle. A side of the equilateral triangle can be formed by connecting the ends of two adjacent sides of the regular hexagon. (See Figure 10.3.) The side subtends an angle of 120°, so half of that line is equal to the sine of 60°. Since that side bisects the radius from the center to the point where the two adjacent hexagon sides meet, we can calculate its length from the Pythagorean Theorem. We have $\sin^2 60° + (0.5)^2 = 1^2$. Therefore, $\sin^2 60° = 1 - (0.5)^2 = 1 - 0.25 = 0.75$. It follows that sin 60° is the square root of 0.75, namely 0.866025404.

This value is the same as the cosine of 30°, calculated just before. But this is not surprising, given formula 3. That equation tells us that

$$\sin 60° = \sin(90° - 30°) = \cos 30° = 0.866025404.$$

Then formula 4 tells us that

$$\cos 60° = \sin 30° = 0.5.$$

Finally, formula 5 says that the tangent of 60° is the reciprocal of the tangent of 30°. Therefore,

$$\tan 60° = \frac{1}{\tan 30°} = \frac{1}{0.577350269} = 1.732050808.$$

Record the values of the three functions at 60° in your table.

Exercises

1. We now determine the side of a square inscribed in a circle of radius 1. (See Figure 10.4.) What angle do the two radii from one side of the square make at the center of the circle?

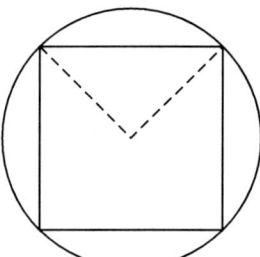

Figure 10.4.

2. Using the result of 1 and the Pythagorean Theorem, calculate to as many decimal places as possible the length of the side of the square inscribed in the circle.

3. Draw a perpendicular from the center of the circle to the midpoint of one of the sides of the square. What angle does this perpendicular make with a radius drawn to the endpoint of that side?

4. Given that the sine of the angle calculated in 3 is half the length of the side of the square calculated in 2, determine the sine of that angle.

5. Determine the cosine and tangent of the angle whose sine you calculated in 4.

10.3. In the Classroom

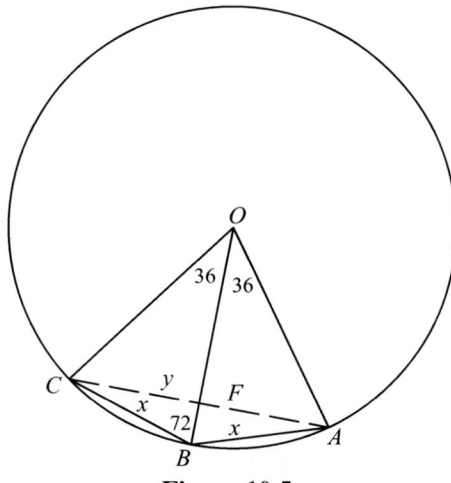

Figure 10.5. **Figure 10.6.**

The results of the exercises should be that

$$\sin 45° = \cos 45° = 0.707106781$$

and that

$$\tan 45° = 1.$$

Record these values in your table.

We now consider the decagon, which is a bit harder than the previous three polygons. Note that the side of a regular decagon inscribed in a circle subtends an angle of $\frac{360}{10} = 36°$. So draw an isosceles triangle OBA with vertex at the center O of the circle of radius 1, with vertex angle equal to $36°$. (See Figure 10.5.) Because the two base angles of this triangle are equal, they are both equal to $72°$, and this triangle is a 36-72-72 triangle. We will let x be the length of the base BA of this triangle, or the length of the side of the regular decagon. To calculate x, we draw the line BD bisecting one of the base angles of this triangle. (See Figure 10.6.) Note that this line divides our original triangle into two isosceles triangle. The triangle at the bottom of the figure, triangle BAD, is a 36-72-72 triangle, while the other triangle, triangle BDO, is a 108-36-36 triangle. Since triangle BAD is isosceles, we know that $BD = BA = x$. Since triangle BDO is also isosceles, we know that $DO = BD = x$. Finally, since $OA = 1$, we can write $DA = 1 - x$.

We now use the basic principle of similarity to determine x. Since triangles BAD and OBA are both 36-72-72 triangles, they are similar. So the ratio of a leg of the first triangle to the leg of the second is equal to the ratio of base of the first triangle to the base of the second. In other words, $BA : BO = DA : BA$, or $x : 1 = (1-x) : x$. We rewrite this ratio as an equation:

$$x^2 = 1 - x \quad \text{or} \quad x^2 + x - 1 = 0$$

We can now solve this quadratic equation for the positive solution:

$$x = \frac{-1 + \sqrt{1^2 + 4}}{2} = \frac{-1 + \sqrt{5}}{2} = \frac{1.236067978}{2} = 0.618033989.$$

We now know the length of the base of the isosceles triangle OBA. Since this base is equal to the side of a decagon and therefore subtends an angle of $36°$, we know from our previous discussion that it is twice the sine of $18°$. Therefore,

$$\sin 18° = \frac{0.618033989}{2} = 0.309016994.$$

We can now calculate that

$$\cos 18° = 0.951056516 \quad \text{and} \quad \tan 18° = 0.324919696.$$

Record these values in your table, then calculate the sine, cosine, and tangent of $72°$ and record these as well:

$$\sin 72° = 0.951056516 \quad \cos 72° = 0.309016994 \quad \tan 72° = 3.077683537.$$

The last of Copernicus' polygons to determine is the pentagon. The simplest way to do this is to consider two adjacent 36-72-72 triangles OAB and OBC as in Figure 10.5. The line AC connecting the vertices A and C then subtends an angle of $72°$ at the center of the circle, so is the side of a regular pentagon inscribed in the circle. To calculate $y = AC$, we note that it is bisected by radius OB. We consider right triangle AFB, where $\angle FBA = 72°$. The hypotenuse of this triangle is $AB = x$, which we have calculated above. The side opposite angle FBA has length $\frac{y}{2}$. By the definition of the sine, we know that the sine of angle FBA is equal to the length of AF divided by the length of BA, That is,

$$\sin 72° = \frac{y/2}{x}.$$

But we know the sine of $72°$ and we also know x. Therefore,

$$\frac{y}{2} = x \sin 72° = 0.618033989 \cdot 0.951056516 = 0.587785252.$$

Thus, not only do we know that the length of the side of a regular pentagon inscribed in this circle is $y = 1.175570505$, but also we know that the sine of $36°$ is equal to $y/2$. That is,

$$\sin 36° = 0.587785252.$$

Exercises

1. Using formulas 1 and 2, calculate $\cos 36°$ and $\tan 36°$.

2. Use the appropriate equations to calculate $\sin 54°$, $\cos 54°$, and $\tan 54°$.

3. It is handy to have values for the sine, cosine and tangent of $0°$ and $90°$. We cannot, however, use the original definitions, because there is no right triangle with an acute angle of $0°$. But if in Figure 10.1 we let angle θ get closer and closer to zero, we see that side a also gets closer and closer to zero. In addition, side b approaches side c in length. Therefore, make a reasonable definition for $\sin 0°$ and $\cos 0°$. Check that your definitions satisfy formula 1.

4. Using your values for $\sin 0°$ and $\cos 0°$ and formula 2, determine $\tan 0°$.

5. Use formulas 3 and 4 as well as the result of 5 above to determine values for $\sin 90°$ and $\cos 90°$. What can you now say about $\tan 90°$?

6. Make sure you now have a table of values, calculated to 8 or 9 decimal places, for sine, cosine, and tangent. There should be values for $0°, 18°, 30°, 36°, 45°, 54°, 60°, 72°$, and $90°$.

Having now developed a table for sine, cosine, and tangent of nine different angles, we want to complete the table so that it gives values for the sine, cosine, and tangent of every angle of an integral number of degrees from 0 to 90. We will not actually work out every angle, but we will develop the tools that would enable you to complete the table if you wanted to. Once this is done, we will be more comfortable with allowing you to use your calculators to determine the values of sine, cosine, and tangent when you need these to solve other problems.

We begin with Copernicus' second theorem:

If a quadrilateral is inscribed in a circle, the rectangle comprehended by the diagonals is equal to the two rectangles which are comprehended by the two pairs of opposite sides.[1, p. 534]

This theorem is actually due to Ptolemy, and Copernicus's proof follows that of the ancient astronomer. But first, we need to be sure of what the theorem says. In Figure 10.7, we have a quadrilateral $ABCD$ inscribed in a circle, with the two diagonals AC and BD drawn. The theorem then says that the "rectangle comprehended by the diagonals," that is, the rectangle whose length and width are AC and BD, respectively, is equal to "the two rectangles which are comprehended by the two pairs of opposite sides," that is, the sum of the rectangle whose length and width are AB and CD and the rectangle whose length and width are AD and BC. In other words, Copernicus claimed that the product $BD \cdot AC$ is equal to the sum of the two products $AB \cdot CD$ and $AD \cdot BC$.

The proof begins by drawing segment BE in such a way that $\angle ABE = \angle CBD$. If we add $\angle EBD$ to both sides of this equation, we get that $\angle ABD = \angle EBC$. But also we know that $\angle ACB = \angle BDA$, because both angles cut off

10.3. In the Classroom

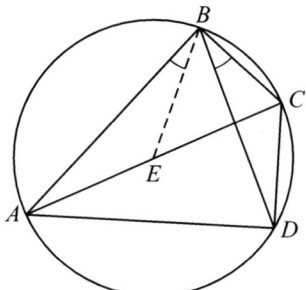

Figure 10.7.

the same arc AB on the circumference of the circle. Because in triangles BCE and BDA, two pairs of corresponding angles are equal, it follows that all three pairs of corresponding angles are equal and the triangles are similar. Therefore $BC : BD = EC : AD$, and so $AD \cdot BC = BD \cdot EC$. That is, the product of one pair of opposite sides of the quadrilateral is equal to one diagonal times a piece of the second diagonal.

To finish the proof, we need to show that the product of the other pair of opposite sides is equal to the product of that same diagonal with the other piece of the second diagonal. We can do that by using the similarity of triangles ABE and CBD. That these triangles are similar follows from the equality of angle ABE with angle CBD, which is true by construction, and of angle BAC with angle BDC, both of which cut off the same arc BC on the circumference. We then have $AB : BD = AE : CD$, so $AB \cdot CD = BD \cdot AE$.

If we add the two equations we derived from the similarity conditions, we have $AD \cdot BC + AB \cdot CD = BD \cdot EC + BD \cdot AE$. But since BD is a factor of each term on the right, we can rewrite the right side of this equation as $BD(EC + AE) = BD \cdot AC$. We have, therefore, $AD \cdot BC + AB \cdot CD = BD \cdot AC$, as desired.

We could now prove Copernicus' third theorem:

> *Hence if straight lines subtending unequal arcs in a semicircle are given, the chord subtending the arc whereby the greater arc exceeds the smaller is also given.*[1, p. 534]

However, this theorem as it is written deals with calculating the chord subtending an arc which is the difference of two given arcs. Since we are dealing with sines and cosines rather than chords, and with angles rather than arcs, it will be better to prove the analogous result enabling us to calculate the sine of the difference of two angles if we know the sine and the cosine of the two angles themselves. Once we know this result, we can, for example, calculate the sine of 12°, because we know the sine and the cosine of both 30° and 18°.

To derive the so-called difference formula, we will use a circle of diameter 1 as in Figure 10.8. Let AD be the diameter of the circle, $\angle BAD = \alpha$, and $\angle CAD = \beta$. Then $\angle BAC = \alpha - \beta$. If we now connect BD, we have a quadrilateral inscribed in a circle, along with its two diagonals. We can therefore apply Ptolemy's theorem (Copernicus's second theorem) to this situation. Before we do so, however, we need to determine what each of the line segments in this diagram represents in terms of sines and cosines.

First, since triangle ABD is a triangle inscribed in a semicircle, it is a right triangle. Therefore $\sin \alpha = \frac{BD}{AD}$. But $AD = 1$. So $BD = \sin \alpha$. Analogously, $AB = \cos \alpha$. Second, triangle ACD is also a triangle inscribed in a

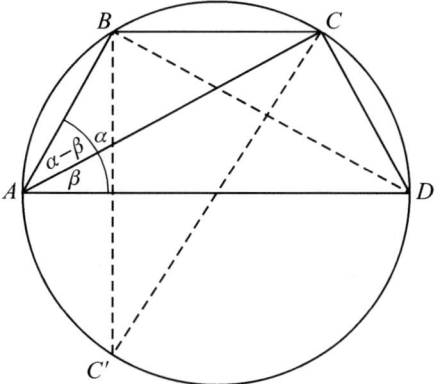

Figure 10.8.

semicircle, so is a right triangle. Therefore $CD = \sin\beta$ and $AC = \cos\beta$. We now have identified five of the six line segments in our figure. The only one missing is the side BC of the quadrilateral. This side is opposite $\angle BAC$, which is equal to $\alpha - \beta$. But since triangle ABC is not a right triangle, we can not use the same argument as above to identify BC with $\sin(\alpha - \beta)$. We need a construction to make BC the leg of a right triangle. We thus draw the diameter CC' and connect BC'. Since now triangle $BC'C$ is a triangle inscribed in a semicircle, it is a right triangle. So, as before, $BC = \sin\angle BC'C$. But angles $BAC(=\alpha - \beta)$ and $BC'C$ cut off the same arc on the circumference of the circle, namely, arc BC. Therefore, those two angles are equal. So finally, we have $BC = \sin(\alpha - \beta)$, and all six line segments in Ptolemy's theorem have been identified.

We now just substitute the values for the six line segments in Ptolemy's theorem. Since $AD \cdot BC + AB \cdot CD = BD \cdot AC$, we get $1 \cdot \sin(\alpha - \beta) + \cos\alpha \sin\beta = \sin\alpha \cos\beta$ or

$$\sin(\alpha - \beta) = \sin\alpha \cos\beta - \cos\alpha \sin\beta, \tag{6}$$

the difference formula for the sine.

Let us now use this formula to determine, as promised, the sine of 12°:

$$\sin 12° = \sin(30° - 18°) = \sin 30° \cos 18° - \cos 30° \sin 18°.$$

But we know the sine and cosine of both 30° and 18°. We therefore calculate

$$\sin 12° = \sin 30° \cos 18° - \cos 30° \sin 18°$$
$$= 0.5 \cdot 0.951056516 - 0.866025404 \cdot 0.309016994$$
$$= 0.475528258 - 0.267616567$$
$$= 0.207911691.$$

We can now calculate the cosine and tangent of 12°: $\cos 12° = 0.978147601$ and $\tan 12° = 0.212556562$.

Of course, now that we know the sine, cosine, and tangent of 12°, we also know the sine, cosine, and tangent of $90° - 12° = 78°$, by use of formulas 3, 4, and 5. We get

$$\sin 78° = \cos 12° = 0.978147601$$
$$\cos 78° = \sin 12° = 0.207911691$$
$$\tan 78° = \frac{1}{\tan 12°} = \frac{1}{0.212556562} = 4.70463011.$$

Exercises

1. Given that you know the sine and cosine of 36° and of 45°, calculate, using formula 6, the sine, cosine, and tangent of 9°.

2. Using formulas 3, 4, and 5 and the results of 1, calculate the sine, cosine, and tangent of 81°.

3. Calculate the sine, cosine, and tangent of 6° and 84°, using values already calculated.

4. Given the angles for which you have already calculated the sine, cosine, and tangent, for what other angles can you calculate these same functions using formulas 1 through 6? Determine all the possibilities, including any that become possible once you calculate other new ones. (It is not necessary actually to calculate the sine, cosine, and tangent; just list the angles for which it is possible using the formulas.)

To fill in the table further, Copernicus stated and proved his fourth theorem:

Given a chord subtending any arc, the chord subtending half of the arc is also given. [1, p. 535]

This result enabled Copernicus to calculate the chord of half of an arc, assuming he knew the chord of a given arc. As before, we will adapt Copernicus's procedure for sines. In other words, we will develop a formula that will enable us to calculate the sine of half of an angle, assuming we know the sine and the cosine of the angle. Thus, because we can calculate the sine and cosine of 15°, we will be able to calculate the sine of $7\frac{1}{2}°$. Or, because we can calculate the sine of 3°, we can also calculate the sine of $1\frac{1}{2}°$.

10.3. In the Classroom

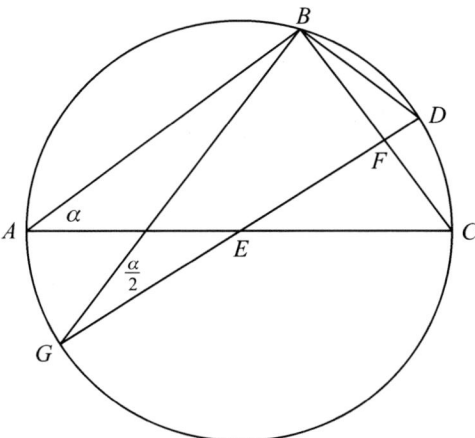

Figure 10.9.

You may wonder why we are calculating sines and cosines of non-integer angles, given that we wanted to calculate the sines and cosines of all angles of an integral number of degrees. Your answer to question 4 of the exercises should have told you that the calculations we have already done enable us to find the sines and cosines of every angle that is an integral multiple of 3°. By taking halves of such angles, we can only reach either other angles that are multiples of 3° or else non-integral angles. (Why?) So to find the sine and cosine of 1°, for example, we will need a different technique. That technique will, however, require the knowledge of the sine and cosine of angles such as $1\frac{1}{2}°$.

To prove what is now known as the half-angle formula, we begin with a circle of diameter $AC = 1$ with an inscribed angle $\alpha = \angle BAC$. (See Figure 10.9.) Connect BC. As before, we know that $BC = \sin \alpha$ and $AB = \cos \alpha$. Choose D to be the midpoint of arc BC cutting off the angle α, and draw the diameter DG and the line GB. Since the arc BD is half the arc BC, the angle DGB is half the angle BAC. That is, $\angle DGB = \frac{1}{2}\alpha$. Therefore, by the definition of sine, we know that $BD/DG = \sin \frac{1}{2}\alpha$, or, since $DG = 1$, that $BD = \sin \frac{1}{2}\alpha$.

To develop a formula for $\sin \frac{1}{2}\alpha$, we want to find a relationship among the line segments BD, BC, and BA. Let F be the point of intersection of DG with BC. Since GD bisects arc BC and therefore line BC, we know from properties of circles that GF is perpendicular to BC. That is, triangle EFC is a right triangle with right angle at F. But then triangle ABC is similar to triangle EFC, because these two right triangles share a common angle. Therefore, since $CF = \frac{1}{2}CB$, it follows that $EF = \frac{1}{2}AB = \frac{1}{2}\cos \alpha$, and, since $ED = \frac{1}{2}$, that $DF = \frac{1}{2} - \frac{1}{2}\cos \alpha$. We know also that BF is the altitude from the right angle of triangle GBD to the hypotenuse. Therefore triangle BDF is similar to triangle BDG, so $GD : BD = BD : DF$, or $BD^2 = GD \cdot DF$. If we substitute for each of these line segments the values we have determined, we get

$$\sin^2 \frac{1}{2}\alpha = 1 \cdot \left(\frac{1}{2} - \frac{1}{2} \cos \alpha \right).$$

We rewrite this in the form

$$\sin \frac{1}{2}\alpha = \sqrt{\frac{1 - \cos \alpha}{2}}. \tag{7}$$

Formula 7 is referred to as the half-angle formula for the sine function. We can calculate a similar half-angle formula for the cosine by applying formula 1:

$$\cos^2 \frac{1}{2}\alpha = 1 - \sin^2 \frac{1}{2}\alpha = 1 - \left(\frac{1}{2} - \frac{1}{2} \cos \alpha \right) = \frac{1}{2} + \frac{1}{2} \cos \alpha.$$

This result can be rewritten in the form

$$\cos \frac{1}{2}\alpha = \sqrt{\frac{1 + \cos \alpha}{2}}. \tag{8}$$

Examples

1. We calculate the sine and cosine of 9° via the half-angle formulas, even though you have already calculated these values via the difference formula. We first apply formula 7:

$$\sin 9° = \sin \frac{1}{2} \cdot 18° = \sqrt{\frac{1 - \cos 18°}{2}}$$

$$= \sqrt{\frac{1 - 0.951056516}{2}} = \sqrt{0.024471742}$$

$$= 0.156434465.$$

We now use formula 8 to calculate the cosine of 9°:

$$\cos 9° = \sqrt{\frac{1 + \cos 18°}{2}}$$

$$= \sqrt{\frac{1 + 0.951056516}{2}}$$

$$= \sqrt{0.975528258}$$

$$= 0.987688341.$$

2. Let us now calculate the sine and cosine of 3°, given that we have already calculated (in exercise 3 in the previous set) the sine and cosine of 6°. Those values are $\sin 6° = 0.104528463$ and $\cos 6° = 0.994521895$. Then

$$\sin 3° = \sqrt{\frac{1 - \cos 6°}{2}}$$

$$= \sqrt{\frac{1 - 0.994521895}{2}}$$

$$= \sqrt{0.002739052}$$

$$= 0.052335956.$$

Similarly,

$$\cos 3° = \sqrt{\frac{1 + \cos 6°}{2}}$$

$$= \sqrt{\frac{1 + 0.994521895}{2}}$$

$$= \sqrt{0.997260948}$$

$$= 0.998629535.$$

3. We now go one step further and calculate the sine and cosine of $1\frac{1}{2}°$. We get

$$\sin 1\frac{1}{2}° = \sqrt{\frac{1 - \cos 3°}{2}}$$

$$= \sqrt{\frac{1 - 0.998629535}{2}}$$

$$= \sqrt{0.000685233}$$

$$= 0.026176948.$$

10.3. In the Classroom

Also,

$$\cos 1\frac{1}{2}° = \sqrt{\frac{1+\cos 3°}{2}}$$
$$= \sqrt{\frac{1+0.998629535}{2}}$$
$$= \sqrt{0.999314768}$$
$$= 0.999657325.$$

Exercises

1. Calculate the tangents of $3°$ and $1\frac{1}{2}°$.

2. Use the half angle formulas and formula 2 to calculate the sine, cosine, and tangent of $7\frac{1}{2}°$.

3. Use the half angle formulas and formula 2 to calculate the sine, cosine, and tangent of $\frac{3}{4}°$.

4. Determine the sine, cosine, and tangent of $87°$ and $88\frac{1}{2}°$.

5. Give an argument as to why you cannot determine the sine of $1°$ or of $2°$ by use of any of the formulas we have developed so far. For what values close to $1°$ or $2°$ can you calculate values for the sine? Calculate the value of the sine of an angle not already calculated that is less than $\frac{1}{2}°$ from $1°$, using the half angle formulas and values you have already calculated. Calculate the value of the sine of an angle within $\frac{1}{2}°$ of $2°$, one you have not already calculated.

Before we attempt to figure out the sine of $1°$, we will continue with Copernicus and determine how to calculate the sine of the sum of two angles, supposing we know the sine and the cosine of each of the angles. This result would enable us to fill in our table in steps of $1\frac{1}{2}°$, or even in steps of $\frac{3}{4}°$, for we know the sines and cosines of both of those angles.

> *When chords are given subtending two arcs, the chord subtending the whole arc made up of them is also given.*[1, p. 535]

As before, what Copernicus claimed is that you can calculate the chord of the sum of two arcs if you know the chords of each of them. We will adapt his procedure for sines. So in Figure 10.10, we will again assume that the diameter AC is equal to 1. We let $\angle CAD = \alpha$ and $\angle CAB = \beta$. Then $\angle BAD = \alpha + \beta$. We connect BC, CD, and BD and now have a quadrilateral $ABCD$ to which we can apply Ptolemy's Theorem. To use Ptolemy's Theorem, we first need to identify all the line segments in quadrilateral $ABCD$. As before, we know that $BC = \sin\beta$, $AB = \cos\beta$, $CD = \sin\alpha$, and $AD = \cos\alpha$. Also, of course, $AC = 1$, To determine BD, we draw a diameter through B, meeting the circle again

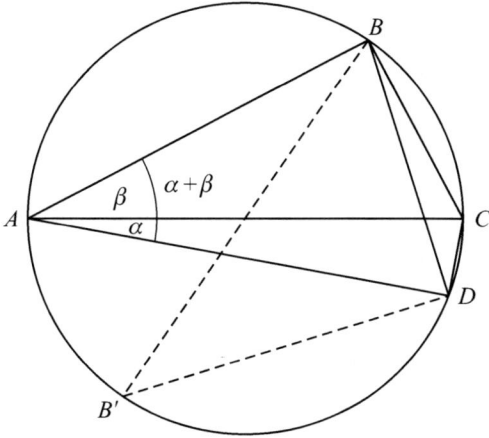

Figure 10.10.

at B' and connect $B'D$. Then $\angle BB'D = \angle BAD = \alpha + \beta$, because both of those angles cut off the same arc BD on the circumference. It follows that $BD = \sin(\alpha + \beta)$. By Ptolemy's Theorem, $AC \cdot BD = AB \cdot CD + AD \cdot BC$. By substituting values for each of the line segments, we get $1 \cdot \sin(\alpha + \beta) = \cos\beta \sin\alpha + \cos\alpha \sin\beta$ or

$$\sin(\alpha + \beta) = \sin\alpha \cos\beta + \cos\alpha \sin\beta. \tag{9}$$

Formula 9 is generally called the sum formula for the sine, because it enables us to calculate the sine of the sum of two angles, given the sine and the cosine of the two angles themselves.

Before using this formula in any calculations, it is handy to determine analogous formulas for the cosine, namely the formulas for cosine of the sum and difference of two angles. We could do this by derivations similar to those of formulas 6 and 9, but it is simpler to do this using formulas 3 and 4. We have

$$\begin{aligned}\cos(\alpha + \beta) &= \sin(90 - (\alpha + \beta)) \\ &= \sin((90 - \alpha) - \beta) \\ &= \sin(90 - \alpha)\cos\beta - \cos(90 - \alpha)\sin\beta \\ &= \cos\alpha \cos\beta - \sin\alpha \sin\beta.\end{aligned}$$

We write the sum formula for the cosine, then, as

$$\cos(\alpha + \beta) = \cos\alpha \cos\beta - \sin\alpha \sin\beta. \tag{10}$$

Similarly, we calculate

$$\begin{aligned}\cos(\alpha - \beta) &= \sin(90 - (\alpha - \beta)) \\ &= \sin((90 - \alpha) + \beta) \\ &= \sin(90 - \alpha)\cos\beta + \cos(90 - \alpha)\sin\beta \\ &= \cos\alpha \cos\beta + \sin\alpha \sin\beta.\end{aligned}$$

Thus, we have the difference formula for the cosine:

$$\cos(\alpha - \beta) = \cos\alpha \cos\beta + \sin\alpha \sin\beta. \tag{11}$$

Examples

1. We use the sum formula for the sine to calculate $\sin 10\frac{1}{2}°$, given that we know the sine and cosine of both $9°$ and $1\frac{1}{2}°$. We get

$$\begin{aligned}\sin 10\frac{1}{2}° &= \sin\left(9° + 1\frac{1}{2}°\right) \\ &= \sin 9° \cos 1\frac{1}{2}° + \cos 9° \sin 1\frac{1}{2}° \\ &= 0.156434465 \cdot 0.999780683 + 0.987688341 \cdot 0.026176948 \\ &= 0.182235525.\end{aligned}$$

2. We use the sum formula for the cosine to calculate $\cos 10\frac{1}{2}°$, even though we could also use formula 1 along with the result just calculated. We get

$$\begin{aligned}\cos 10\frac{1}{2}° &= \cos\left(9° + 1\frac{1}{2}°\right) \\ &= \cos 9° \cos 1\frac{1}{2}° - \sin 9° \sin 1\frac{1}{2}° \\ &= 0.987688341 \cdot 0.999780683 - 0.156434465 \cdot 0.026176948 \\ &= 0.983254908.\end{aligned}$$

3. We can use the sum formula for the sine to develop the double angle formula for the sine. Namely, if we want to calculate the sine of 2α, we can think of this angle as being the sum of α and α. Therefore, we get

$$\sin 2\alpha = \sin(\alpha + \alpha) = \sin\alpha \cos\alpha + \cos\alpha \sin\alpha.$$

We rewrite this in the form

$$\sin 2\alpha = 2\sin\alpha \cos\alpha. \tag{12}$$

Exercises

1. Use the sum formula for the sine to calculate $\sin 13\frac{1}{2}°$.

2. Use the sum formula for the cosine to calculate $\cos 13\frac{1}{2}°$.

3. Use the sum formula for the cosine, with $\beta = \alpha$ to derive the double angle formula for the cosine:

$$\cos 2\alpha = cos^2\alpha - \sin^2\alpha. \tag{13}$$

4. Use formulas 13 and 1 to derive two alternative formulas for the cosine of twice an angle:

$$\cos 2\alpha = 1 - 2\sin^2\alpha \tag{14}$$

$$\cos^2\alpha = 2\cos^2\alpha - 1 \tag{15}$$

5. Use the formulas for sine and cosine of the sum of two angles to derive the formula for the tangent of the sum of two angles:

$$\tan(\alpha + \beta) = \frac{\tan\alpha + \tan\beta}{1 - \tan\alpha \tan\beta}. \tag{16}$$

6. Derive an analogous formula for $\tan(\alpha - \beta)$.

7. Derive a double angle formula for the tangent, that is a formula for $\tan 2\alpha$.

We return now to the problem of calculating the sine of $1°$. Note that once we calculate this and the cosine of $1°$, we can use the double angle formula, for example, to calculate the sine and cosine of $2°$. We could then proceed to calculate the sine and cosine of all angles of an integral number of degrees. We could even do angles in steps of $\frac{1}{2}°$ or $\frac{1}{4}°$.

We have already seen that it is not possible to calculate the sine of $1°$ by use of the formulas we have developed. So we need another method. Copernicus, following Ptolemy, used an approximation procedure based on a theorem about the ratios of arcs and chords. It will be easier, however, to look at this problem in a slightly different, though equivalent way, by studying the ratios of arcs and sines.

Let us look at a piece of the table we have already constructed for the sines of various angles. In particular, we will look at the sines of some very small angles. You may not have calculated all of these values, but you should understand how they all could be calculated from the procedures we have worked out.

Angle	Sine
$6°$	0.104528463
$4\frac{1}{2}°$	0.078459096
$3°$	0.052335956
$2\frac{1}{4}°$	0.039259816
$1\frac{1}{2}°$	0.026176948
$1\frac{1}{8}°$	0.019633692
$\frac{3}{4}°$	0.013089596

Exercises

1. The ratio of 6° to 3° is 6 : 3 = 2. Calculate the ratio of sin 6° to sin 3°. What is the relationship between these two ratios?

2. Calculate the ratio of $4\frac{1}{2}°$ to $2\frac{1}{4}°$. Calculate the ratio of $\sin 4\frac{1}{2}°$ to $\sin 2\frac{1}{4}°$. What is the relationship between these two ratios? How does the ratio of sines in this case compare to the ratio of sines in the previous case?

3. Calculate the ratio of 3° to $1\frac{1}{2}°$ and also the ratio of sin 3° to $\sin 1\frac{1}{2}°$. What is the relationship between these ratios? How does the ratio of sines in this case compare to the ratio of sines in the previous two cases?

4. Calculate the ratio of $2\frac{1}{4}°$ to $1\frac{1}{2}°$ and also the ratio of $\sin 2\frac{1}{4}°$ to $\sin 1\frac{1}{2}°$. How do the ratios compare in this case?

5. Pick any other pair of angles in the table that we have not used so far and compare the ratio of the angle measures to the ratio of the sines of those angles. How do the ratios compare?

6. We have worked here with small angles. Let us try some large angles. The ratio of 30° to 15° is 30 : 15 = 2. What is the ratio of sin 30° to sin 15°? What is the relationship of these two ratios? How does this relationship compare to the relationships already worked out?

7. Pick any other pair of angles, each larger than 30°. Compare the ratio of the measures of the angles to the ratio of the sines of the angles. What is the relationship here? How does this relationship compare to the relationships already discovered?

8. Conjecture a statement about the relationship of the ratio of the angle measures of two (small) angles to the ratio of the sines of those two angles. The statement should fill in the final blank of: "if the ratio of angle α to angle β is r, then the ratio of $\sin \alpha$ to $\sin \beta$ is approximately _____."

The results of the previous exercises should lead you to conclude that the ratio of $\sin \alpha$ to $\sin \beta$ is very nearly equal to the ratio of α to β, when α and β are "small." For our purposes, "small" means angles less than 6°. But the important point is that the ratios become more and more nearly equal, the smaller the angles are.

We can write this fact in another way. If the ratio of α to β is r, then $\alpha = r\beta$. Assuming both α and β are small, we then know that the ratio of $\sin \alpha$ to $\sin \beta$ is very close to r, or that $\sin \alpha \approx r \sin \beta$. Since $\alpha = r\beta$, we can write this in the form $\sin r\beta \approx r \sin \beta$, for small values of r and β, with this approximation becoming closer and closer to an equality the smaller both β and r are. In technical language, we say that the sine function is approximately a **linear** function for small angles.

We can now use this linearity of the sine function to approximate the sine of 1°. We will do this in several different ways to check on how accurate our approximation is. First, we know that 1° is $\frac{2}{3} \cdot 1\frac{1}{2}°$. Therefore,

$$\sin 1° = \sin\left(\frac{2}{3} \cdot 1\frac{1}{2}°\right)$$
$$\approx \frac{2}{3} \sin 1\frac{1}{2}°$$
$$= \frac{2}{3} \cdot 0.026176948$$
$$= 0.017451299.$$

Second, we know that $1° = \frac{4}{3} \cdot \frac{3}{4}°$. Therefore,

$$\sin 1° = \sin\left(\frac{4}{3} \cdot \frac{3}{4}°\right)$$
$$\approx \frac{4}{3} \sin \frac{3}{4}°$$
$$= \frac{4}{3} \cdot 0.013089596$$
$$= 0.017452795.$$

Third, we know that $1° = \frac{8}{9} \cdot 1\frac{1}{8}°$. Therefore,

$$\begin{aligned} \sin 1° &= \sin\left(\frac{8}{9} \cdot 1\frac{1}{8}°\right) \\ &\approx \frac{8}{9} \sin 1\frac{1}{8}° \\ &= \frac{8}{9} \cdot 0.019633692 \\ &= 0.17452171. \end{aligned}$$

We now have three different approximations to the sine of 1°. These are 0.017451299, 0.017452795, and 0.017452171. All three approximations agree on the first five decimal places, and two of the three agree on the sixth place as well. In addition, since those two approximations make use of smaller angles than the first approximation, we have slightly more confidence in the values we get. Thus, we can conclude that to six decimal places, $\sin 1° = 0.017452$. Copernicus himself was satisfied with a five decimal place answer: 0.01745.

Now that we have calculated the sine of 1°, even by an approximation, we can use that value, along with our other values, to calculate the sine, cosine and tangent of every angle of an integral number of degrees by use of our various formulas. We can even use the half angle formula to calculate angles one-half degree apart.

For example, given the sine of 1°, we calculate the cosine of 1°:

$$\begin{aligned} \cos 1° &= \sqrt{1 - \sin^2 1°} \\ &= \sqrt{1 - 0.017452^2} \\ &= \sqrt{0.999695} \\ &= 0.999848. \end{aligned}$$

Note that we have only calculated this to six places, because that is all the accuracy we have on the sine value. We also calculate the sine of 2° using the double angle formula 12:

$$\begin{aligned} \sin 2° &= 2 \sin 1° \cos 1° \\ &= 2 \cdot 0.017452 \cdot 0.999848 \\ &= 0.034899. \end{aligned}$$

Finally, we calculate the sine of $\frac{1}{2}°$ by use of the half angle formula 7:

$$\begin{aligned} \sin \frac{1°}{2} &= \sqrt{\frac{1 - \cos 1°}{2}} \\ &= \sqrt{\frac{1 - 0.999848}{2}} \\ &= \sqrt{0.000076} \\ &= 0.0087. \end{aligned}$$

In this case, since we were taking the square root of a number with only two significant figures, we can only be confident of two significant figures in our answer.

Although in theory, we could now complete our table in this manner, there is no necessity to do so. After all, we do have a calculator available. Everywhere we need to determine sines, cosines, and tangents later, we will use the calculator. But now, at least, we have some idea as to how people were able to calculate the values before such a device was available.

10.4 Conclusion

There is much to be learned about trigonometry by considering how Copernicus or other astronomers presented the subject. By using what we presented here, students will learn how trigonometric values were calculated and why the sum, difference, and half-angle formulas are important. But one can (and should) go further. Trigonometry was developed to enable astronomers to "solve" triangles, both plane triangles (as is usually done in a trigonometry course) and spherical triangles, triangles on a sphere whose sides are arcs of great circles. In particular, knowing how to solve spherical triangles enabled astronomers to predict various heavenly phenomena, such as when and where the sun would set, or when an eclipse of the moon would occur. Solving spherical triangles is also useful in determining distances between points on the surface of the earth or in determining the direction to travel to get from one point to another by the most direct route. Today, it is uncommon to find techniques for solving spherical triangles in a trigonometry course. In my opinion, it would be worthwhile to experiment with reintroducing this material, whether or not one wants to prove all the basic spherical trigonometric identities. Not only does doing this show why people were interested in trigonometry in the first place (and it was *not* to find the distance across a lake or the height of a tree), but also it presents students with the opportunity to solve lots of interesting problems, while at the same time learning some astronomy. For more suggestions on how to use history in the teaching of trigonometry, consult [2].

Bibliography

[1] Nicolaus Copernicus, *On the Revolutions of the Heavenly Spheres*, Charles Glenn Wallis, trans., in *Great Books of the Western World*, vol. 16, pp. 505–838, Encyclopedia Britannica, Inc., Chicago, 1952.

[2] Victor Katz and Karen Dee Michalowicz, eds., *Historical Modules for the Teaching and Learning of Mathematics*, Mathematical Association of America, Washington, 2005. (This is a CD containing eleven different modules; in particular, consult the Trigonometry module.)

11

Cusps: Horns and Beaks

Robert E. Bradley
Adelphi University

11.1 Introduction

This is the mathematical tale of a cusp in the shape of a bird's beak. Although precalculus and calculus courses must stress the idea of function over that of equation, they nevertheless include a number of important topics concerning polynomial equations in two variables, including implicit differentiation and the study of conic sections. Whereas polynomial functions of one variable have very simple graphs, the graphs of polynomial equations in x and y — even those of relatively low degree — can exhibit wonderfully exotic features.

The story of the bird's beak can be used to enrich a course in analytic geometry, precalculus or calculus. For students who know some calculus, it also provides insight into continuous nondifferentiable functions. There is also a connection to power series representations, although this will not be discussed in this chapter (Euler treats them in §5–9 of [1, 2]).

For further reading on these topics, see [3, 4].

11.2 Historical Background

In the 18th century, calculus and the related branches of mathematics gradually changed their perspective from the geometric to the algebraic. When René Descartes (1596-1650) and Pierre de Fermat (1601–1665) invented analytic geometry, for example, mathematicians were already familiar with a large assortment of curves, given by a variety of geometric constructions. Analytic geometry gave them a means of associating equations with these curves. With passing time, the study of equations took primacy, so that the graph came to be seen as an attribute of the equation. Boyer chronicles this shift in his *History of Analytic Geometry* [5, esp. chapter VII]. Perhaps nowhere is this change of perspective better illustrated than in the case of the cusp of the second kind, a kind of point where a curve doubles back on itself, as in Figure 11.5.

The Marquis de l'Hôpital (1661–1704) gave a geometric construction for such a point, but did not give an equation for it. He was a French aristocrat who learned calculus from Johann Bernoulli (1667–1748) in 1691 and 1692. He wrote the first differential calculus textbook *Analyse des infiniment petits pour l'intelligence des lignes courbes* [6] in 1696. The title means "Analysis of the infinitely small for the understanding of curved lines." This is where the rule we call L'Hôpital's Rule first appeared in print, although it was Bernoulli who actually discovered the rule. In the *Analyse des infiniment petits*, l'Hôpital gave the first definition and example of a cusp of the second kind.

Jean Paul de Gua de Malves (ca. 1712–1786) made the next major contribution to the study of cusps. He was the son of Jean de Gua, the baron of Malves, and was born in the south of France, in or around 1712. He became a member of the Paris Academy of Sciences in 1741, a year after the publication of his book *Usages de l'analyse de Descartes* ... [7]. His long-winded title means "Uses of Descartes' analysis to discover, without the aid of differential calculus, the principal properties of geometric lines of every degree."

Although Gua de Malves did not use differential calculus *per se* in his book, like many authors of his era he freely considered the cases in which variables take infinitely small or infinitely large values. This sort of free-wheeling reasoning sometimes led 18th and 19th century authors to false conclusions and Gua de Malves was no exception. In his book, he believed he had proven that "it is impossible to encounter in curves, such a type of point as Mr. le Marquis de l'Hôpital, who first spoke of them, has called a cusp of the second kind."[7, pp. 69–70] He argued that if a curve seemed to exhibit such a cusp, it was in fact an illusion [7, pp. 81–83]. He held "that every time one finds oneself with such a cuspidal point in a curved line, one is nonetheless mistaken, and if one completes the description following the equation that expresses its nature" then the illusion would be resolved [2, §4]. This quotation illustrates the supremacy of algebra by the mid-18th century: where there appears to be a conflict between a geometric construction and an algebraic description, *it is the equation that expresses the true nature of the curve.*

Although Gua de Malves' book was influential, it was a few years before other mathematicians realized that he was wrong on this point: an algebraic curve *can* have a cusp of the second kind. Leonhard Euler (1707–1783) was the first person to discover such a curve, although it was Jean le Rond d'Alembert (1717–1783) who first got an example of such an equation into print.

In 1744, Euler discovered an algebraic equation whose graph exhibits a cusp of the second kind. He wrote about it in a letter to Gabriel Cramer (1704–1752) on October 20 [8], but the equation didn't appear in print until the summer of 1748, when his book *Introductio in analysin infinitorum* [9] was published. Even in the *Introductio* it was only mentioned in passing.[1] It wasn't until Euler wrote a research paper on the problem [1], which appeared in 1751, that he gave a full account of the theory of cusps and an explanation of why Gua de Malves' proof was incorrect.

In 1746, d'Alembert sent the manuscript of his paper "Research on the Integral Calculus" [11] to the Berlin Academy, where Euler was the Director of the mathematics section. The paper was included in the volume of Academy's journal for 1746, which was actually published in 1748. In this paper, he included an even simpler-looking algebraic equation whose graph exhibits a cusp of the second kind.

Jean d'Alembert was competitive and quarrelsome; see [4]. He was feuding with Euler when the latter's article [1] appeared in print. When he read Euler's article, which contained mention of both his own example and Euler's, he believed that Euler was remiss in not giving him credit for being the first to discover an equation exhibiting a cusp of the second kind. Not knowing that Euler had actually made his own discovery in 1744, he sent an angry letter to the Berlin Academy [12, pp. 337–346], demanding that Euler acknowledge his priority in the discovery. Rather than engage in an ugly public debate with d'Alembert, Euler simply ceded priority to him in a brief notice in the next volume of the Berlin Academy's journal [13]. For more on this dispute, see [3]. Not only did Euler truly deserve priority for this result, but his example is superior to d'Alembert's in the sense that it is an example of lowest possible degree, as we will see at the end of this chapter.

11.3 In the Classroom

11.3.1 Cusps

Those who know some calculus may know a cusp as a special kind of point on the graph of a function, where the function is continuous but not differentiable. But even if you don't know calculus, you probably know the word. It appears in the names of two of types of human teeth: the cuspids, which have one sharp point, and the bicuspids, which have two. The word has its origins in the Latin word *cuspis*, or point. Sometimes it means a pointed projection, like the ones on your teeth. Other times it means a point of transition. Both of these meanings are reflected in the mathematical sense of the word.

---

[1] When Euler wrote the *Introductio* in 1743–44, he repeated Gua de Malves' argument that there are no cusps of the second kind. He only discovered his counterexample after he had sent the manuscript of the book to his publisher. As a consequence, his book is self-contradictory, stating in one place that there are no equations with such a cusp and then giving his example of an equation that has one; see [9, 10, §333–334]. For more on this mix-up, see [3].

11.3. In the Classroom

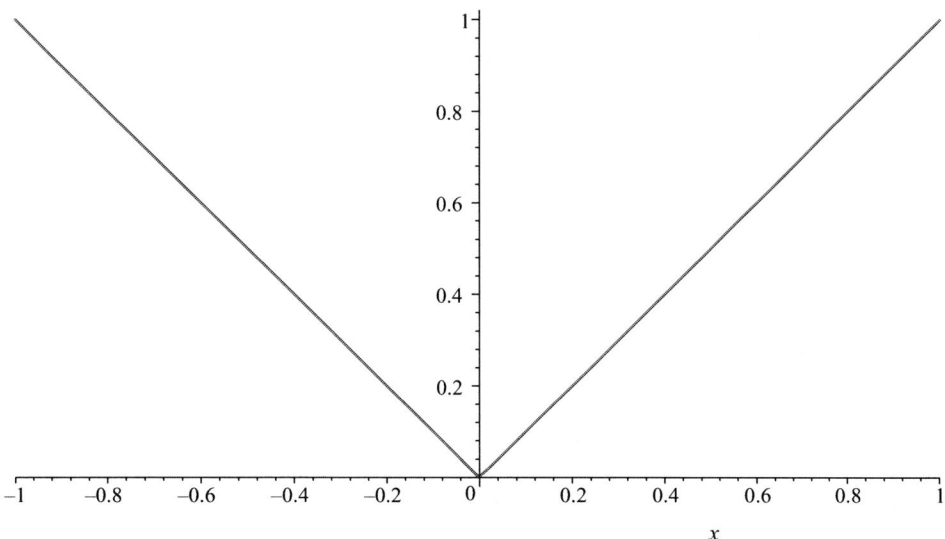

Figure 11.1. The Absolute Value Function

The most familiar example of a graph that is continuous but not differentiable is the graph of $y = |x|$, which has a *corner point* at the origin; see Figure 11.1. Since there is no well-defined tangent line there, the absolute value function $f(x) = |x|$ has no derivative at the origin. But since there are no breaks or holes in the graph, the function is continuous everywhere. The graph has two *branches*. The left branch in follows the line $y = -x$, while the right branch follows the line $y = x$. The two branches meet at the origin, making a right angle.

The corner point in Figure 11.1 doesn't quite count as a cusp, at least not according to the usual definition. Probably the simplest example of a cusp is the origin on the graph of $y = x^{2/3}$; see Figure 11.2. The situation here is quite similar to Figure 11.1: there are two branches meeting in a sharp point at the origin. The crucial difference is that the angle between these two branches is, in the words of Euler, "infinitely small." [1, §2]

Rather than following Euler with an appeal to our informal notion of the infinitesimal, let's just adopt the standard modern definition: a *cusp* is a point where two branches of a graph meet, at which the tangents to the branches coincide. It turns out that if the curve in question is the graph of some function f, then that common tangent line has to be vertical, as in the case of Figure 11.2. That's why some calculus books define a cusp to be a point $(c, f(c))$ at

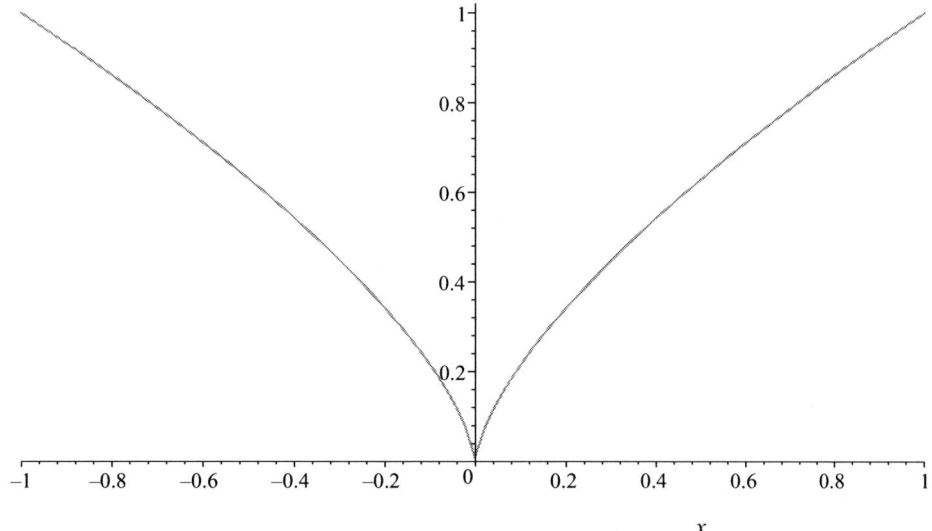

Figure 11.2. The graph of $y = x^{2/3}$.

which the function f is continuous, but for which

$$\lim_{x \to c^-} f'(x) = +\infty \quad \text{and} \quad \lim_{x \to c^+} f'(x) = -\infty,$$
$$\text{or}$$
$$\lim_{x \to c^-} f'(x) = -\infty \quad \text{and} \quad \lim_{x \to c^+} f'(x) = +\infty.$$

A curve certainly can have a cusp at which the tangent line is horizontal (see Figure 11.4, for example) or oblique, but in that case, the curve always fails the familiar "vertical line test." So such a curve is never the graph of a function.

11.3.2 Activities

- Carefully graph the equation $y = |x|$. By also graphing $y = |2x|$ or $y = |3x|$ or $y = |(1/4)x|$, observe that we can make the angle between the two branches take any value between 0 and π radians. The limiting case $y = 0$ gives an angle of π radians, but we observe that the angle between the two straight line branches of such a function can never actually be equal to 0.

- Graph the familiar curve $y = x^3$. Reflect this curve in the line $y = x$ to get the graph of the curve $y = x^{1/3}$. Now reflect the portion of this graph in the third quadrant in the x-axis to get the graph of $y = \sqrt[3]{|x|} = |x|^{1/3}$, which looks quite similar to Figure 11.2.

11.3.3 Functions and Equations

The function concept is one of the two critical ideas behind the modern understanding of calculus, the other being the limit. Any function f that can be expressed as a formula in one variable is naturally associated with an equation $y = f(x)$. Although we often speak about the graph of a function, what we really mean is the graph of this associated equation $y = f(x)$.

Conversely, an equation in x and y may represent a function, if the equation can be solved uniquely for y. The vertical line test expresses this condition in a graphical way: if every vertical line intersects the graph of the equation in at most one point, then the equation implicitly gives y as a function of x. The equation $x^2 + y^2 = 1$ of the unit circle is a good example of a particularly simple equation that does not represent a function. Since the vertical line $x = b$, where $b \in (-1, 1)$, intersects the circle at two points, the graph fails the vertical line test. But since no vertical line intersects the circle in more than two points, the circle can be expressed as the union of the graphs of two functions:

$$y = \sqrt{1 - x^2}, \quad \text{and}$$
$$y = -\sqrt{1 - x^2}.$$

Of course, we usually abbreviate this representation as $y = \pm\sqrt{1 - x^2}$.

Simlarly, ellipses always fail the vertical line test, as do some hyperbolas and parabolas. But in no case does a vertical line — or any other straight line for that matter – intersect any of these graphs in more than two points. This is because all of these curves — called the *conic sections* — are of degree 2. Our next step is the investigation of the connection between the degrees of equations and the intersections of their graphs.

11.3.4 Activities

- Give an example of a parabola that passes the vertical line test and an example of a parabola that fails the vertical line test. Do the same for hyperbolas.

- Draw any hyperbola, parabola, ellipse or circle. Draw a straight line that intersects the curve in two places. Find another straight line that intersects the curve in one place. Find another straight line that doesn't intersect the curve at all. Convince yourself that no straight line can intersect the curve in three places.

11.3.5 Degrees of Equations and Bézout's Theorem

An algebraic equation in x and y has a degree, defined analogously to the degree of a polynomial. Let's consider an equation whose right side is 0 and whose left side consists of a sum of terms, all of which have the form $ax^j y^k$. Here $a \neq 0$ is a real coefficient, while j and k are non-negative integers. The *degree of the term* is defined to be $j + k$ and the *degree of the equation* is the maximum of the degrees of all all the terms. For example, the equation

$$x^5 - x^4 + 2x^2 y - y^2 = 0,$$

which we will encounter later in this chapter, has four terms. Their degrees are 5, 4, 3 and 2, reading from left to right, so the degree of this equation is 5.

An arbitrary algebraic equation, which may involve quotients and fractional powers, can always be written in *standard form*: as a sum of terms of the form $ax^j y^k$ on the left side and the single term 0 on the right side. Thus, every algebraic equation has a degree. For example, the equation

$$y = \frac{1}{x}$$

of the hyperbola can be put into the standard form $xy - 1 = 0$, showing that it has a degree of 2.

Mathematicians in the 18th century were aware that the graphs of two algebraic equations of degrees m and n can intersect in at most mn points. The result is called Bézout's Theorem, in honor of Étienne Bézout (1730–1783), who gave the first acceptable proof of this result. (There is a stronger statement of the theorem: the number of points of intersection is exactly mn, when one counts real and complex roots, as well as roots at infinity, in all their multiplicities. This immediately implies that there are $\leq mn$ real points of intersection.)

In particular, when $m = 1$, Bézout's Theorem says that a line can intersect an algebraic curve of degree n in at most n points. Conversely, if a line intersects the graph of an algebraic equation in n points, then the degree of the equation is at least n.

Let's apply this result to the graph of $y = x^{2/3}$. By trial and error, we can find a line, such as $y = x + 0.1$, which intersects this graph in three distinct points; see Figure 11.3. This tells us that the degree of the equation must be at least 3. Of course, this comes as no surprise, since we can put the equation $y = x^{2/3}$ into standard form by first cubing both sides. The standard form of the equation is $y^3 - x^2 = 0$, so the degree is precisely 3.

11.3.6 Activities

- Graph the familiar equation $y = x^3$, which of course has degree 3. Draw a straight line that intersects it in three places. Convince yourself that no straight line can intersect it in four or more places.

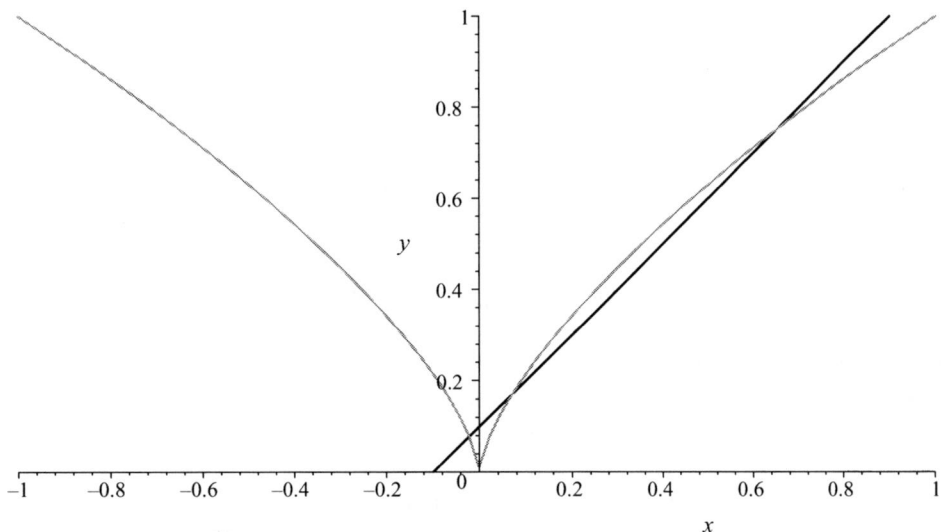

Figure 11.3. The graphs of $y = x^{2/3}$ and $y = x + 0.1$.

- Draw a parabola, then draw an ellipse that intersects it in 4 places. This happens because both have equations of degree 2, so $m = n = 2$ in Bézout's Theorem.

- Do the same thing with a hyperbola and an ellipse. Can you arrange to have all 4 points of intersection fall on the same branch of the hyperbola? Two on each branch? Three on one branch and one on the other?

11.3.7 Involutes and Inflections

In his *Analyse des infiniment petits*, l'Hôpital defined two kinds of cusps. A *cusp of the first kind*, like the one in Figure 11.2, is one where the two branches have opposite concavity with respect to the common tangent line. In Figure 11.2, the common tangent is the y-axis, so we might say that the left branch is concave to the left and the right branch is concave to the right, although this is not standard terminology. Alternately, if we interchange the variables x and y in the equation $y = x^{2/3}$, we get the equation $y^2 = x^3$ or $y = \pm x^{3/2}$; see Figure 11.4. The shape is exactly the same as Figure 11.2, but the branches are now concave up and concave down, in the sense that these terms are usually used in a calculus course.

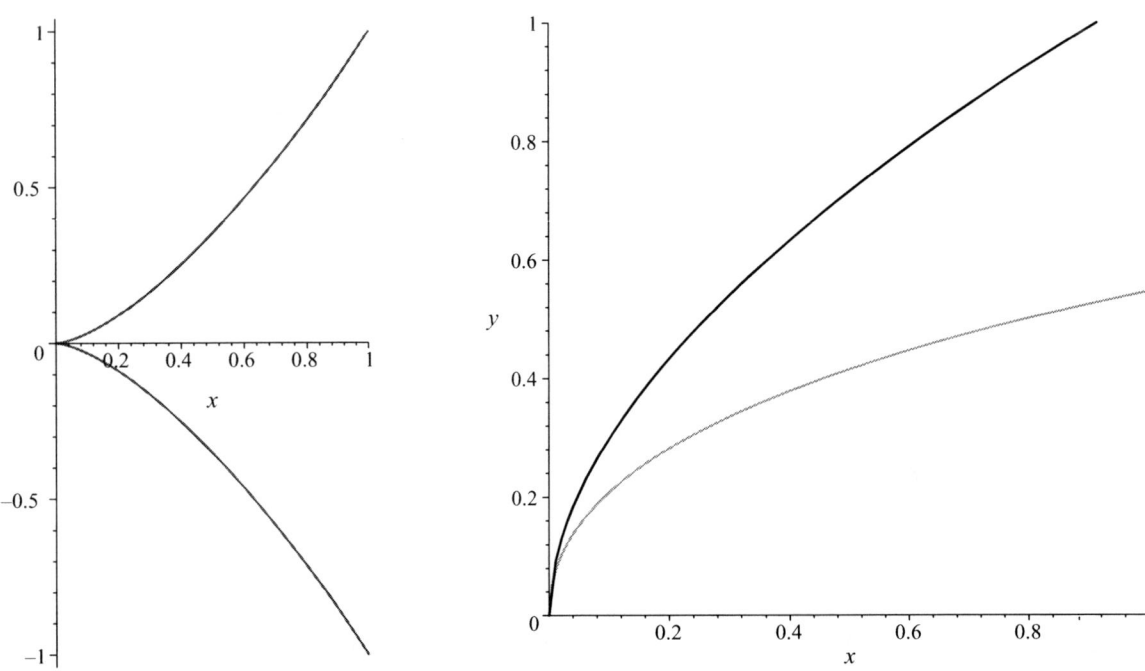

Figure 11.4. The graph of $y^2 = x^3$.

Figure 11.5. A cusp of the second kind, or ramphoid cusp.

L'Hôpital also conceived of a *cusp of the second kind*, in which the two branches are concave in the same direction, as in Figure 11.5, where the common tangent is once again the y-axis. The two kinds of cusps have flowery, if obscure, alternate names in English. A cusp of the first kind is sometimes called a *keratoid* cusp, meaning horn-like. A cusp of the second kind is a *ramphoid* cusp, meaning beak-like. Euler described it in his 1751 paper as "a certain kind of cuspidal point, resembling the beak of a bird, and formed from two branches of a curve, whose concavities turn in the same direction." [2, §1]

It might seem strange to us that l'Hôpital didn't give an equation for a curve with a cusp of the second kind. Instead he gave something that was just as convincing to his late-17th century audience, if not more so: a geometric construction. His construction was based on the process of *involution*, a standard technique at that time for defining a new curve based on an old one. A modern definition of the *involute* usually begins by defining the *evolute* of a given curve as the locus of its centers of curvature. Involution is then defined as the inverse operation: that is, C is the involute of D if and only if D is the evolute of C.

L'Hôpital's definition of the involute was based on a mechanical model. He imagined the given curve as being solid and rigid, with a fine string or thread stretched along its length. He supposed that the thread was peeled off gradually, being held taut from its endpoint as it was pulled away. A new curve is thereby defined by tracing the path of the

11.3. In the Classroom

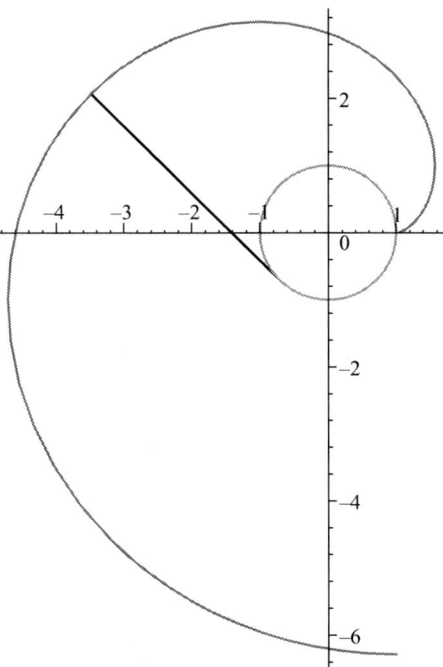

Figure 11.6. The involute of the unit circle.

endpoint. Figure 11.6 is an illustration of this process when the given curve is the unit circle. The thread was initially wrapped clockwise around the circle, just as real thread might be wound around a spool. The endpoint of the thread was initially at the point $(1, 0)$, where the involute (the spiral curve) meets the cirle. As the thread is peeled away in the counterclockwise direction, it defines straight line segments of ever increasing length, tangent to the circle, from the *point of contact* on the circle to the endpoint of the thread, which is a point on the involute. The length of this line segment is the the arc length of the circle from the point of contact to the initial endpoint $(1, 0)$. The straight line in Figure 11.6 illustrates this line segment at the instant when the point of contact is $(-1/\sqrt{2}, -1/\sqrt{2})$, which is $5\pi/4$ radians (225°) counterclockwise from $(1, 0)$. Therefore, its length is $5\pi/4$.

To construct a cusp of the second kind, l'Hôpital starts with any curve that has an inflection point. When constructing the involute of such a curve, the endpoint of the thread will double back on itself the instant that the point of contact coincides with the inflection point. This is illustrated in Figure 11.7, which comes from Euler's article [1]. In this figure, ABM is the given curve, with an inflection point at B. The straight lines BD and MN represent line segments described by the thread at the instants when the points of contact are B and M, respectively.

Here is Euler's description of l'Hôpital's construction:

> Let ABM be an arbitrary curve which has a point of inflection at B. Suppose also that we wrap a thread around this curve, which we then unwrap by pulling it successively from the point A, until it just becomes

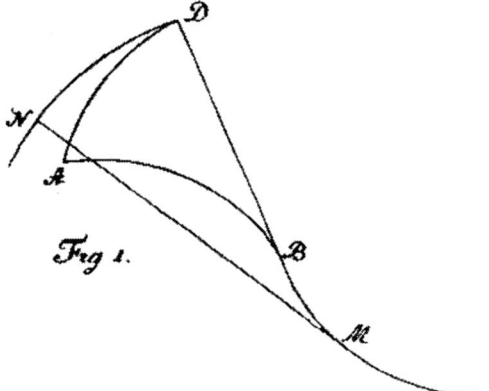

Figure 11.7. Euler's Illustration of L'Hôpital's Construction

detached at the inflection point B, and the extremity A will describe, by this evolution, the arc AD, to which the detached thread BD will be perpendicular so will be its radius of curvature, as we know from the theory of evolutes. But if we continue this evolution beyond the point B, the thread BD will turn back and, coming to the position MN, its extremity will describe the arc DN, which is, as a consequence, the continuation of the arc AD. Now these two arcs AD, DN, which at the point D form an infinitely small angle, are concave in the same direction. [1, §2]

11.3.8 Activity

For this activity, you will need a plastic cup or similar round object, a piece of string longer than the largest circumference of the cup, paper, pencil and tape. Tie a small loop in one end of the string and tape the other end to the lip of the cup. Place the cup upside down in the middle of the piece of paper. Wind the string tightly around the cup in the clockwise direction and put the tip of the pencil in the loop of the string. With the pencil point always in contact with the paper, unwind the string from around the mouth of the cup. As the string unwinds, the pencil point will trace out the involute of the circle, as illustrated in Figure 11.6.

11.3.9 Euler's Example

Euler's equation exhibiting a ramphoid cusp is most easily understood in the form $y = \sqrt{x} \pm x^{3/4}$. The graph is given in Figure 11.8. The dotted line represents the graph of the square root $y = \sqrt{x}$. The solid lines lying above and below it are the two branches of Euler's curve, the result of adding and subtracting the term $x^{3/4}$.

To determine the degree of Euler's example, we'll have to do some algebra to get rid of the square and fourth roots. This process is usually called "rationalizing" the equation. We start by rewriting the equation as

$$y - \sqrt{x} = \pm x^{3/4}.$$

Squaring both sides, we have

$$y^2 - 2\sqrt{x}y + x = x^{3/2}.$$

Rearranging the terms, this gives

$$y^2 + x = x^{1/2}(x + 2y).$$

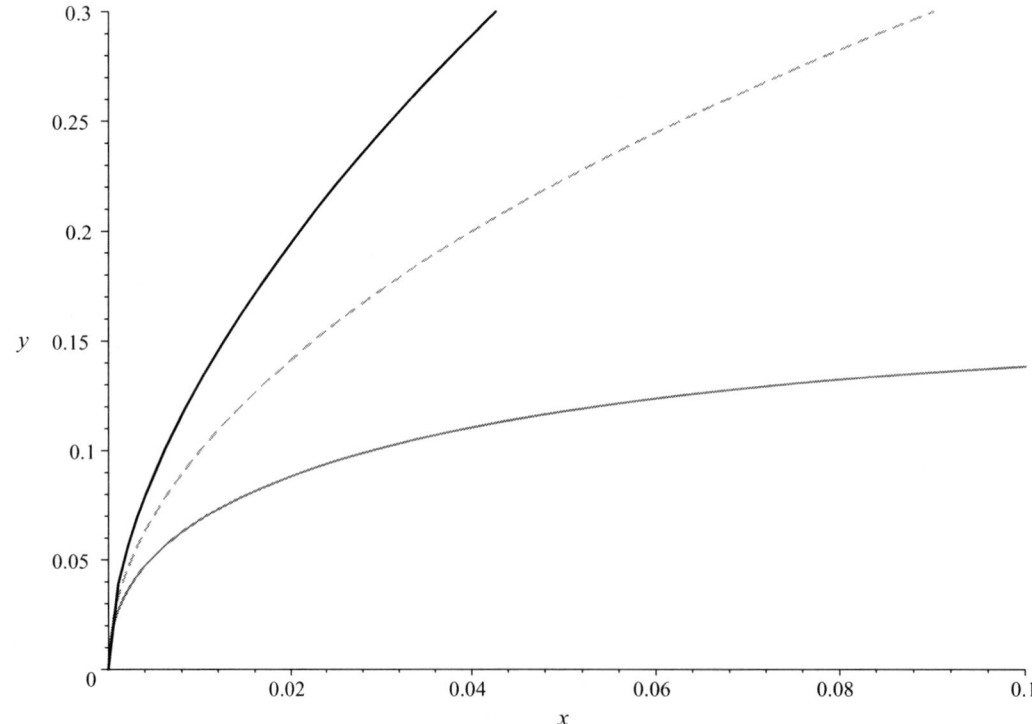

Figure 11.8. Euler's Example

11.3. In the Classroom

Squaring a second time, we get

$$y^4 + 2xy^2 + x^2 = x^3 + 4x^2y + 4xy^2.$$

So the standard form is

$$y^4 - 4x^2y - 2xy^2 - x^3 + x^2 = 0,$$

making this an equation of degree 4, sometimes called a quartic equation.

11.3.10 Activities

- Euler also showed that the equation $y = \sqrt{x} \pm x^{5/4}$ has a ramphoid cusp at the origin. Produce a graph of this equation similar to the one in Figure 11.8. Note that the two branches are squeezed closer together than the ones in Figure 11.8.

- Follow the same steps used in this section to rationalize the equation $y = \sqrt{x} \pm x^{5/4}$. The standard form of this equation should turn out to be $-x^5 + y^4 - 4x^3y - 2xy^2 + x^2 = 0$, an equation of degree 5.

11.3.11 D'Alembert's Example

In his 1748 paper "Research on the Integral Calculus" [11], d'Alembert made passing reference to the problem of cusps. He included an algebraic equation that looks simpler than Euler's example, whose graph has a ramphoid cusp.

The equation is most easily understood in the form $y = x^2 \pm x^{5/2}$. The graph is given in Figure 11.9, where the dotted line represents the parabola $y = x^2$. The solid lines are the two branches of d'Alembert's curve, the result of adding and subtracting the term $x^{5/2}$.

D'Alembert's example is an equation of degree 5. To see this, we rewrite the equation as

$$y - x^2 = \pm x^{5/2}.$$

Squaring both sides, we have

$$y^2 - 2x^2y + x^4 = x^5,$$

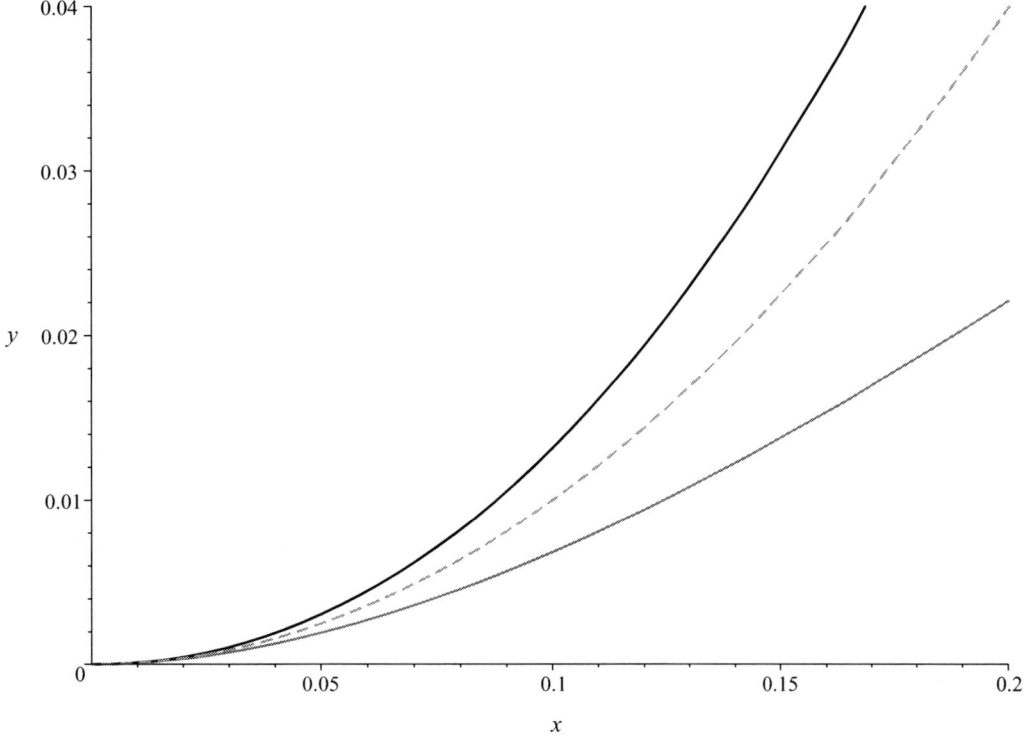

Figure 11.9. D'Alembert's Example

So the standard form for this equation is

$$x^5 - x^4 + 2x^2y - y^2 = 0.$$

11.3.12 Activities

- Euler showed that the equation $y = x^2 \pm x^{7/2}$ also has a ramphoid cusp at the origin. Follow the same steps used in this section to rationalize this equation. Put the equation into standard form to show that it has degree 7.

- The equation $y = x^2 \pm x^{3/2}$ also has a cusp at the origin. Graph this equation to show that this cusp is a keratoid cusp. Rationalize this equation to show that it has degree 4.

11.4 Conclusion

Not only was Euler the first person to find an algebraic example of a ramphoid cusp, his example is the best possible, in the following sense. Euler observed that given any ramphoid cusp, one can always draw a straight line that intersects it in four places. Figure 11.10 is an example illustrating this. It's worth sketching a few other examples of ramphoid cusps and convincing oneself that this can always be done. Consequently, by Bézout's Theorem, any equation with such a cusp must have a degree of at least 4. Since Euler's equation is of degree 4, his example has the lowest possible degree. On the other hand, d'Alembert's example of degree 5 is suboptimal.

Many more examples of cusps can be found in [2], which is surprisingly accessible to modern readers. For example, $y = \sqrt[3]{x} \pm \sqrt{x}$ is a curve of degree 6 with a bird's beak at the origin. Euler even gives a family of examples in which the tangent line of the cusp is oblique to the axes, the simplest one being $y = x + x^2 \pm x^{5/2}$. As with many of his mathematical papers, one get the sense that Euler is enjoying himself immensely as he investigates this delightful topic on the boundary between algebra and geometry.

11.5 Notes on Classroom Use

There is plenty of material in this chapter and to cover all of it would take 1-2 hours of class time. However, most instructors will probably wish to pick and choose topics, depending on the level of the course being taught and the

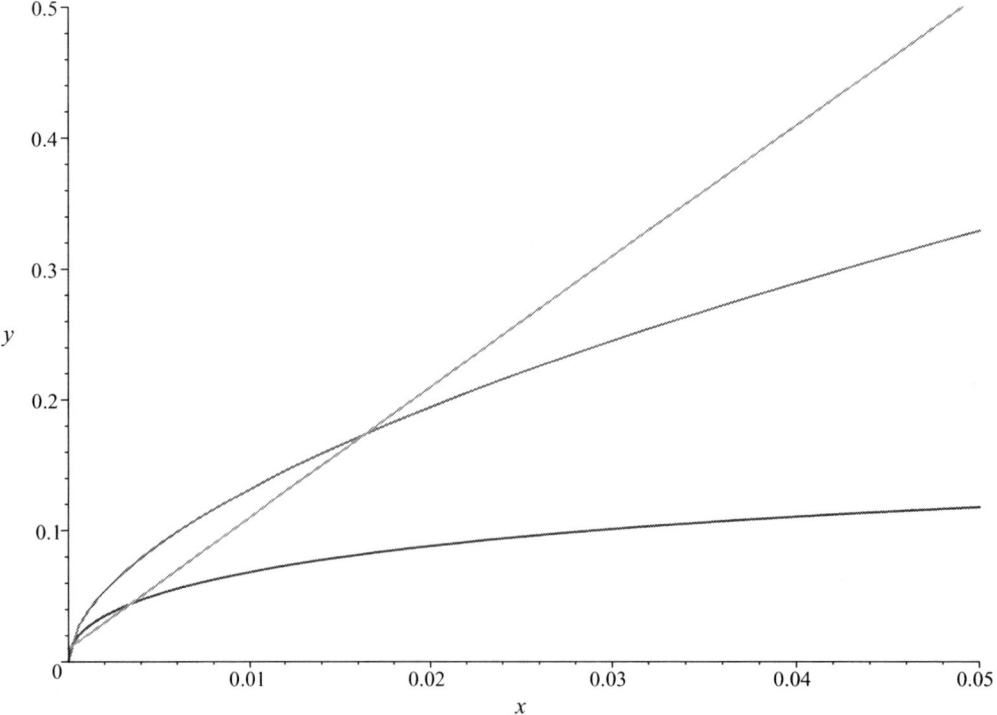

Figure 11.10. Euler's curve and the line $y = 10x + 0.01$

type of activity that would be suitable for their students. The Classroom section of the chapter is divided into six subsections. In many courses, the students will be familiar enough with "Functions vs. Equations" to skip that subsection, although the connection between the well-known vertical line test and the less well-known theorem of Bézout makes this a valuable segue to the next part. The subsection on "Involutes & Inflections" can be omitted without loss of continuity, but the hands-on activity in this portion, which could be done either as a student activity or an instructor's demonstration, provides a wonderful example of an unfamiliar curve that can be constructed without reference to an equation.

The activities included in each subsection are intended to provide insight into the concepts involved in this mathematical narrative. Some of the activities stress curve sketching as an active learning tool in the study of intersections of lines, conics and cubic curves. Others, especially the ones in the core sections of "Euler's Example" and "D'Alembert's Example" provide exercises in algebraic techniques of rationalizing equations, which are necessary to the understanding of the cusp points.

Bibliography

[1] Leonhard Euler, "Sur le point de rebroussement de la seconde espece de M. le Marquis de l'Hôpital," *Mémoires de l'académie des sciences de Berlin*, **5** (1749/1751), 203–221. Reprinted in Leonhard Euler, *Opera Omnia*, Series I, vol. 27, 236–252.

[2] L. Euler, "On the Cuspidal Point of the Second Kind of Monsieur le Marquis de l'Hôpital," translation of [1] by Robert E. Bradley, retrieve via the link in the article [4], or at the eulerarchive.org page for E169.

[3] Robert E. Bradley, "The Curious Case of the Bird's Beak," *International J. Math. Comp. Sci.*, **1** (2006) 243–268.

[4] Robert E. Bradley, "The Nodding Sphere and the Bird's Beak: D'Alembert's Dispute with Euler," *Convergence*, mathdl.maa.org, January 2006.

[5] Carl B. Boyer, *History of Analytic Geometry*, Scripta Mathematica, New York, 1956. Reprinted by Dover, New York, 2004. See pp. 139-191 and esp. p. 180ff for Euler's *Introductio*.

[6] Guillaume François Antoine, Marquis de L'Hôpital, *Analyse des infiniment petits, pour l'intelligence des lignes courbes*, l'Imprimerie royale, Paris, 1696. See section V, esp. pp. 102–103, for the cusp of the second kind.

[7] Jean Paul de Gua de Malves, *Usages de l'analyse de Descartes pour découvrir, sans le secours du Calcul Différentiel, les Propriétés, ou affectations principales des lignes géometriques de tous les ordres*, Briasson, Paris, 1740. See esp. pp. 69–85 for the discussion of cusps.

[8] Leonhard Euler, Letter of October 20, 1744, apparently to Gabriel Cramer, in Leonhard Euler Papers (MSS 490A), Dibner Library of the History of Science and Technology, Smithsonian Institution Libraries.

[9] Leonhard Euler, *Introductio in Analysin Infinitorum*, 2 vols., Bousquet, Lausanne, 1748. Reprinted in Leonhard Euler, *Opera Omnia*, Series I, vols. 8–9. See vol. II, §332-335.

[10] Leonhard Euler, *Introduction to Analysis of the Infinite: Book II*, translated by John D. Blanton, Springer-Verlag, New York, 1989. See esp. pp. 208–211.

[11] Jean d'Alembert, "Recherches sur le calcul intégral," *Mémoires de l'académie des sciences de Berlin*, **2** (1746/1748) 182–224.

[12] Leonhard Euler, *Opera Omnia*, Series IVA, vol. 5 (correspondence with Clairaut, d'Alembert and Lagrange), eds. A. P. Juškevič, R. Taton, Birkhäuser, Basel, 1980. See esp. pp. 19–20, 263, 348.

[13] Leonhard Euler, "Avertissement au sujet des recherches sur la précession des équinoxes," *Mémoires de l'académie des sciences de Berlin*, **6** (1750/1752), 412. Reprinted in Leonhard Euler, *Opera Omnia*, Series I, vol. 31, 124.

12

The Latitude of Forms, Area, and Velocity

Daniel J. Curtin
Northern Kentucky University

12.1 Introduction

Long before the calculus arrived a medieval philosopher, Nicole Oresme, developed what he called the *latitude of forms*, a graphical representation that sheds light on the fundamental connection between area and what we now call the integral. In a calculus course, the latitude of forms can be used to introduce the idea of the integral as area, while simultaneously introducing the idea that the distance traveled is the integral of velocity. Of course the two ideas can be addressed separately, if you prefer. In that case, the latitude of forms might be used to connect the two. In any event, you will be reviewing some simple geometry that students have often forgotten.

At the risk of being untrue to the original, I have modernized my presentation. The Commentary section will attempt to partially correct this distortion.

12.2 Historical Background

Scholastic philosophers, following Aristotle, were greatly interested in explaining the workings of the natural world. In this sense they appear to our eyes as scientists. They also were interested in precise definitions, careful distinctions between cases, and rigorous logical deduction. To us they appear to be mathematicians and analytic philosophers. Yet when we read their works, we can see they were also trying to explain *why* things work, and seeing how well their explanations fit their theology. Thus to us they appear to be trying to tackle everything at once.

This article focuses on Nicole Oresme (c. 1323–1382), who was born in Normandy and studied at the University of Paris. Immediately upon receiving his doctorate he became grand master of the University of Navarre. He served as secretary to the King of France and quite likely as his chaplain and counselor, too, positions of great influence. Oresme had numerous other clerical appointments at various times and taught at the University of Paris. The last five years of his life were spent as bishop of Lisieux. Despite his heavy duties, he was a prolific writer, producing treatises ranging from mathematics to music, physics, philosophy, and more.

The latitude of forms appears in his *Treatise on the configuration of qualities and motions* (*Tractatus de configurationibus qualitatum et motuum*) [1], which a later writer condensed and called the *Treatise on the Latitude of Forms*. Oresme took ideas from earlier scholastics, such as those at Merton College, Oxford, but put in the mix a new geometric idea of *configuration* — what we might call the graph of a function. This is the connection we want our students to see.

12.3 In the Classroom

I offer two ways to present the latitude of forms. The first is entirely modernized, using current terminology such as "constant acceleration," and can be used to get the ideas across quickly. The second uses Oresme's own definitions. Many students find this look at historical ideas interesting, and it really doesn't take much more time. It does, however, require the instructor to do a little more preparation.

12.3.1 Area and the Latitude of Forms

In an era in which analytic geometry is one of the fundamental ideas of mathematics, it may be hard to appreciate how many centuries of work went into its development. One crucial piece of the puzzle is the idea of expressing one quantity in one direction and the other in a different direction, that is, the modern xy plane. Of course sailors and astronomers had been using a very special case of this, latitude and longitude, for a long time.

This was what Nicole Oresme had in mind when he defined his latitude of forms. In studying motion he viewed time as the latitude (our x-axis) and velocity as longitude (our y-axis).

If a body is moving with a constant velocity v over a time period t, Oresme's *configuration* is given by representing the moments of time on a horizontal base segment, while the velocity at each moment is represented by a vertical line segment at that point. The result is a rectangle (Figure 12.1).

As Oresme describes, "Every velocity endures in time. And so time or duration will be the longitude of the velocity and the intensity of the same velocity will be its latitude." [1, p. 289] His terminology is consistent with his picture, since at this time it was common for East to be upward on a map, rather than our modern convention of North. (Hence the term *orientation*.) We see that the area of the rectangle is v multiplied by t and that is also the distance traveled.

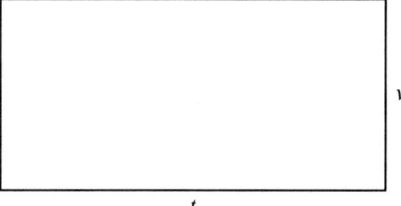

Figure 12.1. Constant velocity.

Here's the crucial idea, in its simplest form: *The area of the configuration, a geometric concept, is the distance traveled, a physical concept.* Note that for Oresme no numbers are actually used for v and t, so the picture gives the general rule directly. We would say that for a constant velocity v, the distance s traveled in time t is given by $s = vt$.

Next Oresme looked at a motion where the velocities start at 0 and increase in such a way that the tops lie on a line segment. We recognize this as motion with a constant acceleration and initial velocity 0. Figure 12.2 represents his diagram. As before he takes the area to represent the distance traveled.

We might use a numerical area formula for the triangle, but Oresme proceeded geometrically; see Figure 12.3. Consider the rectangle $ABCD$ with height half of the segment BE. Since triangles ADF and ECF are congruent,

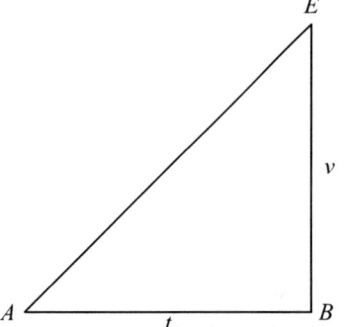

Figure 12.2. Constant acceleration from rest.

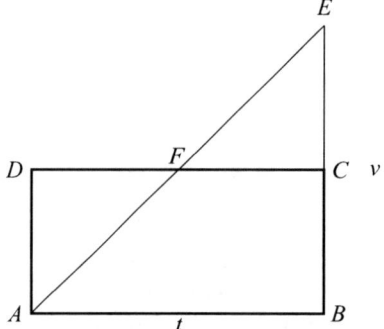

Figure 12.3. Constant acceleration from rest compared with constant velocity at half the final velocity.

the area of the triangle ABE is the same as that of rectangle $ABCD$. Thus the distance traveled under the original motion is the same as a distance traveled at a constant velocity that is half the final velocity (BE). In fact, this result had been discovered already by scholars at Merton College, Oxford, but their derivation was entirely in words!

In our notation, what Oresme has shown is that if a body at rest at time $t = 0$ moves at a constant acceleration to a final velocity v at time t, the distance traveled is $vt/2$.

Although Oresme did not do this, we can draw the configuration for the constant acceleration a over the same time period. In other words, we can make the vertical segment be the acceleration at each time. The area will be at; see Figure 12.4. The area must be the final velocity v, so we get the distance traveled as $at^2/2$. Thus in a sense Oresme has worked out the general formulas for motion under constant acceleration.

Diagrams much like Figure 12.3 appear two hundred years later in Galileo's notes as he works out his theory of falling objects near the surface of the earth, which he correctly inferred involved motion under constant acceleration. Did some of Galileo's ideas come directly from Oresme? We don't know.

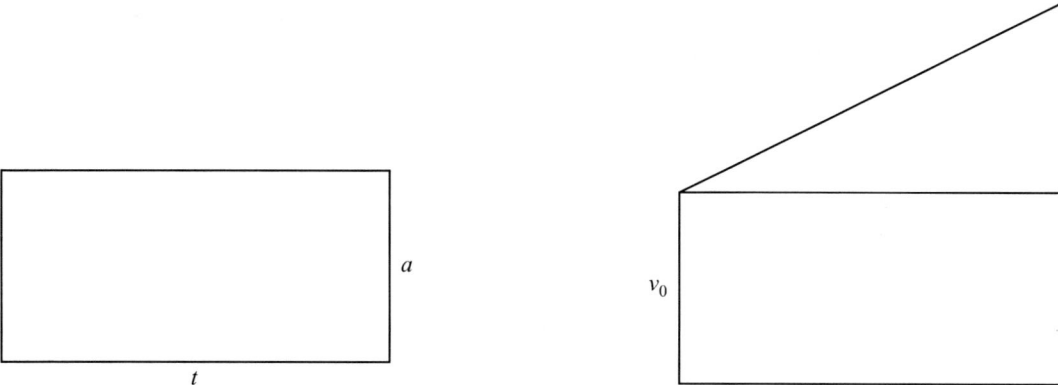

Figure 12.4. Constant acceleration.   **Figure 12.5.** Constant acceleration with non-zero initial velocity.

What if the initial velocity is not 0? Then the picture is as in Figure 12.5. Work out the result Oresme-style and in modern terms. In particular, notice how the picture makes it clear that the total motion may be considered as the sum of two motions, one with constant velocity, one with constant acceleration.

Of course this idea can be applied to any figure whose area can be worked out. For the three cases shown in Figures 6–8, each of which is based on a diagram by Oresme, derive the distance traveled from the area. Also take a shot at describing what the motion would be like for the given velocity configuration. Note that the numerical values are purely arbitrary and the diagrams suggest general results. Oresme would have seen no need for specific values.

More complicated examples are available for the particularly interested student, who could be directed to flip through Clagett's translation [1] and interpret some of the diagrams found there.

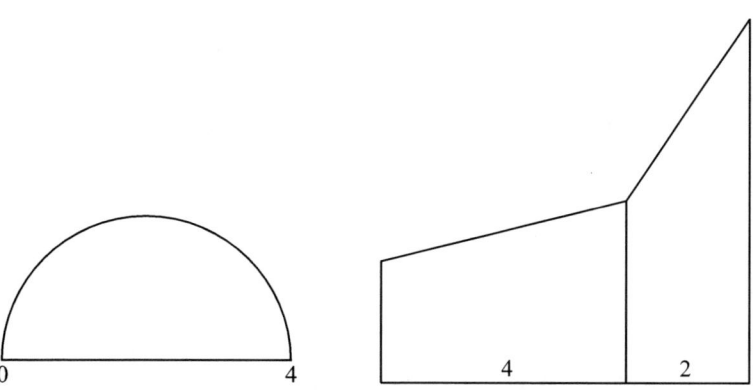 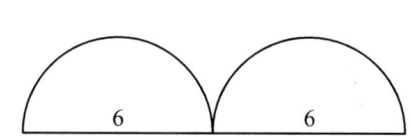

Figure 12.6. A semicircle.   **Figure 12.7.** A motion that combines different motions sequentially.   **Figure 12.8.** More combined motions. (Not quite a figure 8.)

12.4 Taking It Further

Oresme did not actually make the distinctions we do based on our understanding of velocity and acceleration. Our concepts come from much later work by Galileo, Newton, and others. For Oresme the questions were much more philosophical. His latitude of forms was to apply to anything in which one could speak of greater and lesser — not just velocity, but heat and cold, intensity of color, even human goodness. Since he mostly worked with ratios, he was not really interested in specific values but rather the nature or quality of the form.

When I teach this I talk only about motion, and I use Oresme's own definitions, along with the modern interpretation. Historically-minded students find this fascinating — and the rest seem to tolerate it good-naturedly.

12.4.1 Uniform and Difform Motion

Let's start with what we would call motion with constant velocity. Following early writers, Oresme called this *uniform motion*, since there was one (*uni*) form of motion; that is, all the vertical segments were the same.

If the motion isn't uniform, then the segments are different, so it was called *difform motion*. Scholastics are sometimes accused of splitting hairs, but their careful attention to definition and small distinctions often paid great dividends and pointed the way to our modern precision with concepts and definitions. Oresme distinguishes between *uniformly difform motion*, in which the segments change, but in a uniform way, and *difformly difform motion*, in which the segments change in any other way.

What is happening in uniformly difform motion? The segments increase uniformly — that is, in proportion to their time — so that twice the time, for example, leads to twice the increase in velocity. Oresme points out that this means the tops of the segments all lie in a line. We can recognize this as motion with a constant acceleration.

Then Oresme's work shows that the distance traveled under a uniformly difform motion is the same as the distance traveled under a uniform motion over the same time with a velocity equal to half the final velocity of the first motion.

Finally, difformly difform motion covers all the rest, including the semicircle (Figure 12.6) and even something like Figure 12.9.

Figure 12.9. The Wandering Minstrel?

12.5 Conclusion

In principle, the latitude of forms can be applied to any configuration, and Oresme draws many examples beyond the few here.

But how do we find the area in order to know the distance traveled? That's where the calculus comes in. The integral calculus is designed to attack exactly this problem and to solve it in a wide variety of cases. Oresme would have vastly enjoyed it.

12.6 Comments

Oresme did not have any algebraic notation at his disposal, so while the shapes are his, all the formulas above are anachronisms. He never assigned a specific numerical value to any velocity or time. In fact, for the example on constant acceleration (Figure 12.2) he proved that any right triangle can be used to represent it. He went on to say, "if some quality is designated by one triangle, another quality of similar but double intensity must be designated by a triangle that is twice as high, and similarly for proportionally greater [intensities]...." [1, p. 187]

Oresme noted that the semicircle example is different, for if you double all the heights you get half an ellipse, which is not a geometrically similar shape. In effect he stated a formal problem about difformly difform motion, which he left for us to consider.

In his work, Oresme also used the ideas of latitude of forms to sum infinite series and to consider how the quality of curvature differs in kind from that of velocity. The interested reader will enjoy pursuing it further in [1]. For other aspects of Oresme's mathematical works, a good place to start is [2].

Bibliography

[1] Marshall Clagett, *Nicole Oresme and the Medieval Geometry of Qualities and Motions: A Treatise on the Uuniformity and Difformity of Intensities Known as Tractatus de configurationibus qualitatum et motuum*, The University of Wisconsin Press, Madison, Wisconsin, 1968.

[2] Victor J. Katz, *A History of Mathematics: An Introduction*, Second Edition, Addison-Wesley, Reading, Massachusetts, 1998.

13

Descartes' Approach to Tangents

Daniel J. Curtin
Northern Kentucky University

13.1 Introduction

While the modern version of tangents is central to the ideas of the differential calculus, I find students can profit from seeing an earlier and different approach. This minor detour also has the amusing aspect of using quite modern technology to help with an old problem. I use this material at the beginning of Calculus 2, when the students are fairly comfortable with the modern definition of derivative. One class period is used to present Descartes' approach, then students receive a take-home assignment.

13.2 Historical Background

In *La Géometrie* (1637) [2] René Descartes presents his general method of drawing a straight line to make right angles with a curve at an arbitrarily chosen point upon it. He praises his own approach as solving "not only the most useful and most general problem in geometry that I know, but even that I have ever desired to know" [2, p. 95]. In our terms, he sought the normal line to a curve at a given point, from which the tangent line can easily be found as well.

Descartes' approach is quite different from the modern one, which raises the question: If his method was as important as he thought, why did it not prevail? I will describe his method and suggest some exercises that can help a student to understand what Descartes was doing and to see why other approaches won out. Using calculators or computer algebra systems allows us to remove much of the drudgery of this historical reenactment.

13.3 In the Classroom

13.3.1 Descartes' Approach to Tangents

Figure 13.1 is based very loosely on Descartes' own diagram [2, pp. 94 and 98]. The most important anachronism is the vertical axis. Descartes does draw the axis AC. He then labels AM as y and PM as x, thus establishing coordinates, but not coordinate axes. You see Descartes' convention for x and y is the reverse of ours. I tell my students the discrepancy is due to Descartes' well-known habit of working on mathematics while reclining in bed, but they never seem to believe me. We will use the modern convention.

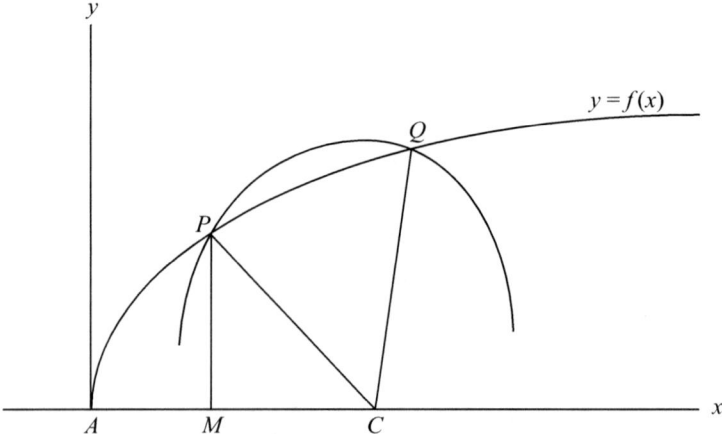

Figure 13.1. For us: $AM = x$, $PM = y$, $AC = v$, $CP = s$.

Descartes observes that a circle with center C and radius CP usually will meet the curve in another point Q besides P. For just the right choice of C, however, the circle will meet the curve only at P, so that CP becomes normal to the curve (and the tangent to the circle at P becomes the tangent to the curve).

Thus the problem of normals is reduced to solving simultaneously the equations of the curve and of the circle, then finding the values of the parameters that yield a solution with a double root. Descartes describes the process thus:

$$s^2 = y^2 + (v-x)^2 = y^2 + v^2 - 2vy + x^2, \quad \text{so} \quad y = \sqrt{s^2 - v^2 + 2vx - x^2}$$

(generally he considered only positive values). He instructs the reader to substitute this expression for y in the equation of the curve. Now think of the result as an equation in the unknown x, with the other letters as parameters. Find the value of v that causes the equation to have a double root.

It is a bit easier for us to follow if we adjust the notation using modern conventions. The notion of defining a curve as $y = f(x)$, with f a function, drawing the y-axis, and writing points as ordered pairs came much later. Let the specific x value of point M be a, so $M = (a, 0)$ and $P = (a, f(a))$.

The circle with center $C = (v, 0)$ and radius s has equation $(x - v)^2 + y^2 = s^2$. From the Pythagorean Theorem, $s^2 = (v - a)^2 + [f(a)]^2$, so $(x - v)^2 + y^2 = (v - at)^2 + [f(a)]^2$. The points P and Q are on the circle and also on $y = f(x)$. For them, therefore,

$$(x - v)^2 + [f(x)]^2 = (v - a)^2 + [f(a)]^2. \tag{13.1}$$

We seek the value of v for which this has a double root (and P and Q coincide). That is, for an appropriate function $g(x)$, we want to solve

$$(x - v)^2 + [f(x)]^2 - (v - a)^2 - [f(a)]^2 = (x - a)^2 g(x). \tag{13.2}$$

For example, let $f(x) = x^2$ and $a = 2$. Then (13.2) becomes

$$(x - v)^2 + x^4 - (v - 2)^2 - 16 = (x - 2)^2 g(x). \tag{13.3}$$

Since $(x - v)^2 + x^4 - (v - 2)^2 - 16$ is a monic fourth-degree polynomial, $g(x)$ must be a monic quadratic, so set $g(x) = x^2 + px + q$. Note: The following calculations can be done by hand, but they are tedious and not very illuminating. This is a good place to call on technology.

We solve (13.3) by expanding and equating coefficients to get the system

$$4q = 4v - 20, \quad 4p - 4q = -2v,$$
$$4 - 4p + q = 1, \quad -4 + p = 0.$$

The solution is $v = 18$, $p = 4$, and $q = 13$. From this the tangent line or normal line can readily be obtained.

This approach entails a bit more algebraic work than our modern methods require, and it gets rapidly worse as the function gets more complicated. Note, however, that the idea of limit or of infinitesimals is not used in the calculation.

13.3. In the Classroom

It is purely algebraic. Descartes felt this conceptual clarity made his method far superior to Fermat's, which is the origin of our modern approach by difference quotients. Later mathematicians preferred the simplicity of Fermat's approach, and when challenged they were able to overcome the philosophical difficulties involved in the ideas of limit and infinitesimals. This story is told in many books on the history of mathematics, e.g., [1, chapter 12].

13.3.2 For the Student

We can use a graphing tool and computer algebra capabilities to explore Descartes' ideas. The following problems can be modified to fit various systems with the appropriate level of detail left to the student. With the exception of solving the system of equations in exercise 3b, almost any graphing calculator can be used for these problems. Programs such as *Mathematica* or *The Geometer's Sketchpad* can be used to create animations. In the spirit of Descartes' times we will confine our attention to the first quadrant.

Above we saw that the circle with center $(v, 0)$ and radius s is given by $(x - v)^2 + y^2 = (v - a)^2 + (f(a))^2$, so the function

$$circle(v, x) = \sqrt{f(a)^2 + (v - a)^2 - (x - v)^2}$$

gives the part of this circle that is in the first quadrant.

The function $f(x)$ and a are usually fixed in each exercise, so defining *circle* as a function of v and x is convenient. On most systems the function *circle* can be defined, perhaps under a shorter name, and stored in the general form above. Be sure to define $f(x)$ and a explicitly before trying to use *circle*. Also, a specific numerical value for v must be entered to graph *circle* on most systems. Defining v as a variable makes it easy to draw graphs on the same screen with different values for v.

Exercises

1. Starting from the modern idea that $f'(x)$ is the slope of the tangent line at P show that the value we seek is $v = a + f'(a)f(a)$; or if you prefer,

$$f'(a) = \frac{v - a}{f(a)}.$$

2. (a) Let $f(x) = \sqrt{x}$ and $a = 1$. Graph the function and $circle(v, x)$ together for $v = 3, 2, 1.7, 1.5$, and 1.1. Note that $v = 1.5$ gives the double root, and $v = 1.1$ shows the second root on the other side of P from the rest. Besides the graph, a "solve" utility can be used to find the roots.

   (b) Now proceed algebraically as for $f(x) = x^2$ above. Identify $g(x)$ and solve for v.

3. (a) Let $f(x) = x^{5/2}$, and $a = 4$. Graph the function and $circle(v, x)$ together for $v = 800$. It may take some tinkering to get a window in which it is clear that the circle meets the curve twice. Then view them for $v = 644$, which gives the double root.

   (b) Identify $g(x)$ as a monic polynomial of degree 3, say $g(x) = x^3 + px^2 + qx + r$. Expand (13.2), identify corresponding coefficients, and list the equations that need to be solved. If you decide to use your CAS to solve these, be aware there may be many complex solutions (in fact pages of them!) that are not relevant to the problem. This gives some insight into the increasing difficulty of Descartes' approach.

4. Let $f(x) = \sin x$, and $a = \pi/4$. Try values of v, starting with $v = 1.6$, and use graphs to approximate the value that gives the double root. (From above we know it is $\pi/4 + \frac{1}{2} \approx 1.285$.)

5. If your students ask for more, count yourself lucky to have such students! They could try analyzing Descartes' first example: the ellipse $y^2 = rx - (r/q)x^2$. I have reversed his x and y to suit modern conventions. Note he puts the left end of the ellipse, not the center, at the origin.

   After this it might be appropriate to have them look at *La Géométrie* itself [2, pp. 94–115].

6. For those with a bent for programming, a nice project might be to animate the process for various curves with a slider to adjust v and display its value, perhaps as a Java Applet.

13.4 Conclusion

After this excursion, a student may conclude that mathematics was right to take the road it did — and yet still be grateful for a brief tour of another path, never fully traveled.

Bibliography

[1] Victor J. Katz, *A History of Mathematics: An Introduction*, Second Edition, Addison-Wesley, Reading, Massachusetts, 1998.

[2] David Eugene Smith and Marcia L. Latham, *The Geometry of René Descartes, with a Facsimile of the First Edition*, Dover Publications, New York, New York, 1954.

14

Integration à la Fermat

Amy Shell-Gellasch
Beloit College

14.1 Introduction

Move over Riemann and make room for Fermat! Most textbooks on the integral calculus focus heavily on the Riemann integral when introducing integration. This method is very effective in transitioning students from the finite (or macro) world of finding area geometrically to the infinite (or micro) world of finding area by integration. Once the notation and abstract idea of an area made up of an infinite number of infinitely thin slices is mastered, most textbooks move directly on to integration techniques. Finding areas using rectangles is usually not mentioned again except in review, to help students visualize a more difficult example, or when transitioning to finding volumes using double integrals.

Riemann used rectangles of uniform width. This is very handy when letting the width, dx, tend to zero. It also corresponds nicely to the definition of the derivative presented in most textbooks, in which the width $h = x_{i+1} - x_i$ in the denominator approaches 0. I still advocate introducing integration in this manner. However, there is no reason to stop there and move directly on to integration techniques.

Prior to Riemann, even prior to Newton and Leibniz, Fermat and others were finding areas using the sum of thin rectangles. However, Fermat's rectangles were not of uniform width. The width of Fermat's rectangles decreased based on a geometric series. Looking at Fermat's method directly after introducing the Riemann integral broadens the student's perspective on the integral calculus. Also, his techniques can be presented in an analysis course to provide depth to the material. In either course it can be used as a point of departure for discussions of convergence as well as discussions of finite areas under infinite curves. Fermat's method also nicely incorporates other topics such as sums of series, factoring, limits and, of course, history.

14.2 Historical Background

The first modern formulation of the integral was given by Augustin-Louis Cauchy (1789–1875), in his 1823 *Resume des leçons donnes a l'École Royale Polytechnique sur le calcul infinitesimal*. In this work Cauchy defined the integral as

$$\int_{x_0}^{X} f(x)\,dx = \lim_{\delta \to 0} \sum_{i=1}^{n} f(x_{i-1})(x_i - x_{i-1})$$

where $f(x)$ is a continuous function on the interval $[x_0, X]$ and δ is the maximum of the rectangle widths δ_i. Thus his rectangles varied in width and he used left-hand end points for his rectangles. [3]

In this important work, Cauchy gave a definition of the definite integral using the mean value of the function within each subinterval. This work is also where he gave us our modern notion of continuity and uniform continuity, as well as explicitly stating the Fundamental Theorem of the Calculus.

In his 1854 "Habilitationschrift", Georg Riemann (1826–1866) expanded on Cauchy's work to make rigorous the integral calculus. Riemann generalized Cauchy's integral by defining the integral as

$$\int_a^b f(x)\,dx = \lim_{\delta \to 0} \sum_{i=1}^{n} f(\overline{x})(x_i - x_{i-1})$$

where $\overline{x_i} = x_{i-1} + \varepsilon_i \delta_i$ is an arbitrary point in the subinterval $[x_{i-1}, x_i]$ [3]. As with Cauchy, δ is the maximum of the rectangle widths δ_i.

The definition of the integral that most current textbooks use to introduce integration and some attribute to Riemann is actually less general than either his or Cauchy's definition.[1] However, this approach has the benefit of being conceptually easier and allows for a convenient discussion of error.

Before Cauchy and Riemann brought us modern definitions and rigor, and even before Newton and Leibnitz gave us the calculus, many mathematicians were working to find a satisfactory method of quadrature; finding the area of a given region. Notable among the investigators of the mid-1600s were John Wallis, Isaac Barrow, Evangelista Torricelli and Fermat [3]. Their work resulted in methods for finding areas under curves that, apart from the use of limits, are equivalent to the latter methods of Cauchy and Riemann.

Though Pierre de Fermat (1601–1665) is most widely known for his work in number theory, he also contributed to early probability, describing spirals, and physical questions such as falling bodies and refraction. However, many feel that his work on maxima and minima and tangents, as well as quadratures, was pivotal to the development of the calculus. The problem of quadratures was one of the hot topics of the day, and helped spur Newton and Leibnitz to invent the calculus. Carl Boyer went so far as to state that, "no mathematician, with the possible exception of Barrow, so nearly anticipated the invention of the calculus as did Fermat." [2, p. 164]

Fermat published little, letting his discoveries be known through his extensive correspondence. For this reason, dating when he made his discoveries is difficult. Most of Fermat's results on quadrature are found in one of his few published works, his *Treatise on Quadrature*, published in 1658 or 1659[2] [5].

14.2.1 Fermat on Quadrature

From his correspondence, we know that Fermat was familiar with the works on quadrature of Wallis and Torricelli, and by about 1636 Fermat likely had solved for the area under curves of the form $y = 1/x^n$ on the interval $[1, \infty]$. By about 1644, Fermat had a method of finding the area under $y = x^n$ for rational exponents (excluding -1) [1, 4]. His method, apart from notation and language, is strikingly similar to our current approach in all but one respect; his rectangles are not of constant width. We will first consider the case of integral powers, $n \geq 1$. The case of negative and rational exponents will then be briefly outlined.

As we do now, Fermat started his investigation of curves of the form $y = 1/x^n$ by dividing the area into rectangles of uniform width, see Figure 14.1.

As with most mathematicians of his time, Fermat was working in the tradition of Archimedes. Thus when finding circumferences and areas, he planned to inscribe and circumscribe his curve with a finite number of rectangles. [5] For an object with finite dimensions, this method works well. However, the curve in Figure 14.1 has infinite length, yet finite area! Fermat realized he would need to have an infinite number of rectangles to cover an infinite length. But how to sum the areas of this infinite series of areas? One easy solution is to convert it in to a series with known closed form sum. Fermat's insightful solution was to use a geometric progression to determine the width of his rectangles. In Figure 14.2 the widths of the rectangles increase in geometric progression.

---

[1] For more detail on the evolution of the integral as a summation of rectangles, see the author's *"The Integration Techniques of Fermat* in the *Proceedings of the Canadian Society for the History and Philosophy of Mathematics*, 2008.

[2] The full translated title of this undated work is (take a deep breath), *On the transformation and alteration of local equations for the purpose of variously comparing curvilinear figures among themselves or to rectilinear figures, to which is attached the use of geometric proportions in squaring an infinite number of parabolas and hyperbolas.*

14.3. In the Classroom

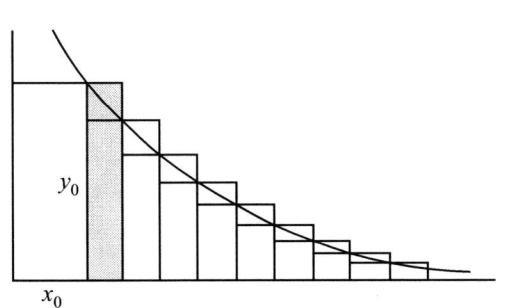

Figure 14.1. Quadrature of $y = 1/x^n$ with finite division.

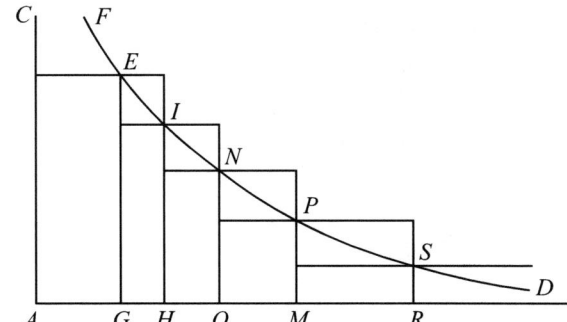

Figure 14.2. Quadrature of $y = 1/x^n$ of infinite division using geometric progression.

Fermat then went on to make the following observation [1]. By reflecting this graph about a line $x = a$ (GE in Figure 14.2) and mapping the interval $[a, \infty]$ onto the interval $[0, a]$, he had, in the limit, the graph of the parabola $y = x^n$ on $[0, a]$ for positive integral n. In this case, the widths decrease from right to left. He then extended his argument to all positive rational powers, $n = p/q$ larger than 1. See Figure 14.3.

In time he was able to generalize this method to all rational powers, as in Figure 14.4, again with decreasing widths.

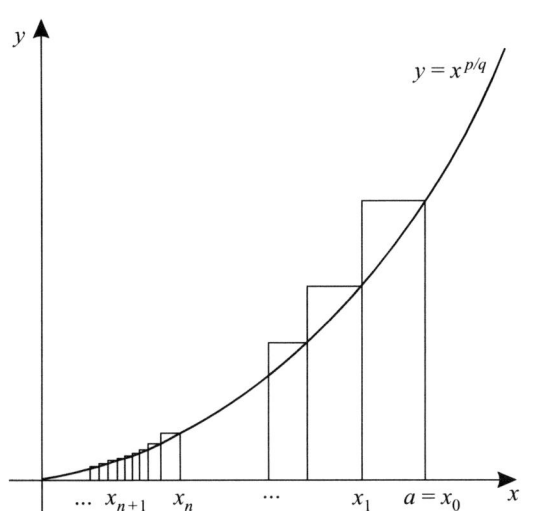

Figure 14.3. Quadrature of $y = x^{p/q}$ with $p/q > 1$.

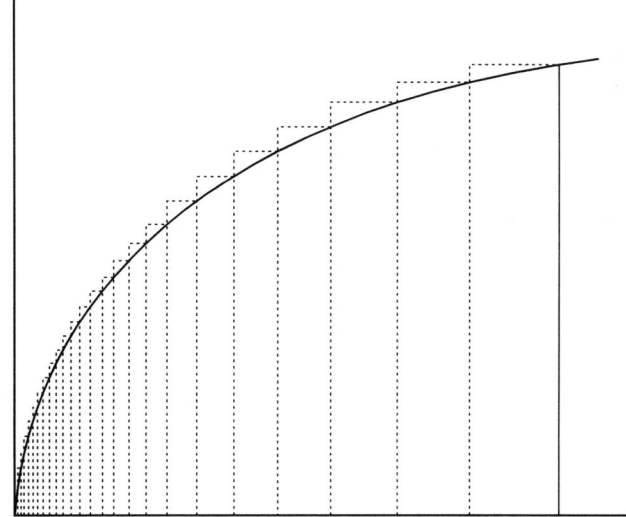

Figure 14.4. Quadrature of $y = x^{p/q}$ with $0 < p/q < 1$.

14.3 In the Classroom

The easiest way to introduce Fermat's method in the calculus class is directly after Riemannian integrals are introduced. Once students are comfortable with that method, simply ask if the rectangles need to be of uniform width. After a few minutes of discussion, Fermat's work can be presented in a Socratic style.

For ease of presentation, I will present two options. First I will present the concrete case of $n = 2$ (the parabola), then I will detail the general case for any positive integer, n; both on the generalized interval $[0, a]$. I have found that working a concrete example in class goes smoothly while still allowing for generalizations to be discovered and discussed. Fermat's general case of n can then be done by the students in groups if time and interest permits.[3]

Keep in mind that what follows is the modern presentation of what Fermat did, and not how it appeared in his work. Since he was working in an era before the use of limits, his arguments relied heavily on the use of proportions, in keeping with the best Greek tradition. However, this approach is foreign to our students and would take an inordinate amount of time to present. The following modern adaptation is sufficient to show students Fermat's work.

---

[3] I started using this in the classroom after reading it in Fred Rickey's as yet unpublished *Historical Notes for the Calculus Classroom*, [6]

14.3.1 Area under $y = x^n$ on $[0, a]$

Let E be any positive value less than 1. Then $a, aE, aE^2, aE^3, \ldots$ is a decreasing geometric sequence approaching 0. Partition the interval $[0, a]$ at these points as shown in Figure 14.3. By creating circumscribing rectangles to the left of the curve, the area can be found in the following manner. Each rectangle has width $aE^i - aE^{i+1}$, starting at the right with $i = 0$. The height of each rectangle is found by evaluating the function on the right.

Case $n = 2$ In the case of exponent 2, the height of each rectangle will be $(aE^i)^2$. Thus the total area is:

$$A = \sum_{i=0}^{\infty} (aE^i)^2 (aE^i - aE^{i+1})$$

$$= a^3 \sum_{i=0}^{\infty} E^{2i} E^i (1 - E)$$

$$= a^3 (1 - E) \sum_{i=0}^{\infty} (E^3)^i$$

$$= a^3 (1 - E) \frac{1}{1 - E^3}$$

$$= \frac{a^3}{1 + E + E^2}$$

By letting E approach 1, $aE^i - aE^{i+1}$ will approach 0, and we get infinitely many infinitely narrow rectangles. Thus the area approaches $a^3/3$, which is the integral of $y = x^2$ evaluated at $x = a$.

General case n In the general case of n any positive integer, the height of the rectangles is given by $(aE^i)^n$. Similarly to the previous case, the total area is:

$$A = \sum_{i=0}^{\infty} (aE^i)^n (aE^i - aE^{i+1})$$

$$= a^{n+1} \sum_{i=0}^{\infty} E^{in} E^i (1 - E)$$

$$= a^{n+1} (1 - E) \sum_{i=0}^{\infty} (E^{n+1})^i$$

$$= a^{n+1} (1 - E) \frac{1}{1 - E^{n+1}}$$

$$= \frac{a^{n+1}}{1 + E + E^2 + E^3 + \cdots + E^n}$$

Again letting E approach 1, the denominator in the last expression above approaches $n + 1$. So Fermat found the area under the curve $y = x^n$ on $[0, a]$ to be $a^{n+1}/(n + 1)$, which agrees with modern integration techniques.

Fermat's investigation of the area under the hyperbola can be incorporated in integral calculus directly after the introduction of the Riemann definition of the integral. A straightforward presentation of Fermat's method would take approximately fifteen minutes of class time. If a more Socratic method is used or if students are asked to evaluate an integral using this method on their own, more time would be needed.

14.4 Taking it Further

All four of Fermat's quadrature problems as depicted in Figures 14.1–14.4 can be developed in class. However, the case of the parabola with positive integer exponent is the easiest and quickest. If time and interest allow, the other three can be done in class or assigned as a project.

14.4. Taking it Further

For the case of negative integer exponents other than -1, E is chosen larger than one. As the graph of the function approaches the x-axis as x tends to infinity, the rectangles get progressively shorter yet longer as in Figure 14.2. By using the same algebraic manipulations as before, as E approaches 1, the area approaches $a^{1-m}/(1-m)$.

In the case of rational powers p/q (other than -1), Fermat used a substitution of the form $E = K^q$ to compensate for the denominator q in the exponent [1, 2]. The calculation is carried out as follows, please refer to Figure 14.4. The area under the curve $y = x^{p/q}$ will be found over the generalized interval $[0, a]$ with $E \in (0, 1)$. Again, starting on the right at a, partition the area by $a, aE, aE^2, aE^3, \ldots$. The area will be:

$$A = \sum_{i=0}^{\infty} a(E^i - E^{i+1})(aE^i)^{p/q}$$

$$= a^{(p/q)+1} \sum_{i=0}^{\infty} (E^i - E^{i+1}) E^{i(p/q)}$$

$$= a^{(p+q)/q}(1-E) \sum_{i=0}^{\infty} E^{i((p+q)/q)}$$

$$= a^{(p+q)/q}(1-E) \frac{1}{1 - E^{(p+q)/q}}$$

Now, to follow the same method as in the previous two cases, we need to compensate for the denominator q in the exponent. So we let $E = K^q$ and we have

$$a^{(p+q)/q} \left(\frac{1 - K^q}{1 - K^{p+q}} \right) = a^{(p+q)/q} \frac{(1-K)(1 + K + K^2 + \cdots + K^{q-1})}{(1-K)(1 + K + K^2 + \cdots + K^{p+q-1})}.$$

As E, and thus K^q, approaches 1, the area under $y = x^{p/q}$ on the interval $[0, a]$ will be

$$a^{(p+q)/q} \left(\frac{q}{p+q} \right).$$

14.4.1 Open Questions for the Classroom

Why is -1 excluded as an exponent?

For $y = 1/x$, how will the function values change? (They will decrease geometrically.) How will the areas of the rectangles change? (The areas of the rectangles will all be the same.)

Since the rectangles all have the same area, how will the total area, or integral, change as you sum the rectangles? (The area increases arithmetically.)

What function has this property of changing multiplication into addition? (Logarithms, so the integral of $1/x$ is the natural log.) [4]

Fermat did not investigate irrational powers. Can this method be adapted to integrate $y = x^n$ for irrational n? Can you get a closed form for the integral or only an approximation?

There are a number of other ways in which Fermat's use of non-uniform rectangles can be presented. The following list gives the most general ones. Each idea can be adapted to the textbook and topics covered in class.

1. Present the problem of finding the area under a curve such as $y = \sqrt{x}$ on the interval $[0, 1]$ using Riemannian rectangles. Start with only a small number of rectangles such as 3 or 4, both circumscribed and inscribed. Discuss the error presented by the large over- or under-estimate of the area near zero. Comparing this to a graph that is less steep near zero will provide a helpful visual. Ask students how to reduce this error. The various techniques such as left endpoint, right endpoint, midpoint and trapezoidal can be reviewed and compared. Then ask the students what other ideas could be used besides altering the top of the rectangles. Where is the approximation more or less accurate? (Less accurate as we approach zero.) If needed, lead the discussion to the idea of altering the rectangles near the area of concern, then move on to the idea of adjusting the widths.

2. Throughout the calculus sequence, short history modules can be incorporated focusing on each of the prominent mathematicians contributing to the development of calculus. Fermat can be presented earlier in the differential calculus[4], and then revisited in the integral calculus. (See chapter "Sharing the Fun" by the editors in this volume.)

3. For a more advanced class, present the underlying idea of rectangles of decreasing width and have students work in groups to rediscover Fermat's techniques. The level of the class will dictate the amount of information provided up front. In an advanced class a hint to the effect that using a geometric progression is all that might be needed. The students can be given a concrete example to tackle, such as $y = x^2$ or $y = \sqrt{x}$ on $[0, 1]$, followed by the general question of $y = x^n$ on the interval $[0, a]$.

14.5 Conclusion

By presenting a little of the history of the standard topics taught in the undergraduate mathematics curriculum, we provide our students with insight into the methods and uses of earlier mathematics. The Riemann integral is the traditional starting place for the teaching of integration. But all too often it is used strictly as a means to present the theory that integration is used to find the area under a curve. As soon as this is mastered, we move on to the various integration rules and techniques without any more attention paid to the physical connection between area, limits and integration.

Presenting the method of Fermat gives students the chance to witness and explore mathematics as a tool for answering challenging problems, not simply as a set of rules used to answer textbook exercises. It is an excellent example of thinking outside the box, or rectangle, as the case may be. Fermat's method of decreasing rectangles shows the flexibility of mathematics, while the method of finding areas by infinitesimal rectangles shows the beauty of mathematics. Letting students experience a variety of historical approaches to mathematics allows for greater comprehension, retention and interest in the topic.

Bibliography

[1] C.B. Boyer, Fermat's Integration of X^n, *National Mathematics Magazine*, 20(1), Oct., 1945, pp. 29–32.

[2] ——, *The History of the Calculus and its Conceptual Development*, Dover, 1959 (Hafner Publishing 1949), pp. 159–164.

[3] C.H. Edwards, *The Historical Development of the Calculus*, Springer-Verlag, 1979, pp. 109–118, 317–328.

[4] S. Hollingdale, *Makers of Mathematics*, Dover, 1994, pp. 142–146.

[5] M.S. Mahoney, *The Mathematical Career of Pierre de Fermat, 1601–1665, 2nd Edition*, Princeton University Press, 1994, pp. 244–254.

[6] V.F. Rickey, *Historical Notes for the Calculus Classroom*, 1996, unpublished, pp. 35–36.

---

[4]See chapter by Vicky Klima in this volume.

15

Sharing the Fun: Student Presentations

Amy Shell-Gellasch and Dick Jardine
Beloit College *Keene State College*

15.1 Introduction

Advocates of incorporating the history of mathematics in teaching mathematics do so believing that providing a human element may spark student interest in mathematics. Incorporating biographical sketches or historical anecdotes into instruction has the potential to enhance student interest, with the hope that interested students will learn more readily and retain the content longer. The learning value of the historical activities can be enhanced when explored and presented *by the student* rather than presented *to the student* by the instructor or textbook. An effective way to have your students deepen their knowledge of mathematics through its history is to have *them* do the historical research and presentations. A student-centered approach to introducing history in a wide variety of undergraduate mathematics courses is an effective teaching tool, in large part because most students like doing the presentations [1].

Invite your students to share in the joy of discovering the "who" and the "why" of the mathematics they are learning, and to take an active role in making the connections between the mathematics they are learning and its historical origins. Student-researched historical presentations can be done in any course, at any level, and require relatively minimal preparation by the instructor. How much time you allow for student presentations in class is up to you. You may limit students to 5 minutes, or require longer presentations. We provide ideas for different approaches to the historical presentations your students can do based on what has worked for us. One approach is general and easy to implement; the other requires more planning on the part of the instructor. Assigned student presentations can be on the historical development of specific topics, such as the origin of e, or on biographies related to the course in general, presented throughout the semester. Another approach, requiring a bit more planning, links each student presentation to a specific lesson in advance. Each student's historical presentation is done during the class period in which the connected mathematical topic is learned.

15.2 Getting Started

Probably the most important factor in deciding how you will include student historical presentations is the amount of class time you choose to make available for their talks. Having included student talks in many different courses, we have found that the benefits outweigh the time taken from instruction. Relatively short presentations help build student confidence in talking about mathematics without a large investment in preparation time.

As an example, have students do individual five-minute presentations starting the second week of classes. In the Appendix are an example schedule and assignment instructions used in a recent Calculus II course. Each student does one presentation as scheduled. The student delivers the talk at an appropriate time in the class period, usually just before or after the mathematical topic to which the mathematician is connected. Improvising the segue for the student presentation is part of the fun for the instructor. Rubrics for grading the essays and presentations are available from the authors.

Longer student presentations can be done if appropriate for each topic. Depending on class size, every student can do one or two presentations at your discretion. Another alternative is to have a team of two or three students do a presentation once a week. Or one or two days of class meetings near the end of the term can be devoted to presentations. You could also incorporate history presentations as part of student-led reviews, either at the end of the semester or prior to an exam, connecting an exam topic with its history, for example the Bernoulli differential equation with Jacob Bernoulli.

Once you decide how many presentations will be conducted and the length of each presentation, compile a list of topics or names of mathematicians, scientists and others who had an impact on the development of the course topic. It is relevant to include topics or names from science, philosophy or other disciplines that had an influence on the course concept to allow students to see the interconnectedness of fields of study. For example, a presentation on William Libby, a Nobel Laureate in chemistry for his work in carbon dating, is relevant when discussing exponential decay. This allows students to see that many mathematicians were also scientists, church figures, politicians, lawyers, etc.—true polymaths. You can make the list of topics a little longer than the number of presentations you expect to allow some choice on the part of the students. Such a list for calculus is provided in the Appendix. Present the list on the first or second day of class and inform students when and how they will choose their topic.

Topic selection can be done as a first-come-first-serve sign-up either in class or by email. Often a student will have already heard a name or topic which is of interest to them. For example, female students are often interested in female mathematicians. When student talks are done in a statistics course, Florence Nightingale always seems to be selected quickly by a female student. We have done the selection in class using the random number generator of *Excel* or a calculator, and included coin flipping or rock-paper-scissors when more than one student wants the same subject. Of course, you can simply assign the topics to students. As mentioned before for both biographical and topical presentations, it makes sense to have the date of each presentation coincide with the lesson or lessons that are relevant. Once students choose or are assigned topics, the presentations are scheduled so that each student knows the date their presentation is due.

To ensure that student work is up to your standards, it is useful to provide students with guidelines for what to include in their presentation: what, if anything, needs to be handed in; information on grading; and a list of resources (web and hardcopy) to get them started. For example, the *MacTutor* website [2] is a good starting place, as is the *Dictionary of Scientific Biography* [3]. It is worthwhile to have a short discussion with your class about the characteristics of good web sources and how to be a discriminating web researcher. To not have students rely too heavily on web sources, you may chose to require a specified number of print sources, including journal articles. This is also an opportunity to introduce your students to reference librarians and all that they can access, including the digital archive *JSTOR*, one of our favorites, if your institution subscribes.

There are measures you can take to reduce the potential for getting presentations of unacceptable quality. We suggest you do the first presentation yourself! That allows you to model what you expect from students, and demonstrates to students that you are willing to do what you expect the rest of the class to do. We have not found that necessary, but it certainly is an option should you want to ensure the quality of what is presented in your classroom. With the arrival of inexpensive and simple to use digital video recorders, student presentations can be recorded and used as examples of good presentations.

15.3 Presentations

PowerPoint, or some other presentation software, is the easiest method of presentation if the technology is available to you. If not, a transparency projector, document reader, or other visual aid can be used for the presentations. You may find it acceptable for the students to just stand up and present without visual aids. Consider offering students assistance in preparing transparencies or handouts they would like to use to support their presentation. Allow time for questions and discussion following the presentations. We have found that students seem to listen much more carefully to their classmates' presentations than to instructor's lectures, so there is potential that wonderful discussions may result.

To ensure student attentiveness to their peers presentations, using the information from the presentations in subsequent quizzes or tests, even as bonus questions, reinforces the importance of the historical information and really makes the students feel that their contribution is significant. If the students know that presentation material is fair game, they will be more involved in the whole process. In order for this to be effective, you have to make sure that any misinformation presented is discussed.

15.4 Assessment

As with many assignments, grading presentations is a subjective classroom assessment activity. Amy requires students to submit a printed copy of their bibliography. She then writes comments about the presentation on that paper as feedback for the student. She grades on the content, how well the students understand what they are presenting, and the style of the presentation itself. Dick employs an oral presentation checklist and an associated analytic rubric. The rubric and checklist are given to the students in advance so they are aware of the grading process. You can also provide students with grading sheets and have them evaluate their peers. Peer grading is a valuable exercise, but be advised that students often grade harder than you might. Presentations can also be used as alternative graded events or as extra credit assignments.

Consider using student submissions of their presentation for program assessment, retaining either electronic or paper copies of their work. These files can become the artifacts necessary to document that students are engaged in activities supporting departmental program goals or state requirements. Many mathematics departments have communicating mathematics as a program goal, and these presentations count toward that objective in addition to any objective your department may have that involves the history of our discipline. In that way, one activity, the student history presentations, is a source of evidence of student work toward two program goals. Additionally, you or your students may want a collection of their presentations in a course, so if you have the electronic files you can easily copy them onto a CD at the end of the term for students who are interested.

15.5 Conclusion

Student presentations provide an alternative learning experience that allows students to practice their talents in the written and spoken word, as well as their interests in history, mathematics, science, the humanities, and other disciplines. Presentations also provide departments with several opportunities for student generated artifacts in support of program goals. The authors have found that end of term student evaluations consistently identify the presentations as a high point in the course. Students comment on how much they learned from doing their own research and presentation, as well as from listening to the other presentations. Students gain a deeper understanding of the subject. More importantly, they learn from the biographical presentations the value of hard work and self-study in enabling almost anyone to achieve success in mathematics or in any other discipline. The topical presentations show that mathematics is a process that evolves over time, with connections to other ideas and fields. Finally, by giving your students the opportunity to do their own research and present their findings, you are sharing the fun, and can sit back and enjoy someone else teaching for a while.

Bibliography

[1] D. Jardine, "Active learning mathematics history," *PRIMUS*, IV (2), (1997), pp. 115–122.

[2] J.J. O'Connor and E.F. Robertson, *The MacTutor History of Mathematics Archive*, `www-history.mcs.st-andrews.ac.uk/index.html`, accessed May 2008.

[3] C. Gillispie, ed., *Dictionary of Scientific Biography*, Scribner, New York, 1978.

[4] The JSTOR digital archive, `www.jstor.org`, accessed May 2008.

Appendix
MATH152 — Calculus II
Mathematical History Essay and Presentation

A graded requirement for MATH152 will be the submission and presentation of a brief essay about a concept or person prominent in the history of mathematics in general and the calculus in particular. The biographical sketch should contain information about the time period in which the person lived and worked, the major contributions of the person, the connection of the person's work to the calculus, and, if possible, any interesting (humorous?) anecdotal information. The essay on the concept should trace the origins of the mathematical idea, identifying key persons responsible for the concept. The written essay (or sketch) should be at least a couple of paragraphs and not more than one page in length, certainly not more than a couple of pages. The oral presentation should take a few minutes of class time, certainly not more than 5 minutes, and should relate the historical figure to the day's lesson or our course. *Important*: both the written essay and the classroom presentation **must** relate the mathematician to the lesson of the day. Remember that your instructor is available for assistance in this endeavor, and full documentation of resources used (at least one print and one web source) is required. The essay and presentation will be worth 30 course points. A maximum grade is possible only if mathematics (e.g., an equation, a definition, a graph, etc.) appears in the reports.

Below is a list of potential subjects for the essays and related lessons:

Lesson	Subject
3	Leibniz, Newton
4	Napier, Seki Kowa, Emilie Du Chatelet
5	Kepler, Hooke
8	Johann Bernoulli, Jacob Bernoulli
10	Archimedes
11	Riemann
12	Thomas Simpson
15	Malthus; Verhulst
16	Galileo, Kovalevskaya
19	Lotka; Volterra
22	Louis Fry Richardson
25	Brooke Taylor, Fermat
27	Laplace
28	Euler
30	Zeno of Elea
31	Colin Maclaurin
32	Lagrange
33	Cauchy
35	L'Hospital
36	Torricelli
37	Cardano, Tartaglia, Bombelli
38	Argand, Wessel, Gauss, Demoivre
39	Agnesi

History Presentation Assignments

Lesson	Mathematician	Student
3	Leibniz	Min
	Newton	Ben
4	Napier	Nicole
	Seki Kowa	Robert
	Emilie Du Chatelet	Stephanie
5	Kepler	Ashley
	Hooke	Amanda
8	Johann Bernoulli	Andy
	Jacob Bernoulli	Joel
10	Archimedes	Matt
11	Riemann	Billy
12	Thomas Simpson	Alyssa
15	Malthus	Michelle
	Verhulst	Mike
16	Galileo	Susan
	Kovalevskaya	Jenna
19	Lotka	
	Volterra	Steve
22	Louis Fry Richardson	Dan
25	Brooke Taylor	Heather
	Fermat	Joe
27	Laplace	Jessica
28	Euler	Ginger
30	Zeno of Elea	Eric
31	Colin Maclaurin	Rob D
32	Lagrange	Mallory
33	Cauchy	
35	L'Hospital	Sean
36	Torricelli	Jeremy
37	Cardano	Silas
	Tartaglia	Duncan
	Bombelli	
38	Argand	Amy
	Wessel	Josh
	Gauss	Justin
	DeMoivre	
39	Agnesi	Nathan

16

Digging up History on the Internet: Discovery Worksheets

Betty Mayfield
Hood College

16.1 Introduction

So you want to include some history of mathematics in your upper-level courses, but you just can't imagine how you can possibly fit anything else in this semester. How will you get to all the topics you want to cover, and still have time for some history?

Instead of giving a lecture on a history topic, or on the name behind a famous theorem, why not let students find the information themselves? Using a discovery worksheet is fun, saves class time, and encourages students to learn things on their own. Some of the answers may be in their own textbooks, or in library books on the history of mathematics, but the activities in this article are designed so that students are encouraged to search the Internet for information. In the process, students will probably be surprised to learn how much material is 'out there' about mathematics and its history and will begin to learn how to separate the online wheat from the chaff. The examples that follow are intended for students in linear algebra and differential equations courses, as indicated, but you can obviously use this idea in any class. This assignment is very flexible: you may assign the worksheets for homework or use them as a class activity; you may have students work independently or in groups.

Students enrolled in a linear algebra class may never have stopped to think that Gaussian elimination was named for someone named Gauss, or that there was a Cramer behind Cramer's Rule. They may be surprised to learn that people have been solving systems of linear equations for thousands of years. They may be equally surprised to learn that the person for whom a theorem or computational method is named is not necessarily the person who first used it. Helping students discover the stories behind the topics they study gives more depth to their understanding and appreciation of mathematics and mathematicians. Many textbooks today include short biographies of ancient and modern mathematicians important in the field. You can use those short articles as a starting point for leading students to learn more about history.

Similarly, students may have little awareness that new mathematics is being discovered every day, that there are famous mathematicians living and working right now. Two of the attached worksheets, designed to be used in a differential equations class, lead students to learn about contemporary mathematicians and their work.

16.2 In the Classroom

In developing a discovery worksheet for your students, you may print the questions on a sheet of paper, with blank space for the students' answers, and hand them out in class; or you may post the questions on the class Web site (using an electronic platform like Blackboard, for instance) and have the students create a document, typing their answers directly into it; or you may do something completely different. If you want your students to work together during a class period, you can give one sheet to each group of students. If you are pressed for time, you may assign the project for homework and have each student submit an individual response. Several worksheets I have used are in the Appendices; in each case, I have just listed the worksheet questions. I generally have students write the answers to these specific questions in complete sentences — but you could also use those questions as a springboard for a more complete written report. In any case, students using discovery worksheets gain experience in reading, writing, searching, and learning about the history of mathematics in a 'real' math course.

16.3 Learning about History — and about the Web

One of the aims of this project is to introduce students to (reputable) online resources in the history of mathematics, and to help them assess the sites they find on their own. You may wish to specify which sources they use, or you can use this experience as an opportunity to help students find and evaluate appropriate resources. If you have not indicated where you want students to look for information, their immediate reaction will undoubtedly be to consult Wikipedia. Our goal is to lead them to more serious, more academic, more trustworthy sources. If you want students to learn about interesting and useful Web sites devoted to the history of mathematics, you will need to offer some guidance on evaluating those sites. Librarians in this country have taken the lead in developing resources to help students figure out whether or not a Web site is biased, or unreliable, or out of date. One example of a helpful tutorial and accompanying checklist for a Web site — the one recommended by our college library — is available from the web site of the Teaching Library at Berkeley [4]. The tutorial includes questions to ask when examining a Web site and techniques for evaluating it, encouraging students to be skeptical and to think critically about sources. You could ask your students to compare two different Web sites on the same topic, using the accompanying checklist [11], for instance. There are, of course, many similar tools from other libraries.

The worksheet activity should always be followed up by a class discussion, however brief. What information did students find? What surprised them? What did they find interesting? Where did they get their facts? Did they find conflicting or contradictory accounts? What could account for that situation? How can we decide who is correct? What further research could we do? If possible, we print out a poster or fact sheet and post it in the classroom, to remind us — and the other classes meeting in that room — of the names behind the math we study.

16.4 Conclusion

Developing discovery worksheets is a fairly easy way to introduce students to the history of mathematics. It encourages independent learning and exploration, reinforces researching and writing skills, and need not take up much class time. You can structure such an experience to emphasize group work, or library research, or evaluating web sites. In any case, the history of a problem can remind us that mathematics is created by people — often very interesting ones.

Bibliography

[1] The American Mathematical Society: www.ams.org

[2] The Association for Women in Mathematics: www.awm-math.org/

[3] Paul Blanchard, Robert L. Devaney, and Glen R. Hall, *Differential Equations, 3rd edition*, Brooks/Cole, Pacific Grove, CA, 2006.

[4] *Evaluating Web Pages: Techniques to Apply and Questions to Ask.* UC Berkeley — Teaching Library Internet Workshops: www.lib.berkeley.edu/TeachingLib/Guides/Internet/Evaluate.html

[5] *The Fields Medal.* The Fields Institute:
www.fields.utoronto.ca/aboutus/jcfields/fields_medal.html

[6] David Lay, *Linear Algebra and its Applications, 3rd edition*, Addison Wesley, Boston, 2003.

[7] The Mathematical Association of America: www.maa.org

[8] The Mathematics Geneology Project. North Dakota State University:
genealogy.math.ndsu.nodak.edu/index.php

[9] John J O'Connor and Edmund F Robertson, *The MacTutor History of Mathematics Archive*,
www-history.mcs.st-andrews.ac.uk/history/index.html

[10] Larry Riddle, *Biographies of Women Mathematicians*,
www.agnesscott.edu/Lriddle/women/women.htm

[11] *Web Page Evaluation Checklist.* UC Berkeley—Teaching Library Internet Workshops Materials:
www.lib.berkeley.edu/TeachingLib/Guides/Internet/
evaluation_checklist_2008_spring.pdf

Appendix 1: Who Was Gauss?

This worksheet was designed for a linear algebra class. Students are encouraged to read the biographical sketch in their textbook [6] first and then search the Internet to find the rest of the answers. There is a link to the MacTutor web site [9] on the Blackboard site for the course, and one to the Mathematics Geneology Project [8], both of which should be helpful. This is the first history assignment in this course, so students are introduced to basic all-purpose sites.

Using your textbook and the web sites listed under External Links on the course Blackboard site, find the answers to these questions. Be sure to list your resources.

For whom is Gaussian elimination named?

When did he live? Where?

What did he do when he was 7 years old that astounded his teacher?

What is a *gymnasium*?

Where did Gauss go to college? Did he have lots of friends? Did he even graduate??

One of Gauss's most famous discoveries is how to construct a particular polygon using only a ruler and compass. How many sides did that polygon have?

What is the name of Gauss's most famous work?

What was the subject of his dissertation? What does that mean?

What method did he use to predict the orbit of Ceres? Have you ever used that method in another class? *[Our students have done so in a calculus lab.]*

What personal tragedies did Gauss experience? Did they affect his work?

What kind of geometry held a special fascination for Gauss?

Who were three famous students of Gauss?

What is the Prime Number Theorem?

Did Gauss ever use the method of Gaussian Elimination himself? In what context? Was he the first person ever to use it?

Appendix 2: History of Matrices and Determinants

This is the second linear algebra worksheet. As with the first assignment, most of the answers can be found at [9], but on another part of the web site: students begin by reading an essay on a particular topic rather than a biography of a particular mathematician. Then they follow links and search around a little to answer further questions. By the end of this assignment, they know their way around this particular web site pretty well. If one Googles 'Cayley-Hamilton Theorem,' all sorts of web sites pop up — some more reputable than others. Students will have an opportunity to choose one that seems appropriate.

In what context did the study of matrices arise?
What is the earliest known occurrence of a system of linear equations?
What is the first known example of the use of a matrix to solve a linear system?
When and where did determinants first appear?
Who first used the term "determinant?" Who first used it in our modern sense of the word?
Who first used the notation of two vertical lines to indicate the determinant of a matrix?
Who first used the term "matrix?"
Who first proved Cramer's Rule for 2×2 and 3×3 systems?
Who proved the rule for the general $n \times n$ case?
What is the Cayley-Hamilton Theorem? Who was Cayley? Who was Hamilton?
Be sure to list all of your sources, and fill out the Web Site Evaluation Check List for any Internet site you used.

Appendix 3: Stephen Smale, a contemporary mathematician

For use in a Differential Equations class. There is a biography of Prof. Smale on [9], and a brief description of his work in [3]; students have to look around a little to answer all of the questions — for instance, on the Fields Medal site [5] and the MAA [7] and AMS [1] sites.

Where did Prof. Smale go to college? Where did he get his graduate degrees? Where is he now?

What is a Fields Medal? How old was Prof. Smale when he received one? What is the maximum age one can be and receive a Fields Medal? Is there a Nobel Prize in mathematics?

Prof. Smale published many papers, and received many honors, for his work in pure mathematics, including his work on the Poincaré conjecture. But then, in the late 1960s, the focus of his work changed. Describe that change.

Every year until 2000, when the American Mathematical Society stopped holding regular summer meetings, the AMS held a series of three Colloquium Lectures on an important topic. When did Stephen Smale give these lectures? What was the topic? Who gave the lectures in 1996? Do you think the room was crowded??

What is the Chauvenet Prize? Who gives it? When did Stephen Smale win it, and what for?

Are there any books in our college library by or about Stephen Smale?

Appendix 4: Nancy Kopell, a female mathematician

Another Differential Equations assignment. Just when students have been lulled into thinking that everything they need to know is just one click away, on the MacTutor web site, they encounter a well-known mathematician who does not appear there. We can have an interesting discussion afterwards about possible reasons why. In this assignment, students are introduced to the Agnes Scott Biographies of Women Mathematicians site [10] and that of the AWM [2]. Students are also reminded that their textbook has biographical sketches of contemporary mathematicians, and they begin to learn about the holdings of our college library in the history of mathematics.

Is this famous mathematician listed on the *MacTutor History of Math* site?

Try the *Biographies of Women Mathematicians* site, hosted by Agnes Scott College. Do you find her there?

Where did Prof. Kopell go to college? to graduate school? Who was her dissertation advisor?

At which institutions has she taught? Where is she now? What kinds of mathematical problems does she work on?

What is the CBD? Where is it?

What honors has Prof. Kopell won?

Every year, at the Joint Mathematics Meetings, the Association for Women in Mathematics sponsors a lecture by a famous woman mathematician, called the Emmy Noether Lectures. Who was Emmy Noether? When did Prof. Kopell give the Noether Lecture? In what city? What was its title?

Does our college library have any books about Prof. Kopell? Say more.

17

Newton vs. Leibniz in One Hour!

Betty Mayfield
Hood College

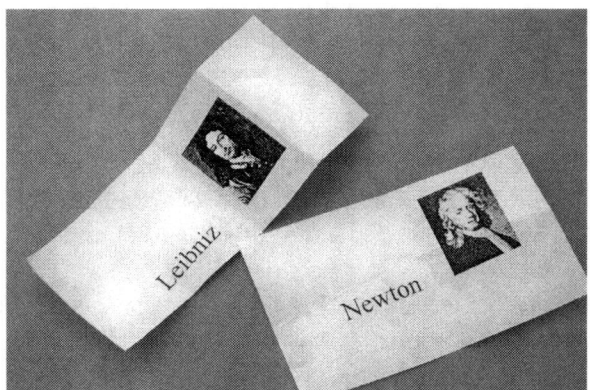

17.1 Introduction

In our college, we teach a quick, one-credit Calculus Workshop course for students who have received credit for taking first-semester calculus elsewhere (in high school or at another college) but who need a brief introduction to some specific topics they may not have seen before. And so we spend one class on using a computer algebra system, one class on Euler's Method, one class on writing about mathematics ... a whirlwind tour of a variety of topics our regular Calculus I students see in more depth. The class meets for one hour and fifteen minutes twice a week during the first half of the semester.

Even in such a condensed course, we want our students to learn something about the history of calculus, and especially about the most famous names associated with its beginnings. They have encountered Isaac Newton's name in solving a murder mystery (Newton's Law of Cooling) and in studying air resistance for a falling body (Newton's Laws of Motion), but few of them know much about his role in the discovery of calculus — and most of them have never even heard of Gottfried Leibniz. Students in our regular calculus sequence read, discuss and write about several articles on the development of the calculus from [2] and [4], but we do not have that luxury in the workshop course. And so on the last day of class, we take a quick dive into the history of the subject. The activity described in this article introduces students to the lives and work of Newton and Leibniz in a way that includes active student participation, collaboration, and the use of technology — in one class period.

17.2 Historical Background

In different parts of Europe, virtually simultaneously, Isaac Newton (1642–1727) and Gottfried Leibniz (1646–1716) discovered the subject we know as calculus. They used different notation, and different words, but they each discovered both differential and integral calculus and the remarkable theorem that links them [8]. The story of Newton and Leibniz and the feud they began is inspiring, instructive, fascinating, and exasperating. It was, indeed, an all-out war between the most brilliant natural philosophers of their time, and it split the mathematical world into Newtonian and Leibnizian camps [5]. This lesson gives students a brief glimpse into the lives and work of these two exceptional mathematicians and their relationship to each other.

17.3 In the Classroom

17.3.1 The Activity

First, divide the class into rival Newton and Leibniz groups. I do that by copying pictures of Newton and Leibniz, with their names attached, onto small pieces of paper, and having students each draw a piece out of a hat. If it is a small class, you may have only two groups; a larger class could be divided into groups of a more manageable size, and the assignment suitably modified by giving different instructions, or lists of resources, to each group. Tell the students that each group will find out as much about their assigned mathematician, especially in relation to calculus, as they can in the time allotted and then will give a report to the rest of the class that very day.

Hand out a list of resources, and give the class about forty-five minutes in which to research their topics and prepare a presentation. That will probably force the groups to divide up the work and combine their findings later to produce a finished product. Then, in the last ten minutes or so of the class period, have the groups present their work to the rest of the class. (See the instructions in the Appendix.)

17.3.2 Resources

An ideal setting for this project is a computer lab, where students have access to several computers and space to work in groups. Or if your classroom is close to a lab, students can move back and forth between the two spaces as needed. If you do not have easy access to computers, this project can still be fun and successful; your students will just rely on print resources and traditional oral reports.

There are many print resources that are ideal for this project; your students may already have access to some of them. Either refer students to books in the classroom, or make handouts available of:

- The "Knighted Newton" (pages 129–142) and "Lost Leibniz" (pages 143–158) chapters of William Dunham's *The Mathematical Universe* [4];

- Eric Temple Bell's "On the Seashore" (pages 36–48), Morris Kline's "The Creation of the Calculus" (pages 49–55), and the excerpt from *Principia Mathematica* (pages 56–59) in *Readings for Calculus* [2];

- The biographies of Newton (pages 131–140) and Leibniz (pages 141–157) from George F. Simmons's *Calculus Gems* [8].

In addition, of course, there are many online resources available; if your students are not already aware of their existence, this is a good time to introduce them to reliable web sites such as:

- The MacTutor History of Mathematics archive [7]: see the biographies of Newton and Leibniz, and the essay on The History of Calculus;

- David Joyce's web site at Clark University [6]: see the article on The Rise of Calculus;

- David Wilkins's site at Trinity College, Dublin [9], for papers by and about Newton.

17.3.3 Student Presentations

If your class meets in a room with an instructor's computer and projector, this is a great project for using software like PowerPoint. Students can gather around one computer and collect all their information together into a short presentation, including text, graphs, and pictures of the great men themselves. Lacking that equipment, you could hand out blank transparencies and colored pens (and perhaps the department code for the photocopy machine), and students can be just as creative. In my experience, people walking past the classroom have stopped and peered in the windows and doors, trying to see what the lively presentations and cheers of "Newton!" and "No, Leibniz!" are all about.

17.3.4 Making it Run Smoothly

The biggest constraint, of course, is time: students can barely scratch the surface of the story of Newton and Leibniz in an hour. On the other hand, as an instructor you must convince yourself that, while rushing to 'cover' all the important topics on the syllabus, it is worthwhile to spend an hour of valuable class time on the history of calculus. Assuming you have answered that question to your own satisfaction, you will have to manage the activity carefully. Make sure that you have all of the materials prepared and in the classroom, and that the students are ready to work, at the beginning of the class period. During the class, keep track of the time, and walk around and see how the groups are doing, reminding them how much time they have left, perhaps answering questions or giving advice, and finally stopping them in time for them to give their oral reports. The presentations are necessarily short (you will have to set limits and stick to them), and necessarily superficial, but students have spent an hour of their lives reading, writing, and listening to information about Newton, Leibniz, and the discovery of calculus — and that can't be bad!

17.4 Taking it Further

If you decide to include this activity in a standard course, you may of course devote more time to it. You could assign the readings for homework and then use class time for discussion and developing the presentations. For students who are intrigued by their introduction to this topic and want to learn more, you might assign an individual project for extra credit. Some possibilities for topics:

- Read A. Rupert Hall's *Philosophers at War* [5], about the Newton-Leibniz feud. Take sides and state the case for either Newton or Leibniz.

- Write a play about the feud and perform it in class.

- Find out who Newton's great antagonist George Berkeley was, and why he was so suspicious of Newton's calculus. (Bishop Berkeley published, in 1734, a treatise called *The Analyst*, in which he famously referred to infinitesimals as 'the ghosts of departed quantities' [9].) Mathematical historian David Wilkins's website, cited above, contains the original text of *The Analyst* as well as contemporaries' reactions to it.

One could, of course, use this approach to introduce students to other controversies in mathematics. Another priority dispute that students find entertaining is the one between Niccolo Fontana (1499–1557), known as Tartaglia, 'the Stammerer,' and his rival Gerolamo Cardano (1501–1576), over which one devised a method to solve a general cubic equation first — a tale that involves public challenges and counter-challenges, gambling, and poisoning by arsenic [1], [3]. Or students could form Fermat and Descartes teams, to debate who really discovered analytic geometry. Many students can correctly identify the mathematician for whom the Cartesian plane is named, and they may have heard of Fermat's Last Theorem, but they do not know much, if anything, about those two men and their work that transformed the ancient Greeks' definition of a circle ('the locus of all points equidistant from a fixed point') into the familiar equation $x^2 + y^2 = r^2$ [8].

17.5 Conclusion

So what do you gain from this project? As a start, your calculus students:

- Begin to see that calculus has a long, rich, interesting history; it didn't just appear in a 600-page textbook in the 21st century;

- Learn something about the lives and work of Isaac Newton and Gottfried Leibniz — from their own work and from their classmates;

- Become aware of the beginnings of calculus, and about the controversy associated with it;

- Collaborate and work in possibly diverse groups;
- Use technology to find and present information about mathematics and its history;
- Give an oral presentation.

And all in one hour!

I have used this activity in our calculus workshop for several years, and it is one of the most successful and fun things we do all semester. I am constantly amazed at how much students can in fact learn in an hour, and how excited they get about the project. The two groups really do compete to see who can do a better job, and their final presentations are instructive and entertaining.

Bibliography

[1] William P. Berlinghoff and Fernando Q. Gouvêa, *Math through the Ages: A Gentle History for Teachers and Others, Expanded Edition*, Oxton House Publishers and the Mathematical Association of America, Washington, DC, 2004.

[2] Underwood Dudley, ed., *Readings for Calculus*, The Mathematical Association of America, Washington, DC, 1993.

[3] William Dunham, *Journey through Genius: The Great Theorems of Mathematics*, Penguin Books, New York, 1990.

[4] William Dunham, *The Mathematical Universe*, John Wiley & Sons, New York, 1994.

[5] A. Rupert Hall, *Philosophers at War: the Quarrel between Newton and Leibniz*, Cambridge University Press, Cambridge, 1980.

[6] David E. Joyce, *History of Mathematics*, aleph0.clarku.edu/~djoyce/mathhist/

[7] John J O'Connor and Edmund F Robertson, *The MacTutor History of Mathematics Archive*, www-history.mcs.st-andrews.ac.uk/history/index.html

[8] George F. Simmons, *Calculus Gems: Brief Lives and Memorable Mathematics*, Mathematical Association of America, Washington, DC, 2007.

[9] David R. Wilkins, *The History of Mathematics*, www.maths.tcd.ie/pub/HistMath/

Appendix 1: Instructions to the class

Today you will learn about two men who were important in the development of calculus: Isaac Newton and Gottfried Leibniz. You will be assigned at random to a group; your group will focus on one of these two mathematical giants. I have posted some resources under Course Documents on the class Blackboard site which you may download and print. Also see the External Links page for some helpful web sites. You are welcome, of course, to search for more information.

Your group will have 45 minutes in which to learn as much as you can about your assigned mathematician and prepare a presentation for the rest of the class. Some questions you should answer:

- Where and when did he live?
- Where did he study? *What* did he study? How did he earn a living?
- What is your mathematician famous for? In what branches of mathematics — or other disciplines — did he do important work?
- Most important, how did your mathematician contribute to the early development of the calculus? What were some of his ideas? When and where did he publish his work?
- What did your mathematician think of the other mathematician?

- How is your mathematician remembered today?
- Had you heard of your mathematician before? What did you know about him before today? What have you learned that surprised you?
- Do you think your mathematician should be given credit for the invention (or discovery) of calculus? Why?

Your group will probably want to split up the reading and report back to each other. When you have developed an outline of your findings, prepare a brief oral report and *PowerPoint* presentation which you will give to the rest of the class. Make your presentation attractive and colorful; include pictures of your mathematician and examples of his mathematics. But do not overwhelm us with animations or objects flying in from above. A simple design is best.

Have fun!

18

Connections between Newton, Leibniz, and Calculus I

Andrew B. Perry
Springfield College

18.1 Introduction

Calculus, like most other well-established branches of mathematics, did not originally appear in the same form as it occurs in modern textbooks. Many mathematicians contributed to the development of calculus over many centuries, using widely varying notation and languages. A proper history of the subject can easily consume a book [1].

Although a thorough study of the history of calculus is completely unnecessary for an introductory calculus student, it is nevertheless of some interest for such students to see an overview of this subject's fascinating and colorful history. Today's calculus students will no doubt consider original papers somewhat cryptic at the very least, and maddeningly cumbersome and obscure in places. Still, there are passages from these writings which will appear comforting in their familiarity. This paper seeks to point out some of these passages and their connections with the modern elementary calculus curriculum. We concentrate on the two mathematicians generally considered to be the fathers of calculus, Sir Isaac Newton (1642–1727) and the German Gottfried Leibniz (1646–1716).

18.2 Historical Background

It would be impossible to say authoritatively when the first ideas of calculus appeared. Arguably, many early mathematicians used a form of integral calculus in approximating the area or volume of irregular objects using finitely many or infinitely many recognizable shapes such as rectangles. In particular, the Greek mathematician Archimedes is famous for estimating the area of circle using the so-called method of exhaustion, and effectively computing the value of π.

One can say with confidence, however, that the English mathematician Sir Isaac Newton (1642–1727) and the German Gottfried Leibniz (1646–1716) are most famous for their discoveries of calculus. The two scholars independently discovered the main ideas of the subject. Because they lived in a time before Cartesian coordinates had become widely used, and because the subject was still in its infancy the two authors' writing looks clumsy and awkward to the modern reader. While neither had put the subject on a firm and rigorous footing, each had effectively "discovered the calculus" according to modern standards.

Much of the drama and the charm of the history of calculus stems from the bitter argument between Newton and

Leibniz as to how the credit for the discovery of calculus should be properly divided. This priority dispute was a major controversy during the mathematicians' lifetimes and for years after their deaths. Newton apparently discovered "fluxions" (derivatives) in 1665 but chose not to publish his work. Leibniz apparently discovered differential calculus independently using vastly different notation and did publish some of his work in 1684 [2]. Newton published a major calculus manuscript in 1704 [3]. At about that time some of Newton's friends, spurred on by Newton himself, accused Leibniz of plagiarism. A major dispute ensued, with British mathematicians rallying around Newton and Continental scholars supporting Leibniz, with each group preferring the notation of their own champion. The British continued to use Newton's notation for at least a century before switching to the more convenient notation of Leibniz which everyone uses today.

18.3 Newton's Work

In part because we currently use Leibniz' notation, Newton's work is difficult to understand for the modern reader, even in translation from the original Latin. However, his computation of the derivative ("fluxion") of the function $f(x) = x^n$ can definitely be understood. I quote below from Evelyn Walker's translation of *Quadratura Curvarum*, published in 1704, and reprinted in David Eugene Smith's *A Source Book in Mathematics* [3, 4].

Newton:

> Let the quantity x flow uniformly, and let it be proposed to find the fluxion of x^n. In the time that the quantity x, by flowing, become $x + o$, the quantity x^n will become $\overline{x + o} \mid^n$, that is, by the method of infinite series,
>
> $$x^n + nox^{n-1} + n^2 - n^2 oox^{n-2} + etc.$$
>
> And the augments o and $nox^{n-1} + \frac{n^2-n}{2} oox^{n-2} + etc.$ are to one another as 1 and
>
> $$nx^{n-1} + \frac{n^2 - n}{2} ox^{n-2} + etc.$$
>
> Now let these augments vanish, and their ultimate ratio will be as 1 to
>
> $$nx^{n-1}.$$

As the reader may have guessed, the notation $\overline{x + o} \mid^n$, means $(x + o)^n$. Newton's construction has the same effect as defining the derivative to be $\lim_{o \to 0} \frac{f(x+o)-f(x)}{o}$. Today we would use either h or Δx in place of the awkward o. He also writes oo instead of the modern o^2, although we can see that he uses exponent notation to write higher powers. Newton's construction is equivalent to following these three steps:

1. Compute $f(x + o)$, that is, $(x + o)^n$.
2. Divide by o.
3. Let $o \to 0$, and take limit.

Instead of "taking a limit", Newton "lets the augments vanish", which has the same effect. He correctly concludes that $f'(x) = nx^{n-1}$.

What we would call the indefinite integral, Newton calls fluents. In his 1704 paper, he uses a nice compact notation, where the nth derivative of a function is denoted by n dots directly above the function, and the nth antiderivative is denoted with n slash marks directly above the function. (Actually, Newton had improved his notation from earlier works in response to criticism from Leibniz.) Below we see Newton's description of his notation system for fluents and fluxions, which is not much like our own.

Newton:

> From the fluxions to find the fluents is a much more difficult problem...
> And after the same manner that $\ddot{z}, \ddot{y}, \ddot{x}, \ddot{v}$ are the fluxions of the quantities $\ddot{z}, \ddot{y}, \ddot{x}, \ddot{v}$ and these the fluxions of the quantities $\dot{z}, \dot{y}, \dot{x}, \dot{v}$ and these last the fluxions of the quantities z, y, x, v; so the quantities z, y, x, v may be considered the fluxions of others which I denote thus: z', y', x', v' and these as fluxions of others z'', y'', x'', v'' and these as fluxions of still others z''', y''', x''', v'''.

18.3.1 Newton's Fluents and Fluxions: Exercises

1. Using the finite expansion $(x+o)^3 = x^3 + 3x^2o + 3xo^2 + o^3$, use Newton's method to find the fluxion of x^3. Change the o to h if you feel more comfortable using the modern notation.

2. Expand $(x+o)^2$ and use Newton's method to find the fluxion of x^2.

3. According to Newton, the fluxion of x^n is nx^{n-1}. Knowing that, what is the fluent of nx^{n-1}?

18.4 Leibniz's Work

Leibniz's pedagogical style is rather more compatible with the modern calculus curriculum than Newton's. Though Leibniz's derivations of the rules of calculus are derived from complex geometric arguments, his rules for calculation of derivatives are very similar to those of a modern Calculus I course.

Leibniz's differential notation may be disconcerting for students used to derivative notation. Rather than writing

$$\frac{d(ax)}{dx} = a,$$

for example, Leibniz writes

$$d(ax) = a dx.$$

Like Newton, Leibniz uses a bar in place of parentheses.

The passages below are quoted from Evelyn Walker's English translation of the Latin paper titled "Nova methodus pro maxima & minimis, itemque tangentibus, qua hec irrationals quantitates moratur, & singulare pro illis calculi genus", originally published in 1684 in the journal *Acta Eruditorum*. This was reprinted in David Eugene Smith's *A Source Book in Mathematics* [4].

In the first passage below, Leibniz describes the product and quotient rules for differentiation and computes the derivatives of various power functions of the form $f(x) = x^n$.

Leibniz:

> Multiplication:
> $$d\overline{vx} = xdv + vdx$$
> or by placing $y = xv$,
> $$dy = xdv + vdx$$
> ... Next, division:
> $$\ldots$$
> $$d\frac{v}{y} = \frac{\pm vdy \mp ydvyy}{yy}$$
> (or z being placed equal to $\frac{v}{y}$)
> $$dz = \frac{\pm vdy \mp ydvyy}{yy}$$
> ... Powers ...
> $$dx^a = ax^{a-1}dx,$$
> for example,
> $$dx^3 = 3x^2dx;$$
> $$d\frac{1}{x^a} = \frac{-adx}{x^{a+1}},$$
> for example, if
> $$w = \frac{1}{x^3},$$
> $$dw = \frac{-3dx}{x^4}.$$
> $$\ldots$$

The above computations should look reasonably familiar to the modern calculus student. Below, we see that Leibniz explains the inverse nature of the integration and differentiation operations.

Leibniz:

> Let the ordinate be x, and the abscissa y, let the interval between the perpendicular and the ordinate...be p; it is manifest at once by my method that
>
> $$p\,dx = x\,dy$$
>
> ...
>
> Which differential equation being turned into a summation becomes
>
> $$\int p\,dy = \int x\,dx$$
>
> But from what I have already set forth in the method of tangents, it is manifest that
>
> $$d\,\overline{\frac{1}{2}xx} = x\,dx;$$
>
> therefore, conversely,
>
> $$\frac{1}{2}xx = \int x\,dx$$
>
> (for as powers and roots in common calculation, so with us sums and differences or $\int$ and d are reciprocals.) Therefore we have
>
> $$\int p\,dy = \frac{1}{2}xx.$$

18.4.1 Leibniz Notation: Exercises

1. How would Leibniz evaluate $\int \frac{-3}{x^4}\,dx$ from his reasoning above?
2. Compute dx^4 as Leibniz would.

18.5 Berkeley's Critique

Leibniz and Newton's works were brilliant and groundbreaking, but also controversial, and not only in the sense of the famous priority dispute. The new science of calculus was sufficiently murky and poorly understood that scholars weren't sure what to make of it. George Berkeley's classic 1734 essay "The Analyst: Or, A Discourse Addressed to an Infidel Mathematician" mocks Isaac Newton's papers on the new subject. (Currently the essay is posted at `www.maths.tcd.ie/pub/HistMath/People/Berkeley/Analyst/Analyst.html`.)

Berkeley:

> Though I am a Stranger to your Person, yet I am not, Sir, a Stranger to the Reputation you have acquired, in that branch of Learning which hath been your peculiar Study; nor to the Authority that you therefore assume in things foreign to your Profession, nor to the Abuse that you, and too many more of the like Character, are known to make of such undue Authority, to the misleading of unwary Persons in matters of the highest Concernment, and whereof your mathematical Knowledge can by no means qualify you to be a competent Judge...
>
> It must, indeed, be acknowledged, that he [Newton] used Fluxions, like the Scaffold of a building, as things to be laid aside or got rid of, as soon as finite Lines were found proportional to them. But then these finite Exponents are found by the help of Fluxions. Whatever therefore is got by such Exponents and Proportions is to be ascribed to Fluxions: which must therefore be previously understood. And what are these Fluxions? The Velocities of evanescent Increments? And what are these same evanescent Increments? They are neither finite Quantities nor Quantities infinitely small, nor yet nothing. May we not call them the Ghosts of departed Quantities?

18.5.1 Questions:

1. Generally, what do you think Berkeley is criticizing in Newton's paper? Try to summarize or paraphrase Berkeley's concerns.
2. *Discussion Question:* In modern notation, we define the derivative to be $f'(x) = \lim_{h \to 0} \frac{f(x+h)-f(x)}{h}$. It would appear that Berkeley at some level is uncomfortable with the very idea of a quantity approaching zero: is it zero or not? What is the truth? Does h equal 0?
3. How would you explain to Berkeley why it might be useful to consider a "limit" as some variable approaches zero?

18.6 Later Developments

Many mathematicians such as the English physician James Jurin and the Irish professor John Walton engaged in public disputes with Berkeley. They each published a vindication of Newton, to which Berkeley replied, and each published a second reply, to which Berkeley replied once more. For an excellent history of some of these disputes, the reader may consult Florian Cajori's 1919 book *A History of the Conceptions of Limits and Fluxions in Great Britain*, which at the time of this writing was available in its entirety on Google Books.

Much subsequent mathematical work strived to put calculus on a firm and rigorous footing. Colin Maclaurin (1698–1746) published an important work "Treatise on Fluxions" in 1742 as a reply to Berkeley's attack. Augustin Louis Cauchy (1789–1857) wrote several rigorous treatises on calculus including "Course d'Analyse" in 1821. Cauchy introduced the notion of limits which made the fundamentals of calculus much easier to understand, and which made the subject seem logically more sound. Mathematicians continue to research the finer points of analysis (the wider branch of mathematics of which calculus is a part) to this day, with no end in sight.

18.7 In The Classroom

An introduction to the stormy historical background of calculus helps to make the subject come alive for introductory calculus students. Teachers wishing to give their students a significant introduction to the origins of calculus may want to assign excerpts from this paper and selected exercises as a homework assignment, with possible class discussion as needed after the students have had a chance to read over this text.

Alternately, in just ten minutes or so of class time, teachers can show students some of the historical passages in class and summarize the controversy. The author of this chapter has most commonly used this "ten minute overview" approach in his Calculus I classes.

18.8 Conclusion

After numerous mathematicians had laid the groundwork over the centuries, Newton and Leibniz discovered the calculus in the late seventeenth century. Newton discovered the subject using clumsy notation in 1665 but neglected to publish his work, leaving the door open for Leibniz to publish his own discovery of calculus in 1684. Mathematicians struggled for decades thereafter to decide who deserved priority, and whether the subject was legitimate. Eventually mathematicians like Cauchy put the subject on a firm footing, and Leibniz' notation became universal.

We feel that a brief overview of this story, and the opportunity to read and discuss samples of Newton and Leibniz' work will enhance any introductory calculus class.

Bibliography

[1] Florian Cajori, *A History of the Conception of Limits and Fluxions in Great Britain from Newton to Woodhouse*, Open Court Publishing Company, Chicago, 1919.

[2] Gottfried Leibniz, "Nova Methodus Pro Maximis et Minimis," *Acta Eridutorum*, 1684, pp. 473–476.

[3] Isaac Newton, "Tractatus de quadratura curvarum," *Optice*, London, 1804.

[4] David Eugene Smith, *A Source Book in Mathematics*, Courier Dover Publications, New York, 1959.

19

A Different Sort of Calculus Debate

Vicky Williams Klima
Appalachian State University

19.1 Introduction

As in most subjects, the historical significance credited to certain events in the development of calculus depends significantly on the historian giving the account. While thinking about how I should interpret selected historical events when presenting them to my first semester calculus classes, I realized that such a decision was unnecessary; my students could determine the appropriate interpretation for themselves through in-class debates. The debate project focuses on two topics: Fermat's method of maxima and minima and Barrow's theorem.

Debates allow students to actively participate in the learning process. David Royse [9] proposes that student learning is at its best when the students have an opportunity to actively engage in an assignment that builds on prior knowledge. The debate assignment was designed to do just that — build upon and shore up the students' understanding of the key concepts of first semester calculus. Bonwell and Eison [1] explain that students are actively learning when they are asked not just to listen, but also to analyze, to synthesize, and to evaluate through active engagement. The debate project requires students to create and analyze arguments, and to actively present these arguments during an in-class debate. As the debate assignment progresses, students begin to take ownership of their arguments and are concerned that they present these arguments well, spending a surprising amount of time in preparation for the debates. During their preparation for and participation in the debates, the students gain a better understanding of some of the fundamental aspects of their beginning calculus course and thus are more likely to remember these fundamental ideas.

19.2 Historical Background

The following gives a brief introduction to both Fermat's method and Barrow's theorem. Worksheets designed to allow small groups, composed of three to four students each, to explore the ideas discussed in these sections can be found in Appendix A (Fermat's method) and Appendix B (Barrow's theorem).

19.2.1 Fermat's method of maxima and minima

Fermat's method of maxima and minima is based on the observation, credited to Kepler, that near a maximum or minimum of a curve, a small change in the independent variable results in an even smaller change in the dependent variable. Fermat used an algebraic interpretation of Kepler's observation to develop an algorithm for finding maxima and minima and used his method to correctly identify maxima and minima of polynomials. Descartes (1596–1650) and Fermat (1601–1665) were among the first to use algebraic techniques to solve geometric problems.

Fermat communicated his algorithm in a letter written around 1638. The following summary of Fermat's method is based on Struik's [11] translation of Fermat's letter. Suppose $f(x)$ has a maximum or minimum value at a. Fermat uses the observation concerning the behavior of functions near their maxima and minima to assert that if $a + e$ is near a then $f(a + e)$ is near $f(a)$. Fermat instructs us to adequate $f(a)$ and $f(a + e)$. The term adequate comes from a process used in ancient Greek mathematics, where it meant "to approximate a certain number as closely as possible" [11, page 220]. Thus, we are to assume that $f(a + e)$ approximates $f(a)$ as closely as possible and form the "pseudo-equality" $f(a + e) \sim f(a)$, where $\sim$ means approximately equal. Most historians use this "pseudo-equality" to describe Fermat's method, although some disagree with the use of this idea (see [7, p. 233]). Then, we are to cancel out terms which appear on both sides of the pseudo-equality. Every remaining term in the pseudo-equality has an e in it. Fermat instructs us to divide all terms by e. Fermat's actual instructions were "to divide all terms by e or by a higher power of e so that we completely remove e from at least one of the terms." [11, p. 223]; however in all of his examples and in his further explanations, he divided by e to the first power. Next, we should suppress e, which means we are to set $e = 0$, and after doing so change our pseudo-equality into an actual equality. Fermat finally explains that solving the resulting equation for a and then substituting the value that you find back into the original equation will yield the maximum or minimum value for $f(x)$.

To clarify Fermat's method we will use the method to identify the minimum value of the function $f(a) = a^2 + 3a + 1$. First we adequate $f(a + e)$ and $f(a)$ to obtain the pseudo-equality

$$a^2 + 2ae + e^2 + 3a + 3e + 1 \sim a^2 + 3a + 1.$$

Next we cancel out terms that appear on both times of the pseudo-equality, yielding $2ae + e^2 + 3e \sim 0$. All of the remaining terms in our pseudo-equality have an e in them and we divide all terms by e to obtain $2a + e + 3 \sim 0$. Finally, we solve $2a + 3 = 0$ for a to obtain the value $a = \frac{-3}{2}$. Substituting $a = -\frac{3}{2}$ into $f(a)$ yields our minimum value $f\left(\frac{-3}{2}\right) = -\frac{5}{4}$.

One may quickly observe that Fermat's method is actually identifying stationary points of the polynomial $f(a)$, and thus only gives necessary but not sufficient conditions to identify maxima and minima. In our modern day calculus classes we teach our students that maxima and minima may occur only when the derivative of a function is equal to zero. Does Fermat's method express this fact for polynomial equations? In his article "History of Calculus", Arthur Rosenthal explains Fermat's method using a relatively simple mathematical formula. Rosenthal [8, p. 79] explains that Fermat's method means that we should "determine a" from the following equation

$$\left[\frac{F(a + e) - F(a)}{e}\right]_{e=0} = 0.$$

This equation should look similar to that created by expressing $f'(a)$ using the limit definition of the derivative and setting this expression equal to zero. However, a very important part of the definition of the derivative is missing. Fermat does not take the limit as e approaches zero. Fermat would not even know what we mean by taking the limit as e approaches zero, for this concept was not developed during his time. Does Fermat's description of his method imply that he is letting e tend towards zero, or is he simply treating e sometimes as a nonzero number and sometimes as zero?

19.2.2 Barrow's theorem

Isaac Barrow (1630–1677) was a well known theologian and mathematician. He was a professor of mathematics at Cambridge University, where his most famous pupil was Isaac Newton. Barrow's theorem, which can be found in his *Lectiones Geometricae* (London, 1670) is re-phrased below. For simplicity we will assume that the curve $y(x)$, given in the theorem, passes through the origin, and that all of the points mentioned in the theorem lie either on the x-axis or in the first quadrant. The figure given below may be useful when reading the theorem.

19.3. In the Classroom

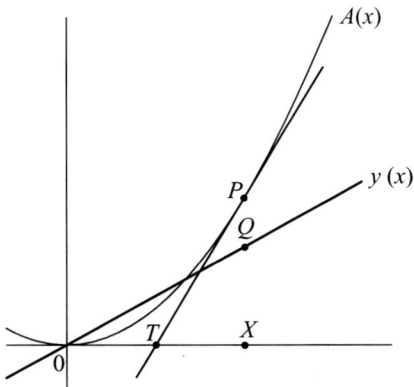

Barrow's Theorem [5]: Suppose that at any point $X = (x, 0)$ on the x-axis, $A(x)$ gives the area of the region enclosed by the curve $y(x)$, the x-axis, and the vertical line containing the point X; Q is the point on the curve $y(x)$ directly above X; and P is the point on the curve $A(x)$ directly above X. If T is chosen on the x-axis so that the length of $\overline{TX}$ times the length of $\overline{XQ}$ equals the length of $\overline{XP}$, then the line $\overleftrightarrow{TP}$ is tangent to the curve $A(x)$ at point P.

If the hypothesis of Barrow's theorem is met, then T is the point on the x-axis such that

$$|\overline{TX}| \cdot |\overline{XQ}| = |\overline{XP}|.$$

Simply rearranging the previous equality yields

$$|\overline{XQ}| = \frac{|\overline{XP}|}{|\overline{TX}|},$$

and thus

$$y(x) = |\overline{XQ}| = \frac{|\overline{XP}|}{|\overline{TX}|} = \text{the slope of the line } \overleftrightarrow{TP}.$$

Barrow concludes $\overleftrightarrow{TP}$ is tangent to $A(x)$, the area curve, at the point P. If we are willing to impose our modern notion of the derivative onto Barrow's statement of his theorem, we see Barrow is claiming that

$$A'(x) = \text{the slope of the line } \overleftrightarrow{TP} = y(x),$$

where $A(x)$ is the area function for $y(x)$. Therefore, if we view Barrow's theorem in light of our modern day notions of the derivative, we might conclude that this theorem is a geometric statement of the fundamental theorem of calculus.

19.3 In the Classroom

By introducing my Calculus I class to Fermat's method and Barrow's theorem, I aim to make them think about what makes the limit definition of the derivative and the fundamental theorem of calculus important. Are these theorems just powerful computational tools or do they have a deeper meaning? In order to encourage my students to tackle this question wholeheartedly, I structure a debate in which the small groups of students work together to debate the significance of the method and theorem. Initially, I break the class into small groups of three to four students each. Half the groups receive a worksheet exploring Fermat's method (see appendix A) and the other half receive a worksheet exploring Barrow's theorem (see appendix B). Each group is responsible for submitting one completed worksheet. When I return the graded worksheets, I announce that the groups will now become debate teams. Each team will be assigned to debate another team that received the same type of worksheet that they did. Those teams who received the Fermat's method worksheet will be debating Resolution 1, given below, and those teams who received the Barrow's theorem worksheet will be debating Resolution 2, also given below.

Resolution 1. *Pierre de Fermat's creation of his method of maxima and minima demonstrates that he understood the concept of the derivative, and thus Fermat should be given credit for the development of this key idea in calculus.*

Resolution 2. *Isaac Barrow's famous theorem (stated at the end of his Geometrical Lectures) is a geometric statement of the fundamental theorem of calculus, and thus Barrow should be given credit for the development of this key idea in calculus.*

Of the groups receiving Resolution 1, half will be arguing for the resolution and half against. However, team members learn which side, affirmative (i.e., taking the position that the resolution is true) or negative (i.e., taking the position that the resolution is false), their team must argue only moments before the debate begins. Thus, they must prepare to argue both sides of their resolution. The debate concerning Resolution 2 is set up in exactly the same manner.

19.3.1 Structuring the debates

The form of debate that I use in my classroom is based on the parliamentary format of debate [4]. However, the classroom debates have considerably shorter time limits than usual debates using the parliamentary format. In addition, unlike debates in the parliamentary format, the classroom debates do not allow for interruptions during the speeches, instead they include a cross-examination period after the constructive speeches.

During the debates, each team member has a specific assigned task. The teams consist of a leader, member, cross-examiner, and rebuttal speaker. The leader and the member give constructive speeches that lay out their team's argument, also taking opportunities to point out weaknesses in their opponents' case. The cross-examiner listens to the other team's argument and asks them a question that challenges this argument. The rebuttal speaker gives counter-arguments for the points made by the constructive speeches given by her opponent, concluding with a summary of her team's case. If a group contains only three students, the group as a whole performs the cross-examination function. Appendix C contains a detailed explanation of the structure of the debates and some hints to help students in preparing for the debate.

19.3.2 Conducting the debates

When I use the debates in my classroom, the students conduct the debates during our regularly scheduled class time. In each of my classes, I have been able to break the class into eight debate teams. Thus, I can schedule the debates over two days. Each day we see two teams debate the historical significance of Fermat's method and two teams debate the historical significance of Barrow's theorem. Those students who are not members of the four teams scheduled for debate on a certain day work on an in-class assignment in a separate room. I have considered holding a debate tournament in the evening and inviting faculty to watch all or part of the tournament. However, I have not yet implemented this idea.

Before each debate the students toss a coin to determine which team argues which side. One team chooses whether they want the coin toss to apply to them or the other team, and the remaining team chooses whether heads represents the affirmative or negative argument. After the coin toss, before a debate officially begins, the debating teams take a few moments to collect their thoughts. During this time, I briefly explain the resolution to the audience, those students who are scheduled to debate the other resolution on the same day, so that they understand the statement of the topic for debate. Then the experts begin their debate and the audience listens and critiques. While listening to the debate, the audience answers a few simple questions about the debate they are watching. "What do you think are the strongest and weakest parts of the affirmative argument? What do you think are the strongest and weakest parts of the negative argument? Which team do you think won? Why?"

After all of the debates have concluded, the class discusses the general themes of the debates. We decide that often at the beginning of a calculus class, the class seems to take the same attitude that the affirmative side took in the debate; they just want to learn a method that works. However, by the end of the semester the class should see the importance of the negative sides' arguments as well; they should understand that calculus is a dynamic subject with infinitesimally small changes at its core.

19.4 Conclusions

Usually, students initially take the side that Fermat's method clearly develops the idea of the derivative and Barrow's theorem is exactly the same as the fundamental theorem of calculus. When they first hear of the debates, students often say that they can only hope that they will be assigned the affirmative side of the argument. Then the students begin their research. This is when they really begin to think about what the definition of the derivative and the fundamental theorem of calculus truly mean. When asked immediately before the debates which side they would prefer to argue, most teams choose the negative side. Their thoughts concerning the ideas that make differentiation and the fundamental theorem of calculus important have moved from a mechanical to a conceptual level.

The debate project requires students to study the derivative and fundamental theorem of calculus in much more detail than when these ideas were first introduced in their calculus course. Thus the students leave their beginning calculus course with a deeper conceptual understanding of these topics. In some sense the students develop this deeper understanding without really intending to do so. Their natural desire to win the debate drives them to prepare and a meaningful understanding of fundamental calculus ideas results. In addition, students begin to develop an appreciation for the foundation that Newton and Leibniz had to build upon when they began their work.

Bibliography

[1] Charles Bonwell and James Eison, "Active Learning: Creating Excitement in the Classroom," *ASHE-ERIC higher education report*, School of Education and Human Development, George Washington University, Washington, DC, 1991.

[2] Florian Cajori and J. M. Child, "Recent Publications: Book Review: Who Was the First Inventor of the Calculus?: The Geometrical Lectures of Isaac Barrow", *American Mathematical Monthly*, 26 (1919), 15–20.

[3] C. H. Edwards Jr., *The Historical Development of the Calculus*, Springer-Verlag, New York, 1979.

[4] Jon M. Erickson, James J. Murphy, and Raymond Bud Zeuchner, *The Debater's Guide, Third Edition*, Southern Illinois University Press, Carbondale, IL, 2003.

[5] Martin Flashman, "Historical Motivation for a Calculus Course: Barrow's Theorem," in *Vita Mathematica*, Ronald Calinger, ed., MAA Notes #40, 1996, pp. 309–315.

[6] Judith V. Grabiner, "The Changing Concept of Change: the Derivative from Fermat to Weierstrass," *Mathematics Magazine*, 56(1983), 195–206.

[7] Michael Sean Mahoney, *The Mathematical Career of Pierre de Fermat*, Princeton University Press, Princeton, NJ, 1973.

[8] Arthur Rosenthal, "The History of Calculus," *American Mathematical Monthly*, 58(1951), 75–86.

[9] David Royse, *Tips for College and University Instructors: a Practical Guide*, Allyn and Bacon, Boston, MA, 2001.

[10] P. Stromholm, "Fermat's Methods of Maxima and Minima and of Tangents. A Reconstruction," *Archive for History of Exact Sciences*, 5(1968), 47–59.

[11] D. J. Struik (ed.), *A Source Book in Mathematics, 1200–1800*, Harvard University Press, Cambridge, MA, 1969.

[12] Robert Trapp, Joseph P. Zompetti, Jurate Montiejunaite, and William Driscoll, *Discovering the World Through Debate: a Practical Guide to Educational Debate for Debaters, Coaches and Judges, Third Edition*, International Debate Education Association, New York, 2005.

Appendix A: Fermat's Method Worksheet

An Observation

Fermat's method of maxima and minima is based on the observation, credited to Kepler, that near a maximum or minimum of a curve, a small change in the independent variable results in an even smaller change in the dependent variable.

1. Suppose $f(x) = -x^2 + 6x - 7$.

    (a) Draw a graph of $f(x)$, and use this graph to explain the observation of Kepler as it applies to $f(x)$.

    (b) Fill in the missing values in the table below, and use this table to explain the observation numerically as it applies to $f(x)$.

e	0.9	0.7	0.3	0.1
$3 + e$				
$f(3 + e)$				
$2 - f(3 + e)$				

    Note $(3, 2)$ is the maximum point of $f(x)$

The Method

Fermat used an **algebraic** interpretation of Kepler's observation to develop an algorithm for finding maxima and minima. Descartes (1596–1650) and Fermat (1601–1665) were among the first to use algebraic techniques to solve geometric problems. Fermat communicated his algorithm in a letter written around 1638. A summary of Fermat's method follows (see [3]).

- Suppose $f(x)$ has a maximum or minimum value at a.
- Fermat uses the observation concerning the behavior of functions near their maxima and minima (introduced in problem one) to assert that if $a + e$ is near a then $f(a+e)$ is near $f(a)$. Fermat only explains this reasoning in a later communication regarding Descartes' challenge of Fermat's method.
- Next, Fermat instructs us to adequate $f(a)$ and $f(a + e)$. The term adequate comes from a process used in ancient Greek mathematics, where it meant "to approximate a certain number as closely as possible" [3, p. 220]. Thus, we are to assume that $f(a + e)$ approximates $f(a)$ as closely as possible (which makes sense in light of the previous observation) and form the "pseudo-equality" $f(a+e) \sim f(a)$, where $\sim$ means approximately equal. Most historians use this "pseudo-equality" to describe Fermat's method, although some disagree with the use of this idea [1, p. 164].
- Then we are to cancel out terms which appear on both sides of the pseudo-equality.
- Every remaining term in the pseudo-equality has an e in it. Fermat instructs us to divide all terms by e. Fermat's actual instructions were "to divide all terms by e or by a higher power of e so that we completely remove e from at least one of the terms" [3, p. 223]; however in all of his examples and in his further explanations, he divided by e to the first power.
- We are then to suppress e, which means we are to set $e = 0$, and after doing so change our pseudo-equality into an actual equality.
- Fermat finally explains that solving the resulting equation for a and then substituting the value that you find back into the original equation will yield the maximum or minimum value for $f(x)$.

2. Use Fermat's method to find the minimum of the equation $f(x) = x^2 + 5x - 3$ and the maximum of the equation $f(x) = -2x^2 + 5x - 1$.

3. In the first account of his method, Fermat used his method to solve the following problem, "To divide the segment AC [of length b] at E so that AE [of length a] $\times EC$ may be a maximum" [2, p. 233]. Use Fermat's method to solve this problem.

19.4. Conclusions

4. In his *American Mathematical Monthly* article "History of Calculus", Arthur Rosenthal explains Fermat's method using a relatively simple mathematical formula. Rosenthal [2, p. 79] explains that Fermat's method means that we should "determine a" from the following formula

$$\left[\frac{F(a+e) - F(a)}{e}\right]_{e=0} = 0.$$

Show that Rosenthal's explanation of Fermat's method coincides with the method as described above. You may need to re-order the steps in our description of Fermat's method to make it coincide with Rosenthal's formula.

5. Give a critique of Fermat's method. Do you think it correctly identifies maxima and minima? Can you find any faults with the method? If so, what are they and why do you consider them faults?

Bibliography

[1] Michael Sean Mahoney, *The Mathematical Career of Pierre de Fermat*, Princeton University Press, Princeton, NJ, 1973.

[2] Arthur Rosenthal, "The History of Calculus," *American Mathematical Monthly*, 58(1951), 75–86.

[3] D. J. Struik (ed.), *A Source Book in Mathematics, 1200–1800*, Harvard University Press, Cambridge, MA, 1969.

Appendix B: Barrow's Theorem Worksheet

Understanding the Theorem

Isaac Barrow (1630–1677) was a well known theologian and mathematician. He was a professor of mathematics at Cambridge University, where his most famous pupil was Isaac Newton. Barrow's theorem, which can be found in his *Lectiones Geometricae* (London, 1670) is re-phrased below. For simplicity we will assume that the curve $y(x)$, given in the theorem, passes through the origin, and that all of the points mentioned in the theorem lie either on the x-axis or in the first quadrant. The figure given below may be useful when reading the theorem.

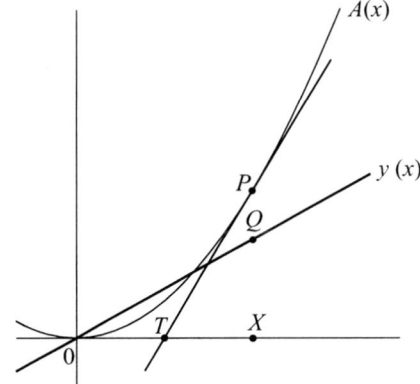

Barrow's Theorem [1]: Suppose that at any point $X = (x, 0)$ on the x-axis, $A(x)$ gives the area of the region enclosed by the curve $y(x)$, the x-axis, and the vertical line containing the point X; Q is the point on the curve $y(x)$ directly above X; and P is the point on the curve $A(x)$ directly above X. If T is chosen on the x-axis so that the length of $\overline{TX}$ times the length of $\overline{XQ}$ equals the length of $\overline{XP}$, then the line $\overleftrightarrow{TP}$ is tangent to the curve $A(x)$ at point P.

1. The function $A(x)$ can be thought of as $\int_0^x y(t)\, dt$; explain why this is true.

2. In this question we will look at an example and check that Barrow's theorem holds true for our example. We will use the curve $y(x) = 2x$ and after finding the algebraic equation for $A(x)$ we will investigate the point $X = (3, 0)$ on the x-axis in detail. Please begin this question by drawing a careful graph of the curve $y(x) = 2x$ and include the point $X = (3, 0)$ on your graph.

   (a) First, we will determine what the hypothesis of Barrow's theorem (the part before the **then**) would look like for our example. Please fill in the blanks and as you do so add each function/point that you determine to the graph that you began earlier. For our specific example the hypothesis would become ...

   Suppose that at any point $X = (x, 0)$ on the x-axis, $A(x)$ gives the area of the region enclosed by the curve $y(x) = 2x$, the x-axis, and the vertical line containing the point X. Then an algebraic equation for $A(x)$ is given by $A(x) = $ _____. As a specific example consider the point $X = (3, 0)$ on the x-axis. Suppose Q is the point on the curve $y(x) = 2x$ directly above $X = (3, 0)$, and P is the point on the curve $A(x) = $ _____ directly above $X = (3, 0)$. Then $Q = ($ ___ , ___ $)$ and $P = ($ ___ , ___ $)$. If the point T is chosen on the x-axis so that the length of $\overline{TX}$ times the length of $\overline{XQ}$ equals the length of $\overline{XP}$, then the length of $\overline{TX}$ is _____ units and $T = ($ ___ , ___ $)$.

(b) Barrow's theorem's concludes that if the hypothesis is met then

"the line $\overleftrightarrow{TP}$ is tangent to the curve $A(x)$ at the point P."

We will check that this is in fact the case for our example. Recall that in part (a) you found the points T and P and the algebraic equation of the curve $A(x)$ for our specific example.

  i. Find the equation of the line $\overleftrightarrow{TP}$ and draw this line on your graph.
  ii. Find the equation of the line tangent to the curve $A(x)$ at the point P. (You will want to use calculus for this question.)
  iii. Note that your answers to part (i) and (ii) should describe the same line.

3. Repeat question two with the function $y(x) = x^2$ and $X = (4, 0)$.

Why the fundamental theorem of calculus?

4. Some mathematicians (see for example, [1]) have offered the following modern day interpretation of Barrow's theorem

If $f(x)$ is a continuous function (like the function $y(x)$ in Barrow's theorem) and $A(x) = \int_0^x f(t)dt$ then $A'(x) = f(x)$.

Explain how one could translate Barrow's theorem into this modern statement. Hint: You may wish to begin by writing the statement "If the point T is chosen on the x-axis so that the length of $\overline{TX}$ times the length of $\overline{XQ}$ equals the length of $\overline{XP}$." as an equation and then solve this equation for the length of $\overline{XQ}$.

5. State the fundamental theorem of calculus as you learned it in class and explain why the modern day interpretation of Barrow's theorem given in question four of this section is actually equivalent to the fundamental theorem of calculus as you learned it in class.

6. Give a critique of the interpretation of Barrow's theorem as an early version of the fundamental theorem of calculus. Do you think this geometric theorem is actually equivalent to the modern day fundamental theorem of calculus? Can you find any faults with the idea that Barrow's theorem conveys the essence of the fundamental theorem of calculus? If so, what are the faults, and why do you consider them to be faults?

19.4.1 Bibliography

[1] Martin Flashman, "Historical Motivation for a Calculus Course: Barrow's Theorem," in *Vita Mathematica*, Ronald Calinger, ed., MAA Notes, 40(1996), 309–315.

Appendix C: The Debates: Roles, Structure, Hints

The Roles

- **Leader**: gives a constructive speach that lays out his or her teams argument, and takes opportunities to point out weaknesses in the opponent's case. (speaks first)

- **Member**: gives a constructive speech that lays out his or her teams argument, and takes opportunities to point out weaknesses in the opponent's case. (speaks second)

- **Cross-examiner:** listens to the other team's argument and asks them a question that challenges this argument.

- **Rebuttal Speaker**: gives counter-arguments for the points made by the constructive speeches given by her opponent, concluding with a summary of her team's case.

If your group contains only three students, the group as a whole will perform the cross-examination function.

The Structure

- The debate begins with the leaders' constructive speeches, each a maximum of three minutes in length. The affirmative leader argues first, using his time to define terms and criteria, give an outline of the team's entire case, and begin to develop the main case for the team. The affirmative leader concludes with a brief summary. The leader of the negative speaks next. Her task is much the same, but because she has already heard the affirmative speech, she should point out where her argument differs from the affirmative, refuting the affirmative points and explaining why the negative argument is the better one.

- The members of the affirmative and the negative give the next set of constructive speeches in the debate, with the affirmative side speaking first. These speeches are also each a maximum of three minutes in length. The affirmative speaker repeats the affirmative case, responding to the challenges that the leader of the negative made during her speech and identifying points that the previous negative speaker failed to refute. Then, the speaker extends the affirmative case, concluding with a brief summary. The member of the negative fulfills mainly the same role, except that he is refuting the points make by the member of the affirmative.

- Two periods of cross-examination follow the constructive speeches. The cross-examination periods consist of posing a question to the opposing team, and then having the opposing team confer and offer their response. The time limits for these actions are thirty seconds, and two minutes, respectively. During this time the cross-examiner must ask a question that builds his team's credibility. Typically these questions focus on exposing some sort of weakness in his opponent's case. During the first cross-examination period the negative side asks the question and the affirmative side responds, with the roles being reversed for the second cross-examination period.

- The debate concludes with two rebuttal periods, each a maximum of two minutes in length. The negative rebuttal speaker speaks first, identifying the key arguments of the opponent's case and refuting those arguments, ending her speech with a summary of her team's case. The affirmative rebuttal speaker follows, using the same plan.

Some Hints

The following tips are based on the advice found in [1].

- Begin by brainstorming for the affirmative and then separately for the negative; write any argument you can think of; don't worry if the argument is a good one. After you are finished look through your arguments and mark out the ones you do not like, put arguments that are interconnected together, and rank your arguments from strongest to weakest.

- Make a T-chart, labeling one side as affirmative and the other as negative. One by one put the arguments from your brainstorming session into the chart. Once you have entered an argument for the affirmative, try to find a related argument for the negative to put across from your affirmative argument in the chart. If you have already listed this negative argument in your brainstorm then cross it off.

- Carefully define terms, and decide upon the correct interpretation of terms that may have ambiguous meaning. Remember, sometimes opposing sides decide to define terms differently and your success in the debate may hinge on whether your definition is seen as the most reasonable one.

- Determine an outline for your argument. What are your main points? Develop support for your main points with evidence (e.g., historical facts, examples, and expert commentary). Research is key at this step.

- You can work to prepare for your cross-examination and rebuttals before the debate begins. Look carefully at your affirmative argument. Think of it as your opponent's argument and try to determine weakness and write cross-examination questions and rebuttal points that expose these weaknesses. Do the same with your negative argument. You will also want to collect evidence and develop arguments that refute the weaknesses that you found; you can use these arguments during your response to the opposing team.

Bibliography

[1] Robert Trapp, Joseph P. Zompetti, Jurate Montiejunaite, and William Driscoll, *Discovering the World Through Debate: a Practical Guide to Educational Debate for Debaters, Coaches and Judges, third edition*, International Debate Education Association, New York, 2005.

20

A 'Symbolic' History of the Derivative

Clemency Montelle
University of Canterbury
New Zealand

20.1 Introduction

How often do you have to tell your students to brush up on their notation? When they have dropped limit notation, forgotten critical modulus signs, mixed up their integrals, muddled up their derivatives, how do you convey to them the importance of recording it right?

What of your exasperation as they fail to appreciate the precision that mathematical notation affords them—notation which has been developed and refined over centuries, and notation that will continue to be improved for centuries more. Indeed, mathematics is the one subject in which they can really express *exactly* what they mean. How can we help them appreciate the symbolism they use?

This paper provides a light-hearted look at various notational conventions concerning the derivative. It will briefly cover the various proposals for its symbolic representation over history and the reasons behind the prevalence of the various manifestations over others. The examination of the history of the development of calculus notations is not only fascinating, but suggests a paradigm for the development of future notations. An examination of the reasons behind the failures and successes of past notation can equip one with a certain foresight regarding the future of newly introduced symbols, as particular areas of mathematics expand.

This capsule is intended for undergraduate calculus classes and should ideally be offered directly before or after covering the derivative and its various applications, or indeed right in the middle when the students need a bit of a break from theory and practice! It may also be suitable for higher level secondary school mathematics.

This capsule is intended to take 30 minutes and contains visual material which may be presented to the students as well as directed questions to encourage them to think critically about notation.

20.2 The Derivative

How many ways to represent the derivative have you come across in mathematics? Are they *really* equivalent? Who developed them and what can history reveal about the success of some notations over others? Among the many disagreements Newton and Leibniz are remembered for, they had a big one over notation. They both independently developed distinctive notation for the derivative when they published their results in calculus. In turn, allegiance in both British and European mathematical communities to strictly one or the other persisted for almost half a century until Leibniz's notation finally prevailed. We will look at this scuffle and its consequences.

20.3 Isaac Newton (1643–1727):

To understand Newton's use of symbolism, we need to be familiar with the basics of his conception of calculus. In Newton's mind, the fundamental notion of calculus was concerned with motion. Each variable in an equation could be ultimately interpreted as a distance reckoned with respect to time. Therefore, central to his account and thus his symbolic devising, was the notion of time.

Throughout his mathematical researches, Newton developed several distinct ways to represent the derivative. The first of these was contained in *De Analysi* 1669 (but not published until 1711), in which among many other things, he developed a systematic method for differentiation. Notation-wise, Newton set a 'little zero', literally o, as a very small interval of time and op and oq as the corresponding small increments over which x and y change in this interval.

The second notational system appeared in about 1671, where p and q were reoriented symbolically as well as mathematically. Newton developed what he called Fluxional Calculus, in which he consciously avoided the use of infinitesimals as he considered these mathematically imprecise. He imagined that a curve was the result of the motion of a particle with respect to time. This curve, which was the path traced by the particle, could be then analyzed via changes in x and in y. In order to consider the various properties of the curve, he defined *fluents*, or flowing quantities, as the variables x and y themselves and *fluxions* the velocity in the x and y direction respectively with respect to time. Thus, Newton's fluxions were equivalent to our first derivative.

Newton represented fluxions notationally as:

$$\dot{x} \quad \ddot{x} \quad \dot{\ddot{x}}$$

which are equivalent to the modern $\frac{dx}{dt}$, $\frac{d^2x}{dt^2}$ and $\frac{d^3x}{dt^3}$ respectively. These are occasionally referred to as 'pricked letters' (for obvious reasons!), but more commonly today the system is called 'dot-notation'. As you can see, by repeating the dots and introducing dashes Newton could present fluxions of fluxions, or fluents of fluents! The natural succession therefore was:

$$\overset{\parallel}{x} \quad \overset{\mid}{x} \quad x \quad \dot{x} \quad \ddot{x} \quad \dot{\ddot{x}}$$

each term being the fluxion of the term preceding.

Historians often connect particular features of Newton's personality (in particular being private and guarded) and his continually evolving account of calculus, as responsible for the poor reception of his notation and indeed its ultimate demise. That he seemed to write for himself, they assume, had the consequence that he tended to adopt the notation he thought of at the time and put little thought into its practicality for wider audiences and broader mathematical function.

English mathematicians of the eighteenth century remained loyal to Newton's notation, but ultimately the tradition lost favor. It is perhaps because of the inability of the notational system to allow for a richer development in the field of calculus, and to incorporate and represent for the equivalents in multi-variate calculus and calculus of variations. Newton's system was also somewhat cumbersome, because of its orientation around the notion of time. Any expression containing a derivative not taken with respect to time was quite involved. For example, in order to express the derivative $\frac{dz}{dx}$, Newton had to write:

$$\dot{z} \; : \; \dot{x}$$

which literally meant $\frac{dz}{dt} : \frac{dx}{dt}$.

Newton's dot-notation, however, didn't disappear completely from the mathematical scene. It is used occasionally when expressing differential equations, particularly in engineering contexts, and can still be seen today in some textbooks.

20.4 Gottfried Wilhelm von Leibniz (1646–1716):

Leibniz, Newton's European contemporary, in contrast took the greatest pains to develop his notation, believing that good notation was of primary importance, not only for representation of mathematical concepts, but for human understanding. In fact, he credited all his mathematical insights to his notation. The symbolic representation of a mathematical concept was, in his opinion, to be designed with the greatest attention—capturing a mathematical idea with carefully selected notation would help mathematicians make progress. The importance of notation seems to have been fostered in his early days as a student, as historians record episodes which reveal Leibniz's teachers instilling in him a reverence for simplicity and clarity in discourse.[1]

Leibniz's account of the calculus[2] was centered, in contrast to Newton, on the differential, a variable dx which was an arbitrarily small finite increment upon the variable x. In a letter dated November 11th 1675, he introduced the

---

[1] See Bardi [1] pp. 86–87.
[2] For more details see for example Boyer [2] pp. 403–405 or Katz [4] pp. 527ff.

symbol dx and dy as the differentials of x and y and something like $\frac{dx}{dy}$ for what we have come to call the derivative, but not in that exact form—rather expressed as:[3]

$$dx \text{ ad } dy$$

or

$$dx : dy$$

He wrote in this letter:

> ... *idem est dx et $\frac{x}{d}$, id est differentia inter duas x proximas*
> ... dx and $\frac{x}{d}$ are the same, that is, the difference between two proximate x's

Higher derivatives were to be denoted as

$$ddx$$

being, in turn, an infinitesimally small increment upon the variable dx, and so on.

Almost immediately after this, he discovered the product and quotient rules. It was as if the symbolism suggested these concepts and manipulations.[4]

> ... *videndum an dx dy idem sit quod $d\overline{xy}$, et an $\frac{dx}{dy}$ idem quod $d\frac{x}{y}$*
> ... let us now examine whether $dx\ dy$ is the same thing as $d\overline{xy}$ and whether $\frac{dx}{dy}$ is the same things as $d(\frac{x}{y})$

These rules were followed shortly by his articulation of the power rule and the rule for roots with respect to the derivative. Other rules fundamental to differentiation, such as the chain rule, seem trivial using his notation. His notation also allowed him in 1684 to solve a problem that he described as "one of the most difficult and most beautiful problems of applied mathematics, which without our differential calculus ... no one could attack with any such ease." This is the problem of finding a curve whose subtangent is a given constant a.[5]

Leibniz's notation offered a simple, yet clear representation of the infinitesimal demands of the calculus, which no doubt facilitated its spread. His use of the 'd' which *acted* upon something emphasized the operator quality of the process. This proved important in later developments. His notation was immediately embraced by the continental mathematical community to their great advantage.

There are many reasons as to why it ultimately prevailed. One of the most important was the fact that it could incorporate further details and features as mathematical developments and insights in calculus deepened. Furthermore, it recorded the right amount of information without obscuring anything, but neither was it too cumbersome.

One of the main distinctions in the approach to notation by Newton and by Leibniz is one of motivation. Newton searched for a means of expression for concepts he already conceived of. Leibniz, on the other hand, developed a platform in order to research new mathematics, with the result that, in many cases, the resulting symbolism proved to be suggestive of new mathematical insights and relations—just as he had intended.[6]

20.5 Joseph-Louis Lagrange (1736–1813)

At the end of the eighteenth century, Lagrange gave a definitive new approach to the calculus, being dissatisfied with earlier accounts. In his mathematical description, he purposefully avoided all references to fluxions and fluents, and to infinitesimals, (and indeed limits for that matter), because he believed they were unable to be defined precisely enough in a mathematical context.

---

[3] The first expression uses the Latin preposition *ad* which literally means 'to', or in keeping with the mathematical context 'by'.

[4] Cajori [3] p. 204.

[5] For details see Katz [4] pp. 528–9.

[6] In fact Leibniz's 'symbolic' ambitions went beyond mathematics. He went so far as to develop a *characteristica universalis* or "universal characteristic", founded on an alphabetic-like indexing of human thought. Through it, each element essential to human thought could be then represented and then a method of combining these symbols to indicate more complex notions. "It is obvious that if we could find characters or signs suited for expressing all our thoughts as clearly and as exactly as arithmetic expresses numbers or geometry expresses lines, we could do in all matters...For all investigations which depend on reasoning would be carried out by transposing these characters and by a species of calculus." (Preface to the General Science, 1677. Revision of Rutherford's translation in Jolley [5] p.234.)

20.6. Louis François Antoine Arbogast (1759-1803)

Lagrange emphasized the centrality of the concept of the function and established a new basis for calculus with recourse to algebraic processes. Any function, Lagrange argued, can be represented as a power series.[7] Given then some function $f(x)$, its related function $f(x + i)$ (with i indeterminate) can be expanded:

$$f(x + i) = f(x) + p(x)i + q(x)i^2 + r(x)i^3 + \cdots$$

where $p(x), q(x), \ldots$ etc., are independent of i. The first coefficient $p(x)$ of i can be identified with the ratio $\frac{dy}{dx}$, and likewise for higher order ratios.[8] He named $p(x)$ the "fonction dérivée" (from which we get 'derivative'), because of the fact that it is 'derived' from the initial function $f(x)$. He denoted all higher order derivatives in turn:

$$f',\ f'',\ f'''\ \text{etc} \ldots$$

to emphasize the fact that the functions p, q, r are all likewise "derived" from the initial function $f(x)$.

The notation of Lagrange is used extensively today, as an important and useful adjunct of other notations current in modern calculus.

> 9. Nous appellerons la fonction fx, *fonction primitive*, par rapport aux fonctions f'x, f''x, etc. qui en dérivent, et nous appellerons celles-ci, *fonctions dérivées*, par rapport à celle-là. Nous nommerons de plus la première fonction dérivée f'x, *fonction prime*; la seconde fonction dérivée f''x, *fonction seconde*; la troisième fonction dérivée f'''x, *fonction tierce*, et ainsi de suite.
>
> De la même manière, si y est supposée une fonction de x, nous dénoterons ses fonctions dérivées par y', y'', y''', etc., de sorte que y étant une fonction primitive, y' sera sa fonction *prime*, y'' en sera la fonction *seconde*, y''' la fonction *tierce*, et ainsi de suite.
>
> De sorte que x devenant $x + i$, y deviendra
>
> $$y + y'i + \frac{y'' i^2}{2} + \frac{y''' i^3}{2.3} + \text{etc.}$$
>
> Ainsi, pourvu qu'on ait un moyen d'avoir la fonction prime d'une fonction primitive quelconque, on aura, par la simple répétition des mêmes opérations, toutes les fonctions dérivées, et par conséquent tous les termes de la série qui résulte du développement de la fonction primitive.
>
> Au reste, pour peu qu'on connaisse le calcul différentiel, on doit voir que les fonctions dérivées y', y'', y''', etc. relatives à x, coïncident avec les expressions $\frac{dy}{dx}, \frac{d^2y}{dx^2}, \frac{d^3y}{dx^3}$, etc.

Figure 20.1. Excerpt from *Théorie des fonctions analytiques* of Lagrange (1797) in which there is the first appearance of the prime notation for the derivative

20.6 Louis François Antoine Arbogast (1759-1803)

Indeed, many other notational systems were proposed for representing various concepts in calculus with varying reception from the mathematical community. We can't give an exhaustive survey here, but will consider one final proposal which is significant because it conceives of the processes in the calculus as operations; the symbols become

---
[7]This was to prove not always the case, see Boyer [2] pp. 489–10.
[8]See Katz [4] pp. 586–8 for more details.

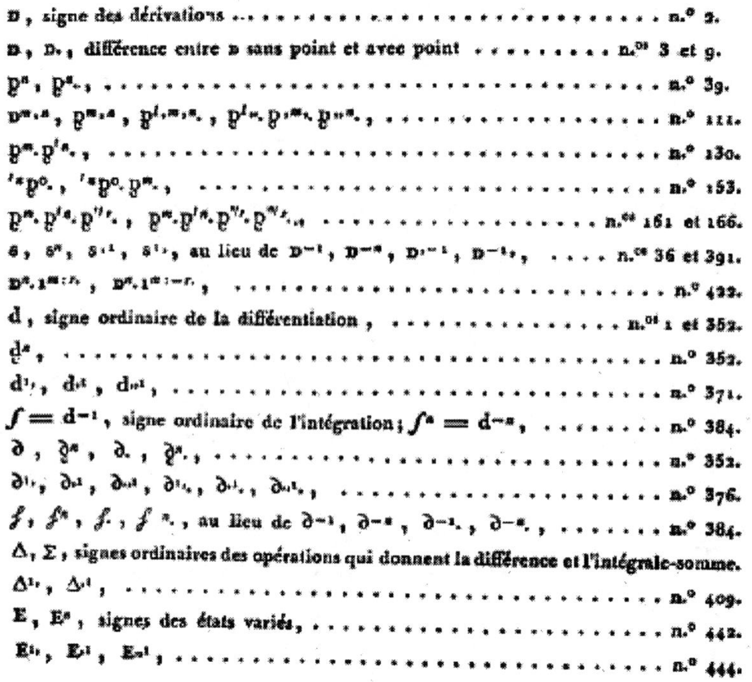

Figure 20.2. Excerpt from the *Calcul des Dérivations* of Arbogast (1800) p. xxi.

operators, which act on one function, to produce another. This notation is less common, but still has currency today in some mathematical contexts, particularly linear differential equations.

The first to champion this orientation was Arbogast in 1800,[9] although such symbolism was used by Johann Bernoulli (1667–1748) earlier. Arbogast's approach to calculus was similar to Lagrange. He used as his main symbol D, to indicate the process of differentiation. Similar to Lagrange, Arbogast supposes that $F(a+x)$ can be represented as the series:

$$a + bx + \frac{c}{1 \cdot 2}x^2 + \cdots$$

in which $a = Fa$. The symbol D then is the operation on Fa such that:

$$DFa = b$$

and all higher derivatives:

$$DDFa = c$$

and so on. Arbogast also developed additional features which were not adopted much beyond their inception, but are historically interesting (see Figure 20.2).[10]

He states his motivations ([3] p. 209):

> To form the algorithm of derivations it became necessary to introduce new signs; I have given this subject particular attention, begin persuaded that the secret of the power of analysis consists in the happy choice and use of signs, simple and characteristic of the things which they are to represent. In this regard I have set myself the following rules:
> (1) To make notations as much as possible analogous the the received notations;
> (2) Not to introduce notations which are not needed and which I can replace without confusion by those already in use;
> (3) To select very simple ones, yet such that will exhibit all the varieties which the different operations require…

---

[9]For further detail see Cajori [3] pp. 208–211.
[10]As cited from Cajori [3] p. 210 Fig. 123, originally a facsimile from Arbogast's *Calcul des Dérivations* (1800), p. xxi.

As well as emphasizing the operator like quality of the process, Arbogast's notation found favor as it avoided reference to infinitesimals, which was appealing to many mathematicians, as succinctly expressed by a mathematical analyst from the early nineteenth century, Christian Kramp (1760–1826) ([3] pp. 210–211.):

> Later researches have convinced me of the absolute inutility of this constant factor or divisor dx, as well as the notion of the infinitely small ...[the adoption of the notation D] banishes all idea of the infinite and causes all this part of analysis to re-enter the domain of ordinary algebra.

20.7 Conclusion

What makes a good notation? Is it how it responds to changes and developments? Its applicability to other fields? Its ability to be incorporated into new technology and discoveries? Should notation somehow reflect its function? Should we be conservative about our mathematical notation, or encourage innovation? Should we be revolutionary or retaining? Notation is pivotal for a mathematician and is often given very little consideration and taken for granted, not just by students, but by the very practitioners themselves!

As we have seen, Newton and Leibniz feuded not only over the authorship of the calculus but also how notationally to express it. Correspondence reveals heavy criticism of the other's approach: "Newton", Newton himself once anonymously wrote, "does not confine himself to symbols"![11] Similar sentiments were returned from Leibniz and his supporters. Newton pursued research into the calculus through ideas of velocity and acceleration; Leibniz, through sums and differences. In turn, the English mathematical community subscribed to Newton's methods and notation, the continent to those of Leibniz. As a result of these incompatible approaches, English and continental mathematicians effectively terminated all mathematical collaboration and exchange. In terms of notation, as the calculus was developed, Leibniz's proposals turned out to be easier to work with and as a result, mathematical analysis on the continent progressed rapidly. To their loss, English mathematicians missed out on a century of stimulating and advancing mathematical activity, until they too adopted Leibniz's notation by the eighteenth century. Indeed, the struggle to establish the expression for the derivative shows us the ways in which notational considerations can make a huge difference in mathematics.

20.8 Reflective Questions

1. Brainstorm some features you deem important to keep in mind when developing or evaluating a particular notational system. Here are some possible criteria to get you going:

   comprehensibility—simplicity—avoidance of ambiguity—uniqueness (i.e., make up symbols or borrow from other languages)—aesthetics—symbols which suggest their meaning—rhetorical potential—ability to represent multiple/additional features—minimisation of the strokes of a pen *etc* ...

   Rank them from the most to the least important.

2. In what ways were the various proposals for the derivative a product of the concept their instigators were trying to emphasize?

3. Mathematical symbols not only represent a specific idea, they are a guide to operation by their careful placement. Select a notation that you use on a regular basis. What are its benefits? What are its limitations? What would you like to change about it?

4. We have looked at an episode from the History of Mathematics in which notation played an important role in facilitating mathematical insights. This is but one of many episodes which raises some very important questions for mathematicians regarding notation:

   > To what extent is our ability to conceptualize and develop mathematics based upon the symbolism we employ?

   Consider this with respect to some specific examples, such as the integral sign, matrices, fractions, function notation, powers and indices.

---
[11] Bardi [1] p. 215

Bibliography

[1] Jason Socrates Bardi, *The Calculus Wars: Newton, Leibniz, and the Greatest Mathematical Clash of All Time*, Thunder's Mouth Press, New York NY, 2006, 86–88, 214–215.

[2] Carl B. Boyer, *A History of Mathematics*, 2nd Ed., John Wiley and Sons, 1991, 391–414.

[3] Florian Cajori, *A History of Mathematical Notations*, 2 Vols., Dover, 1993, reprinted from the work first published as two volumes by the Open Court Publishing Company, La Salle Illinois, in 1928 and 1929, 196–242.

[4] Victor Katz, *A History of Mathematics An Introduction*, 2nd ed., Addison Wesley Longman, 1998, 468–543.

[5] Nicholas Jolley, *The Cambridge Companion to Leibniz*, Cambridge University Press, 1995.

21

Leibniz's Calculus (Real Retro Calc.)

Robert Rogers
State University of New York at Fredonia

21.1 Introduction

To many students, differential calculus seems like a set of rules to be applied for solving problems such as optimization problems, tangent problems, etc. This really should not be surprising as differential calculus literally is a set of rules for calculating differences. These rules first appeared in Leibniz's 1684 paper *Nova methodus pro maximus et minimus, itemque tangentibus, quae nec fractus nec irrationals, quantitates moratur, et singulare pro illi calculi genus* (*A New Method for Maxima and Minima as Well as Tangents, Which is Impeded Neither by Fractional Nor by Irrational Quantities, and a Remarkable Type of Calculus for This*). A translation of this appears in [5, p. 272–80]. As the title suggests, our students' perceptions are not far off. Indeed, Leibniz's differential calculus is very recognizable to modern students and illustrates the fact that this is really a collection of rules and techniques to compute and utilize (infinitesimal) differences. The fact that Leibniz's notation is so modern in appearance, or rather our notation is that of Leibniz, allows these rules to be presented in a typical calculus class. The author has typically done this while covering the differentials section of the course, as the rules are rules for differentials, not derivatives. Doing this reinforces the rules for computing derivatives and introduces the student to the manipulation of differentials that will be necessary in integration.

A bolder approach, which the author has employed, is to replace the typical "limit of difference quotient" derivations with these heuristic arguments and adopt the point of view that $\frac{dy}{dx}$ is a ratio of infinitesimals. In this case, the derivation of the typical derivative rules can be accomplished without reliance on limits of difference quotients. The fact that these can be obtained in a more algebraic manner might be more in the comfort zone of the typical beginning calculus student.

Even in the less bold classroom, students can be made to appreciate that the modern "limit of difference quotient" definition is not how the subject was originally conceived. The fact that these difference rules allowed a systematic approach to problems in tangents and optimization led to a golden period in mathematics during the eighteenth century. Indeed, Leibniz himself applied his differential rules to study classical problems such as Snell's Law of Refraction and his quite modern approach is worth presenting. Furthermore, the exploitation of these calculus rules and the power of differentials can be seen in Johann Bernoulli's 1696 solution of the brachistochrone problem. This masterpiece of mathematical work certainly deserves a place in any calculus class and can easily be included at the end of an AP calculus course.

This article includes these applications along with exercises which will enhance students' heuristic understanding of Leibniz's calculus rules and fluency with differentials. More applications of differentials can be found in [3, p.

9–20]. Just as it is important in any discipline for students to experience some of the development of the subject, here it is important for students to be introduced to the beginning ideas of the calculus. Students can develop a heuristic understanding of the calculus rules and can follow in the steps of eighteenth century mathematicians in exploiting the power of these rules.

21.2 Differential Calculus (Rules of Differences)

Throughout this chapter, I will use modern functional notation though it should be noted that Leibniz's own notation is quite modern in appearance. In the beginning of his paper, Leibniz used the tangent of a curve to define the differential of a quantity $y = y(x)$. Essentially, given an arbitrary change in x (dx), the change in y (dy) must satisfy

$$\frac{dy}{dx} = \text{the slope of the tangent line at } (x, y).$$

There is no mention of limits, as these differentials (differences) represent the momentary (instantaneous) changes in x and y, respectively. In other words, the differential dx represents an infinitesimal change in the quantity x. This coincides with our telling students that $\frac{dy}{dx}$ represents the instantaneous rate of change of y with respect to x. Leibniz next provided the rules for dealing with these infinitesimal differences.

$$\text{If } a \text{ is constant, then } da = 0 \text{ and } d(ay) = a\, dy$$
$$d(y + z - w) = dy + dz - dw$$
$$d(yz) = y\, dz + z\, dy$$
$$d\left(\frac{y}{z}\right) = \frac{z\, dy - y\, dz}{z^2}$$

> Knowing thus the algorithm (as I may say) of this calculus, which I call differential calculus, all other differential equations can be solved by a common method. We can find maxima and minima as well as tangents without the necessity of removing fractions, irrationals, and other restrictions, as had to be done according to the methods that have been published hitherto. The demonstration of all this will be easy to one who is experienced in these matters ... [5, p. 276].

Leibniz did not include these demonstrations, but they provide an opportunity for students in calculus to justify these rules without the burden of using limits. It should be pointed out that these arguments are somewhat heuristic and ignore aspects such as knowing that the derivative of a function exists. In fairness to Leibniz, he was not computing derivatives. The term *fonction dérivé* was coined by Joseph Louis Lagrange (1736–1813) in his *Théorie des fonctions analytiques* of 1797, where it represented the linear term in the power series expansion of a function. Leibniz was relying on the existence of a tangent line to provide a geometric basis for his differential calculus. It can be pointed out to students that these very real concerns over existence contributed to later rigorous formulations using limits. For now, we will allow ourselves to conform to seventeenth century rigor and not encumber ourselves with these issues. With this in mind, the rest of this chapter incorporates exercises to provide an opportunity for students to understand and apply Leibniz's rules heuristically. The solutions to these are included in the appendix. The first exercise is meant to emphasize the fact that a differential is really just a difference, a fact that is often lost in the world of derivatives.

Exercise 1. Let $\Delta y = y_2 - y_1$, $\Delta z = z_2 - z_1$, and $\Delta w = w_2 - w_1$. Show that if $x_1 = y_1 + z_1 - w_1$ and $x_2 = y_2 + z_2 - w_2$ then the difference $\Delta x = x_2 - x_1$ satisfies

$$\Delta x = \Delta y + \Delta z - \Delta w.$$

How is this related to Leibniz's differential rule $d(y + z - w) = dy + dz - dw$?

Whereas the previous exercise is fairly straightforward when one recognizes that differentials are considered as momentary differences, the product rule is not as transparent and requires some guidance. Consider the following geometric argument for Leibniz's product rule $d(yz) = y\, dz + z\, dy$. The product $a = yz$ can be thought of as the

21.2. Differential Calculus (Rules of Differences)

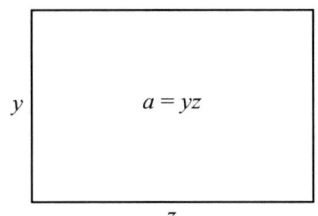

Figure 21.1. Geometric interpretation of the product yz

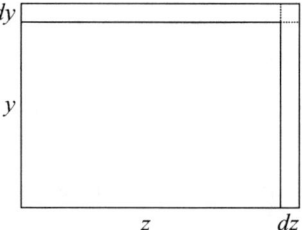

Figure 21.2. $d(yz) = y\,dz + z\,dy + dy\,dz$

area of the following rectangle [Figure 21.1]. With this in mind, $da = d(yz)$ can be thought of as the change in the area when y is changed by dy and z is changed by dz. This can be seen as the L-shaped region in Figure 21.2.

By dividing the L shaped region into 3 rectangles, we obtain $d(yz) = y\,dz + z\,dy + dy\,dz$. Since dy and dz are infinitely small, then the product $dy\,dz$ is infinitely small compared to the other terms and can thus be ignored, leaving $d(yz) = y\,dz + z\,dy$

The idea of ignoring products of infinitesimals, though questionable, was utilized by other mathematicians of the day. In particular, in his method of fluxions, Newton introduced an "infinitely short" time interval o to compute instantaneous changes in his fluents (quantities which change over time). Newton allowed terms containing o to a higher power than one to vanish as they are "infinitely less" than those in which o appears to the first power [2, p. 193–4]. This notion of "ignoring higher powers" may seem strange on the surface but, in fact, is embedded in our modern limit based derivations and in the more recent theory of asymptotics.

Exercise 2. Compare the above derivation of Leibniz' product rule with the following "modern" derivation of $\frac{d(x^2)}{dx}$. In particular, how does this compare with the seventeenth century notion of "ignoring" higher powers of infinitely small quantities?

$$\begin{aligned}
\frac{d(x^2)}{dx} &= \lim_{\Delta x \to 0} \frac{(x + \Delta x)^2 - x^2}{\Delta x} \\
&= \lim_{\Delta x \to 0} \frac{x^2 + 2x \cdot \Delta x + (\Delta x)^2 - x^2}{\Delta x} \\
&= \lim_{\Delta x \to 0} \frac{2x \cdot \Delta x + (\Delta x)^2}{\Delta x} \\
&= \lim_{\Delta x \to 0} (2x + \Delta x) \\
&= 2x.
\end{aligned}$$

Exercise 3. Use Leibniz' product rule to show $d(x^2) = 2x\,dx$. More generally, use induction to show that if n is a positive integer, then $d(x^n) = nx^{n-1}\,dx$.

Exercise 4. Use Leibniz' product rule to derive the quotient rule

$$d\left(\frac{y}{z}\right) = \frac{z\,dy - y\,dz}{z^2}.$$

(Hint: Let $v = \frac{y}{z}$ so that $v \cdot z = y$.)

Exercise 5. Use the quotient rule to show that if n is a positive integer, then

$$d(x^{-n}) = -nx^{-n-1}\,dx.$$

Exercise 6. Let p and q be integers with $q \neq 0$. Show

$$d\left(x^{p/q}\right) = \frac{p}{q} x^{(p/q)-1}\,dx.$$

Thus, it was as Leibniz said in his title, "*... Which is Impeded Neither by Fractional Nor by Irrational Quantities, and a Remarkable Type of Calculus for This*".

Applications

As was stated earlier, Leibniz did not provide derivations of the rules, but he provided applications of his calculus to prove its worth. As an example in the paper, Leibniz derives Snell's Law of Refraction using his calculus.

Given light which travels through medium 1 with a speed of v_1 and through medium 2 with a speed of v_2, the problem is to find the fastest path from point A in medium 1 to point B in medium 2 [Figure 21.3].

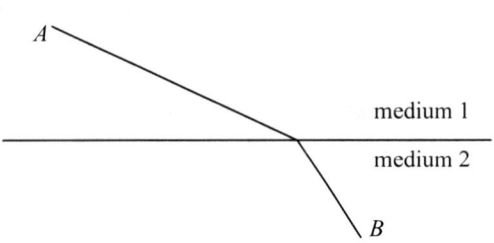

Figure 21.3. Light traveling through two media

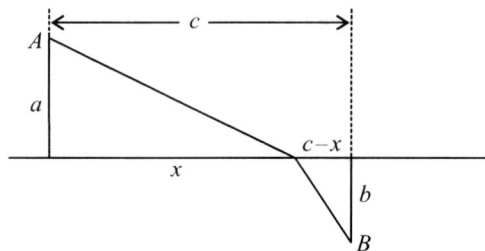

Figure 21.4.

According to Fermat's Principle of Least Time, this fastest path is one that light will travel. Using the fact that *time = distance/velocity* and the labeling in Figure 21.4, we can obtain a formula for the time T it takes for light to travel from A to B, namely

$$T = \frac{\sqrt{x^2 + a^2}}{v_1} + \frac{\sqrt{(c-x)^2 + b^2}}{v_2}.$$

Using the rules of Leibniz' calculus, we obtain

$$dT = \left[\frac{1}{v_1} \cdot \frac{1}{2}(x^2 + a^2)^{-(1/2)}(2x) + \frac{1}{v_2} \cdot \frac{1}{2}[(c-x)^2 + b^2]^{-(1/2)}(2(c-x)(-1))\right] dx$$

$$= \left(\frac{1}{v_1} \frac{x}{\sqrt{x^2 + a^2}} - \frac{1}{v_2} \frac{c-x}{\sqrt{(c-x)^2 + b^2}}\right) dx$$

Notice that in this calculation, we used the chain rule. The fact is that to a mathematician in Leibniz's day, the notion that $\frac{dz}{dx} = \frac{dz}{dy}\frac{dy}{dx}$ would have been as obvious as cancellation is to our students now, and, in fact, is built into the rules. Specifically, if $y = x^2 + a^2$ then

$$d(y^{1/2}) = \frac{1}{2}y^{-(1/2)} dy = \frac{1}{2}y^{-(1/2)} \cdot 2x\, dx.$$

We seem to make a big deal about this rule. Be assured that there are issues about division by zero, but are these appropriate for the beginning calculus class?

Using the fact that at the minimum value for T, $dT = 0$, we have that the fastest path from A to B must satisfy

$$\frac{1}{v_1}\frac{x}{\sqrt{x^2 + a^2}} = \frac{1}{v_2}\frac{c-x}{\sqrt{(c-x)^2 + b^2}}.$$

Inserting angles θ_1 and θ_2 [Figure 21.5], we get that the path that light travels must satisfy

$$\frac{\sin\theta_1}{v_1} = \frac{\sin\theta_2}{v_2}$$

which is Snell's Law. In the words of Leibniz, "Other very learned men have sought in many devious ways what someone versed in this calculus can accomplish in these lines as by magic." [5, p. 279]

21.2. Differential Calculus (Rules of Differences)

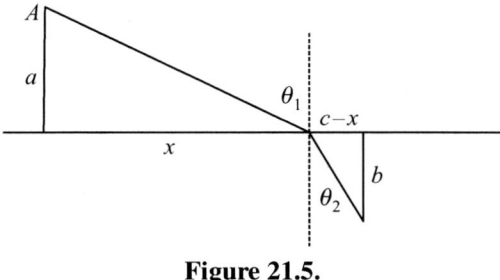

Figure 21.5.

Exercise 7. Why would Leibniz say that at the minimum value for T, $dT = 0$? You may want to explain this graphically.

As another application of the calculus, consider the solution of the brachistochrone problem by Johann Bernoulli (1667–1748) [4, p. 315–17]. In 1696, Bernoulli posed the brachistochrone problem; that is, to find the shape of a frictionless wire joining points A and B so that the time it takes for a bead to slide down under the force of gravity is as

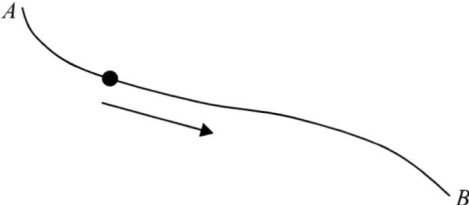

Figure 21.6. Brachistachrone problem

small as possible [Figure 21.6]. Bernoulli posed this "path of fastest descent" problem to challenge the mathematicians of Europe and used his solution to demonstrate the power of Leibniz' calculus as well as his own ingenuity.

> I, Johann Bernoulli, address the most brilliant mathematicians in the world. Nothing is more attractive to intelligent people than an honest, challenging problem, whose possible solution will bestow fame and remain as a lasting monument. Following the example set by Pascal, Fermat, etc., I hope to gain the gratitude of the whole scientific community by placing before the finest mathematicians of our time a problem which will test their methods and the strength of their intellect. If someone communicates to me the solution of the proposed problem, I shall publicly declare him worthy of praise [7].

Five solutions were obtained. Newton, Jacob Bernoulli, Leibniz, and de l'Hôpital solved the problem in addition to Johann Bernoulli [5]. We will present Johann Bernoulli's ingenious solution [4, p. 315–17 or 5, p. 392–6].

Bernoulli's solution began, strangely enough, with Snell's Law of Refraction. If we extend Snell's law to an object whose speed (v) is constantly changing, then along the fastest path, the ratio of the sine of the angle that the curve's tangent makes with the vertical and the speed must remain constant [Figure 21.7]. If we include axes and let P denote the position of the bead at a particular time then we have the following picture [Figure 21.8].

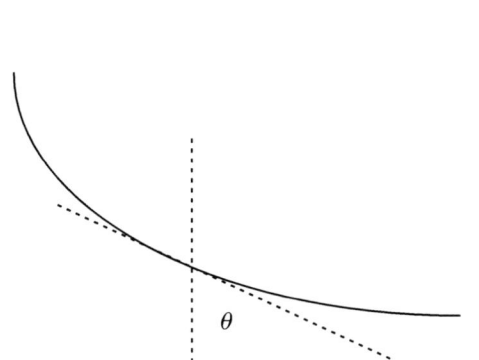

Figure 21.7. $\frac{\sin\theta}{v} = k$, where k is a constant

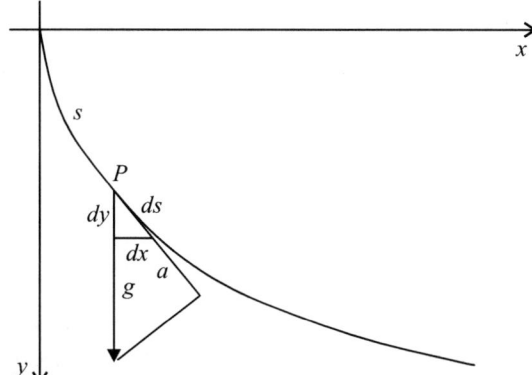

Figure 21.8. Forces in the brachistochrone problem

In Figure 21.8, s denotes the length that the bead has traveled down to point P (that is, the arc length of the curve down to that point) and a denotes the tangential component of the acceleration due to gravity g. Since the bead travels only under the influence of gravity then $\frac{dv}{dt} = a$. By similar triangles we have $\frac{a}{g} = \frac{dy}{ds}$.

Exercise 8. Use the above equations and the fact that $v = \frac{ds}{dt}$ to show that $g\,dy = v\,dv$ and conclude $v = \sqrt{2g\,y}$.

If we insert θ, then we can readily see that $\sin\theta = \frac{dx}{ds}$ [Figure 21.9].

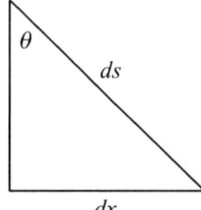

Figure 21.9.

Exercise 9. Substitute the above expressions for $\sin\theta$ and v into Snell's Law to show that the brachistochrone must satisfy the differential equation
$$\frac{dx}{\sqrt{2gy((dx)^2 + (dy)^2)}} = k.$$

Exercise 10. Show that the equations
$$x = \frac{t - \sin t}{4gk^2}, \quad y = \frac{1 - \cos t}{4gk^2}$$
satisfy the differential equation of the brachistochrone
$$\frac{dx}{\sqrt{2gy((dx)^2 + (dy)^2)}} = k.$$

Even without presenting Bernoulli's complete solution, students can explore the claim that the curve given by the parametric equations in exercise 10 actually is the brachistochrone. Specifically, consider the following.

Exercise 11. We know that under the influence of gravity, the speed of the bead sliding down the curve $y = y(x)$ is given by $\frac{ds}{dt} = v = \sqrt{2g\,y}$. Assuming that the positive y axis points down as in Figure 21.8, show that the time it takes for the bead to slide along the curve from the origin to point B with coordinates (x_1, y_1) is given by
$$T = \int_{(0,0)}^{(x_1, y_1)} \frac{ds}{\sqrt{2gy}}$$
and use this to compare the time it takes for the bead to slide along the curve given by
$$x = t - \sin t \quad y = 1 - \cos t$$
with the time it takes to slide along the straight line path from $(0,0)$ to $(\pi, 2)$.

Bernoulli recognized the above solution to be an inverted cycloid, the curve traced by a fixed point on a circle as the circle rolls along a horizontal surface. Interestingly, Christian Huygens (1629–1695) showed in 1659 that the cycloid was also a tautochrone; that is, a curve with the property that the time taken by a particle sliding down the curve under uniform gravity to its lowest point is independent of its starting point. Huygens applied this property

21.3. Conclusion

to the design of a pendulum clock. By designing a clock whose pendulum swung along the path of a cycloid, the period of the oscillation would be independent of the amount of swing. In theory, this would make the clock more accurate than one whose pendulum swung in a circular arc. See [1, p. 377] for a diagram of such a clock design. At the time, designing an accurate timepiece to be used for determining longitude at sea was of paramount importance and cash prizes were offered for successful designs [8]. While Huygens' cycloidal pendulum design was perfectly accurate in theory, it proved no more accurate than a regular pendulum design in practice. Nonetheless, Huygens work on the cycloid was recognized by contemporary and subsequent mathematicians. Bernoulli ended his solution of the brachistochrone problem with these words:

> Before I end I must voice once more the admiration I feel for the unexpected identity of Huygens' tautochrone and my brachistochrone. I consider it especially remarkable that this coincidence can take place only under the hypothesis of Galileo, so that we even obtain from this a proof of its correctness. Nature always tends to act in the simplest way, and so it here lets one curve serve two different functions, while under any other hypothesis we should need two curves ... [7].

Exercise 12. Referring to Figure 21.8, suppose that the initial position of the bead is the point $P_0(x_0, y_0)$. Show that the speed at the point (x, y) is given by $v = \frac{ds}{dt} = \sqrt{2g(y - y_0)}$ and so the time it takes for the bead to slide from P_0 to the bottom of the curve B is given by

$$T = \int_{P_0}^{B} \frac{ds}{\sqrt{2g(y - y_0)}}.$$

Given the cycloid defined by the equations $x = t - \sin t$ and $y = 1 - \cos t$, let $(x_0, y_0) = (x(t_0), y(t_0))$. Show that the time required for the bead to slide to the bottom of the cycloid is given by

$$T = \frac{1}{\sqrt{g}} \int_{t=t_0}^{\pi} \frac{\sin(t/2) dt}{\sqrt{\cos^2(t_0/2) - \cos^2(t/2)}} = \frac{\pi}{\sqrt{g}}$$

which is independent of the starting point [6, p. 695–6]. Hence the cycloid is, in fact, a tautochrone.

21.3 Conclusion

As can be seen by the above examples, the calculus was a powerful tool. In the late eighteenth and nineteenth centuries, mathematicians developed a formal definition of limit to address the foundational issues which Newton and Leibniz could not, but there was no doubting the power of the calculus. Leibniz's rules are very modern in appearance and in how they are utilized. While it is true that students should learn the concept of limits to justify calculus techniques in a rigorous fashion, this rigor sometimes masks the basic concept. Perhaps it would be wise to follow a more historical approach in the classroom and allow students to heuristically understand and utilize these techniques before they address the foundational underpinnings with limits.

Appendix

The following are outlines of solutions to the exercises in the chapter.

Exercise 1. $\Delta x = (y_2 + z_2 - w_2) - (y_1 + z_1 - w_1) = (y_2 - y_1) + (z_2 - z_1) - (w_2 - w_1) = \Delta y + \Delta z - \Delta w$. Since Leibniz applied the same rules to infinitesimal differences, then this reasoning would apply to (infinitesimal) differentials as well.

Exercise 2. In the difference quotient, the numerator represents the area of the L-shaped region consisting of the square of side $x + \Delta x$ minus the square of side x. When we divide by Δx and take the limit, we are essentially discarding the higher powers of Δx that were in the numerator.

Exercise 3. $d(x^2) = d(xx) = x\,dx + x\,dx = 2x\,dx$. In the inductive step of the general case, we have $d(x^{n+1}) = d(x^n x) = x^n\,dx + x(nx^{n-1}\,dx) = (n+1)x^n\,dx$.

Exercise 4. If $v = \frac{y}{z}$, then $dy = d(vz) = v\,dz + z\,dv$. From this we get

$$dv = \frac{dy - v\,dz}{z} = \frac{dy - \frac{y}{z}\,dz}{z} = \frac{z\,dy - y\,dz}{z^2}.$$

Exercise 5. $d(x^{-n}) = d\left(\frac{1}{x^n}\right) = \frac{x^n \cdot 0 - 1 \cdot nx^{n-1}\,dx}{x^{2n}} = -nx^{-n-1}\,dx.$

Exercise 6. Let $y = x^{p/q}$, so $y^q = x^p$ and $qy^{q-1}\,dy = px^{p-1}\,dx$. From this we get

$$dy = \frac{px^{p-1}\,dx}{q(x^{p/q})^{q-1}} = \frac{p}{q} x^{(p-1-p+(p/q))}\,dx = \frac{p}{q} x^{(p/q)-1}\,dx.$$

Exercise 7. If you graph a function such as $T = T(x)$, then you can see that as you approach a minimum (or maximum) point on the graph, then the change in T is approaching zero. If you use the fact that the slope of the tangent line is used to relate dT to dx, then the above assertion is equivalent to saying that the slope of the tangent line is zero at a local extremum.

Exercise 8. We have $g\,dy = a\,ds = \frac{dv}{dt}\,ds = dv\frac{ds}{dt} = v\,dv$. Antidifferentiating, we get $gy = \frac{1}{2}v^2 + c$, for some constant c. Since $v = 0$ when $y = 0$, then $c = 0$. Solving for v, we get $v = \sqrt{2gy}$.

Exercise 9. $k = \dfrac{\sin\theta}{v} = \dfrac{dx}{ds\sqrt{2gy}} = \dfrac{dx}{\sqrt{(dx)^2 + (dy)^2}\,\sqrt{2gy}}.$

Exercise 10. $dx = \dfrac{1 - \cos t}{4gk^2}\,dt = y\,dt$ and $dy = \dfrac{\sin t}{4gk^2}\,dt$. This yields

$$\frac{dx}{\sqrt{2gy((dx)^2 + (dy)^2)}} = \frac{y\,dt}{\sqrt{2gy\frac{(1-\cos t)^2 + \sin^2 t}{(4gk^2)^2}(dt)^2}} = \frac{4gk^2 y}{\sqrt{2gy(2 - 2\cos t)}}$$

$$= \frac{4gk^2 y}{\sqrt{4gy(1-\cos t)}} = \frac{4gk^2 y}{\sqrt{4gy \cdot 4gk^2 y}}$$

$$= \frac{4gk^2 y}{4gky}$$

$$= k.$$

Exercise 11. $T = \displaystyle\int_{(0,0)}^{(x_1,y_1)} dt = \int_{(0,0)}^{(x_1,y_1)} \frac{ds}{\sqrt{2gy}}$. For the curve defined by $x = t - \sin t$ and $y = 1 - \cos t$, we have

$$T = \int_{(0,0)}^{(x_1,y_1)} \frac{ds}{\sqrt{2gy}} = \int_{(0,0)}^{(x_1,y_1)} \frac{\sqrt{(dx)^2 + (dy)^2}}{\sqrt{2gy}}$$

$$= \frac{1}{\sqrt{g}} \int_{t=0}^{\pi} \frac{\sqrt{(1-\cos t)^2 + \sin^2 t}\,dt}{\sqrt{2y}} = \frac{1}{\sqrt{g}} \int_{t=0}^{\pi} \frac{\sqrt{2(1-\cos t)}\,dt}{\sqrt{2y}}$$

$$= \frac{1}{\sqrt{g}} \int_{t=0}^{\pi} dt$$

$$= \frac{\pi}{\sqrt{g}} \approx \frac{3.1416}{\sqrt{g}}.$$

21.3. Conclusion

For the line joining $(0, 0)$ to $(\pi, 2)$, we have $y = \frac{2}{\pi}x$ and so

$$T = \int_{(0,0)}^{(x_1,y_1)} \frac{\sqrt{(dx)^2 + (dy)^2}}{\sqrt{2gy}}$$

$$= \frac{1}{\sqrt{g}} \int_{x=0}^{\pi} \frac{\sqrt{1 + \left(\frac{2}{\pi}\right)^2} \, dx}{\sqrt{2y}}$$

$$= \frac{1}{\sqrt{g}} \int_{x=0}^{\pi} \sqrt{\frac{1 + \frac{4}{\pi^2}}{\frac{4x}{\pi}}} \, dx$$

$$= \frac{1}{\sqrt{g}} \sqrt{\frac{\pi^2 + 4}{4\pi}} \int_{x=0}^{\pi} x^{-(1/2)} \, dx$$

$$= \frac{1}{\sqrt{g}} \sqrt{\pi^2 + 4} \approx \frac{3.7242}{\sqrt{g}}.$$

Exercise 12. As in exercise 8, we have $gy = \frac{1}{2}v^2 + c$. We have $v = 0$ when $y = y_0$, which yields $c = gy_0$ and so $v = \sqrt{2g(y - y_0)}$. Using $x = t - \sin t$ and $y = 1 - \cos t$, we get

$$T = \int_{t=t_0}^{\pi} \frac{ds}{\sqrt{2g(y - y_0)}}$$

$$= \frac{1}{\sqrt{g}} \int_{t=t_0}^{\pi} \frac{\sqrt{(1 - \cos t)^2 + \sin^2 t}}{\sqrt{2((1 - \cos t) - (1 - \cos t_0))}} \, dt$$

$$= \frac{1}{\sqrt{g}} \int_{t=t_0}^{\pi} \frac{\sqrt{(1 - \cos t)^2 + \sin^2 t}}{\sqrt{2((1 + \cos t_0) - (1 + \cos t))}} \, dt$$

$$= \frac{1}{\sqrt{g}} \int_{t=t_0}^{\pi} \frac{\sqrt{2(1 - \cos t)}}{\sqrt{2((1 + \cos t_0) - (1 + \cos t))}} \, dt$$

$$= \frac{1}{\sqrt{g}} \int_{t=t_0}^{\pi} \frac{\sqrt{4 \sin^2 \left(\frac{t}{2}\right)}}{\sqrt{2 \left(2 \cos^2 \left(\frac{t_0}{2}\right) - 2 \cos^2 \left(\frac{t}{2}\right)\right)}} \, dt$$

$$= \frac{1}{\sqrt{g}} \int_{t=t_0}^{\pi} \frac{\sin \left(\frac{t}{2}\right)}{\sqrt{\cos^2 \left(\frac{t_0}{2}\right) - \cos^2 \left(\frac{t}{2}\right)}} \, dt$$

$$= \frac{\pi}{\sqrt{g}}.$$

Bibliography

[1] C. B. Boyer & U. C. Merzbach, *A History of Mathematics*, 2nd ed., Wiley, New York, 1991.

[2] C. H. Edwards, *The Historical Development of the Calculus*, Springer-Verlag Inc., New York, 1979.

[3] R. Rogers, "Putting the Differential Back Into Differential Caclulus," *From Calculus to Computers. Using the Last 200 Years of Mathematics History in the Classroom*, MAA Notes #68, Washington, DC, 2005.

[4] G. Simmons, *Calculus Gems*, MAA, Washington, DC, 2007.

[5] D. Struik, *A Source Book in Mathematics, 1200–1800*, Harvard University Press, Cambridge, MA, 1969.

[6] M. Weir, J. Haas, & F. Giordano, *Thomas' Calculus, Early Transcendentals, 11th ed.*, Pearson, Boston, MA, 2006.

[7] `www-gap.dcs.st-and.ac.uk/~history/HistTopics/Brachistochrone.html`

[8] `www-gap.dcs.st-and.ac.uk/~history/HistTopics/Longitude1.html`

22

An "Impossible" Problem, Courtesy of Leonhard Euler

Homer S. White[1]
Georgetown College

22.1 Introduction

The second semester of calculus may well be the busiest course in the standard undergraduate mathematics curriculum. Between applications of integration, integration techniques, polar coordinates, parametric representations of curves, sequences and infinite series, one usually has no time to give conic sections their due. For quite some time, therefore, I have been looking for interesting things to say about conics that tie in well with students' recently acquired calculus tools.

Recently I got lucky. I happened upon an article[2] published in 1755 by the great Swiss mathematician Leonhard Euler, which considers a problem that fits the bill perfectly. Euler's treatment of the problem synthesizes a number of ideas from elementary calculus: trigonometric identities, techniques of integration including partial fractions, representation of curves by polar equations, and separable differential equations, with a particular conic section—the parabola–leading off the action.

22.2 Historical Setting

Suppose that you are given a parabola, and that you draw an arbitrary line through its focus F, meeting the parabola at points M and N. The tangent lines to these points will always meet at a right angle! One possible approach to a proof is to work from the reflection property of parabolas, as follows:

In Figure 22.1, the points P and Q are chosen so that PN and QM are parallel to the central axis of the parabola. By the reflection property, a ray of light traveling from P to N will bounce off the parabola and head toward F, with PN and NF making equal angles, of measure α let us say, to the tangent line YZ at point N. Similarly QM and MF

---

[1] The author would like to acknowledge the assistance in the French language of his student Katy Thompson, who read [2], on which this paper is based, as part of an Honors project in a second-semester calculus class in the Fall semester of 2006.

[2] The English title of the article is "Reflections on a problem which has been treated by certain geometers but which is, nonetheless, impossible." The original article appeared in French in the *Memoirs of the Berlin Academy of Sciences*, (see [2]) or it may be viewed in the online Euler Archives (http://www.math.dartmouth.edu/˜euler/) as E. 220. Unfortunately it seems that the editors of the *Memoirs* did not include the diagrams to which Euler refers in the text.

169

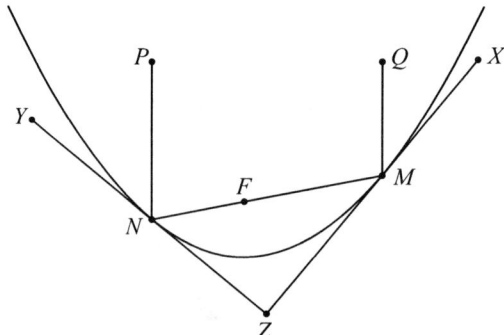

Figure 22.1. The tangent-intersection property of a parabola.

make equal angles β to the tangent XZ at M. Hence

$$180 = m(\angle PNF) + m(\angle QMF)$$
$$= 180 - 2\alpha + 180 - 2\beta.$$

We get $\alpha + \beta = 90$, so the angle at the intersection Z of the tangent lines must be a right angle.[3]

The following question occurs naturally: are there any other curves that share with the parabola the foregoing tangent-intersection property, and how may we go about finding them? In his 1755 article, Euler took up the matter:

Problem 4. Given a fixed point F and a fixed angle α with $0 \leq \alpha \leq \frac{\pi}{2}$, can we find smooth curves so that for any straight line through F, meeting the curve at exactly two points M and N on opposite sides of F, the tangent lines at M and N meet at angle α? (The case $\alpha = 0$ asks that the tangents lines be parallel.)

In the eighteenth century European mathematicians were quite fond of this type of "reciprocal" problem, which involves taking a familiar property of a familiar curve—one that often can be demonstrated by classical geometric methods—and trying to find other curves that share this property.[4] The search for new solution curves allowed mathematicians to test the power of newly-discovered techniques of what they called "Analysis"—essentially calculus and differential equations—along with newly-developed coordinate systems for representing locations on the plane or in space.[5]

Now, we have observed already that any parabola with focus F is a solution, when $\alpha = \frac{\pi}{2}$. Also, any circle with center F would solve the problem when $\alpha = 0$. Eventually we will see that for $\alpha = 0$ there are many other solutions. Euler also demonstrated that there are *no solutions at all* for any α *between* 0 *and* $\frac{\pi}{2}$.

The proof given here is essentially Euler's, although it has been streamlined a bit and the notation has been modernized.

Let's suppose that α is not $\frac{\pi}{2}$. Euler set up the situation using polar coordinates, with the given point F located at the origin. The solution curve, if it exists, may be represented as the graph of some equation $r = f(\theta)$, where f is a continuous function of θ. Also, in order to guarantee that there are just two points along the line through F, one on each side of F, he required that $f(\theta)$ be periodic with period 2π: In other words, when θ has increased 360 degrees, we must return to the same point on the curve as when we started. (Hence, for example, spirals like $r = \theta$ are not permitted.) Euler was able, therefore, to restrict his attention to functions of the form $f(\cos\theta, \sin\theta)$, since

$$f(\cos(\theta + 2\pi), \sin(\theta + 2\pi)) = f(\cos\theta, \sin\theta).$$

Now in the diagram below, let M have polar coordinates $(f(\theta), \theta))$, where θ is the angle made by FM with the positive x-axis, so that N has coordinates $(f(\theta + \pi), \theta + \pi))$. The two tangents at M and N meet at I, where

---

[3] It is also not difficult to show that the point Z of intersection lies on the directrix of the parabola.

[4] Here is another example of such a "reciprocal" problem: a ray of light beginning from one focus of an elliptical mirror will bouce twice off the ellipse and arrive back at its starting point. If for a given curve there exists a point having this double-relfection property with respect to the curve, then that curve is called a *catoptrix*. What other catoptrices, besides ellipses, can we find? Can we find gemoetric ways to construct them? For a discussion of Euler's treatment fo the problem of the catoptrix and other reciprocal problems, see [3].

[5] In some respects Euler was a pioneer in the representation of points by coordinates. Although he did not invent the polar coordinate system, he was a leader in the systematic exploitation of its three-dimensional relatives, the cylindrical and spherical coordinates. For a history of polar coordinates, see [1].

22.3. In the Classroom

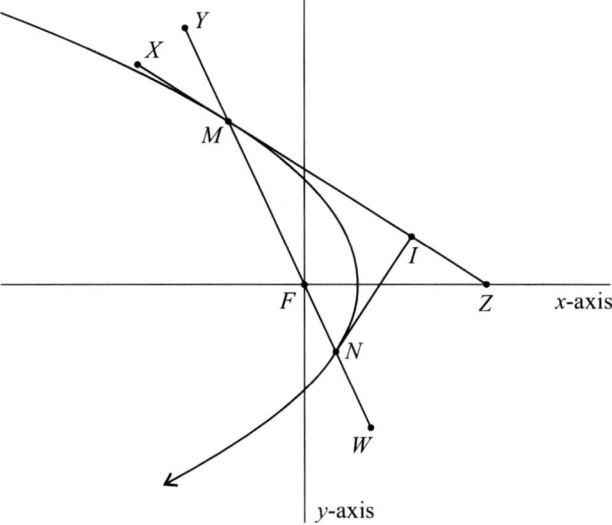

Figure 22.2. Setting up the impossibility proof.

$\angle MIN$ has constant measure α. Let the tangent of angle $\psi = \angle XMY$ be denoted by t, and let the tangent of angle $\psi' = \angle INW$ be denoted by t'. (Note: The $'$ notation is Euler's, and it has nothing to do with derivatives.) Clearly t and t' are functions of θ, and in fact

$$t'(\theta) = t(\theta + \pi)$$

for all θ for which M and N both exist. Of course,

$$t''(\theta) = t'(\theta + \pi) = t(\theta)$$

by the periodicity of our curve.

Considering external angles and employing the identity for the tangent of the difference of angles, we arrive at

$$\tan \alpha = \frac{t'(\theta) - t(\theta)}{1 + t(\theta)t'(\theta)}.$$

Note that the assumption $\alpha \neq \frac{\pi}{2}$ guarantees that the denominator is nonzero.

When the equation in the previous exercise holds for θ, it also holds for $\theta + \pi$, so

$$\tan \alpha = \frac{t'(\theta + \pi) - t(\theta + \pi)}{1 + t(\theta + \pi)t'(\theta + \pi)}$$
$$= \frac{t''(\theta) - t'(\theta)}{1 + t'(\theta)t''(\theta)} = \frac{t(\theta) - t'(\theta)}{1 + t'(\theta)t(\theta)} = -\tan \alpha.$$

We conclude that $\tan \alpha = 0$, so $\alpha = 0$. That α is between 0 and $\frac{\pi}{2}$ is not a possibility.[6]

22.3 In the Classroom

How might Euler's problem be introduced in the classroom? One very appropriate occasion is in the standard second-semester calculus class. As soon as polar coordinates and conic sections have been introduced, one should bring up the tangent-intersection property for parabolas, with which this article began. Students could be assigned the theorem as a homework problem. Although most are unlikely to think of the geometric proof given at the outset, an analytic proof using polar coordinate methods is well within their reach, with perhaps a hint or two on set-up. For instance, if we begin with the equation of a parabola in polar coordinate form, say $f(\theta) = \frac{a}{1+\cos\theta}$, the graph will resemble Figure 22.2 above: the focus of the parabola is at the origin, the vertex is on the positive x-axis, and the directrix is

---

[6]Euler does not say which "geometers" had treated the problem in the past. Perhaps they had given incorrect solutions, and Euler wanted to spare them unnecessary embarrassment.

$y = a$. Letting M be an arbitrary point $\left(\theta, \frac{a}{1+\cos\theta}\right)$ on the parabola, we obtain the diametrically opposed point N as $\left(\theta + \pi, \frac{a}{1+\cos(\theta+\pi)}\right)$. Next, using the well-known identities

$$\frac{dy}{d\theta} = \frac{dr}{d\theta}\sin\theta + r\cos\theta,$$
$$\frac{dx}{d\theta} = \frac{dr}{d\theta}\cos\theta - r\sin\theta,$$

one arrives at an expression for the slope of the tangent line at M:

$$\frac{\frac{\sin^2\theta}{(1+\cos\theta)^2} + \frac{\cos\theta}{1+\cos\theta}}{\frac{\sin\theta\cos\theta}{(1+\cos\theta)^2} - \frac{\sin\theta}{1+\cos\theta}},$$

and for the slope of the tangent at N:

$$\frac{\frac{\sin^2\theta}{(1-\cos\theta)^2} - \frac{\cos\theta}{1-\cos\theta}}{\frac{\sin\theta\cos\theta}{(1-\cos\theta)^2} + \frac{\sin\theta}{1-\cos\theta}},$$

where in the latter expression use has been made of the fact that

$$\sin(\theta + \pi) = -\sin\theta,$$
$$\cos(\theta + \pi) = -\cos\theta.$$

The student now has to employ algebra and familiar trigonometric identities to verify that the two slopes are indeed negative reciprocals of one another.[7]

Alternately one could give the synthetic-style proof in class, and then raise Euler's "reciprocal" question.

Next, Euler's argument that there are no solutions for $0 < \alpha < \frac{\pi}{2}$ should be given in class, just as above.

The next step is to cue the students to Euler's advice on finding other solution curves, beginning with the case $\alpha = 0$. Euler states, but does not prove, that

$$\tan\psi = \frac{r}{dr/d\theta}, \qquad (22.1)$$

a result that today is often given as an exercise in calculus textbooks. If the instructor does not assign this particular exercise, then the in-class proof could be given as follows: in the previous diagram, let ϕ denote the angle made by the tangent at M with the positive x-axis. Of course

$$\tan\phi = \frac{dy}{dx}.$$

We also see that $\tan\psi = \tan(\phi - \theta)$. Employing a trigonometric identity we find

$$\tan(\phi - \theta) = \frac{\frac{dy}{dx} - \tan\theta}{1 + \frac{dy}{dx}\tan\theta} = \frac{\frac{dy/d\theta}{dx/d\theta} - \tan\theta}{1 + \frac{dy/d\theta}{dx/d\theta}\tan\theta}$$
$$= \frac{\frac{dy}{d\theta} - \frac{dx}{d\theta}\tan\theta}{\frac{dx}{d\theta} + \frac{dy}{d\theta}\tan\theta}.$$

The result then follows from the aforementioned identities

$$\frac{dy}{d\theta} = \frac{dr}{d\theta}\sin\theta + r\cos\theta,$$
$$\frac{dx}{d\theta} = \frac{dr}{d\theta}\cos\theta - r\sin\theta.$$

---

[7]Having gone this far, one might as well also set up the equations for the two tangent lines, and discover that their intersection point I always lies along the directrix $y = a$!

22.3. In the Classroom

Now imagine that $r = f(\theta)$ solves our problem for the case $\alpha = 0$. In that case, the tangents at M and N are parallel, so the alternate exterior angles ψ and ψ' are congruent, so

$$\tan \psi = \tan \psi'.$$

which by (1) gives

$$\frac{f(\theta)}{df/d\theta} = \frac{f(\theta + \pi)}{\frac{df}{d\theta}(\theta + \pi)},$$

hence

$$\frac{df/d\theta}{f(\theta)} = \frac{\frac{df}{d\theta}(\theta + \pi)}{f(\theta + \pi)},$$

or

$$\frac{d \ln f(\theta)}{d\theta} = \frac{d \ln f(\theta + \pi)}{d\theta}.$$

Euler says that when we integrate, we get

$$f(\theta) = f(\theta + \pi),$$

so our problem is solved by any curve $r = f(\theta)$ where $f(\theta)$ has period π. The circle, various multi-leaved roses, and other familiar polar curves turn out to be solutions.

The case $\alpha = \frac{\pi}{2}$ is more challenging, and the instructor should provide at least a start in the classroom. Here, we need the function $t = \tan \psi$ to satisfy the condition

$$tt' + 1 = 0.$$

Recall also that t can be considered as a function of $\sin \theta$ and $\cos \theta$, so that the $'$ operation—the operation of increasing θ by π radians—is equivalent to substituting $-\cos \theta$ for $\cos \theta$ and $-\sin \theta$ for $\sin \theta$. Therefore, Euler casts about for expressions $t = t(\cos \theta, \sin \theta)$ so that substituting $-\cos \theta$ for $\cos \theta$ and $-\sin \theta$ for $\sin \theta$ and multiplying the result by the original t yields the product -1. We will call this the *perpendicularity condition for t*. Once a t satisfying the perpendicularity condition is found, since $t = \tan \psi$, by (1) we have only to solve the separable differential equation

$$t(\cos \theta, \sin \theta) = \frac{r}{dr/d\theta}$$

for r as a function of θ, obtaining a polar-coordinate form of a solution curve.

Euler suggests, as one of two initial examples,

$$g(\cos \theta, \sin \theta) = \frac{-\sin \theta}{1 + \cos \theta},$$

which meets the perpendicularity condition, since

$$\frac{-\sin \theta}{1 + \cos \theta} \cdot \frac{--\sin \theta}{1 - \cos \theta} = \frac{-\sin^2 \theta}{\sin^2 \theta} = -1$$

and which leads to the differential equation

$$\frac{dr}{r} = \frac{1 + \cos \theta}{-\sin \theta} d\theta.$$

which is solved by integrating both sides:

$$\ln r = \int \frac{1 + \cos \theta}{-\sin \theta} d\theta + \ln a = -\int \frac{1 + \cos \theta}{\sin \theta} \frac{1 - \cos \theta}{1 - \cos \theta} d\theta + \ln a$$

$$= -\int \frac{\sin \theta}{1 - \cos \theta} d\theta = \ln\left(\frac{1}{1 - \cos \theta}\right) + \ln a.$$

Note that we have written the usual constant of integration as $\ln a$, where a is an arbitrary positive constant. This permits us to write:

$$\ln r = \ln\left(\frac{1}{1-\cos\theta}\right) + \ln a,$$

$$\ln r = \ln\left(\frac{a}{1-\cos\theta}\right),$$

$$r = \frac{a}{1-\cos\theta},$$

which graphs as a parabola with focus at the origin and directrix $x = -a$. It is suggested that the preceding be provided by the instructor.

A selection of other solution curves in Euler's article may be obtained by the students in a series of assigned exercises, which give practice in separable partial fractions expansion, separable differential equations, trigonometric identities, and the even use of symbolic computation packages.

Exercise 22.1. Show that Euler's other initial example,

$$g(\cos\theta,\sin\theta) = \frac{\sin\theta}{1+\cos\theta},$$

results in the family of cardioids $r = a(1-\cos\theta)$ with cusp at the origin.

Exercise 22.2. Verify Euler's next claim: For any odd integer λ, either of

$$g(\cos\theta,\sin\theta) = \pm\frac{\sin\lambda\theta}{1+\cos\lambda\theta}$$

satisfy the perpendicularity condition.

Exercise 22.3. Solve for r, and graphically investigate the solution curve, for several odd values of λ in the cases

$$\frac{dr}{r} = \frac{1+\cos\lambda\theta}{\sin\lambda\theta}\,d\theta$$

and

$$\frac{dr}{r} = -\frac{1+\cos\lambda\theta}{\sin\lambda\theta}\,d\theta.$$

Formulate conjectures about the shape of the curves for general odd λ.

Exercise 22.4. Show that the perpendicularity condition is satisfied by

$$g(\cos\theta,\sin\theta) = \pm\left(\frac{\sin\lambda\theta}{1+\cos\lambda\theta}\right)^{\frac{m}{n}},$$

where m and n are odd integers.

Exercise 22.5. A function h of $\cos\theta$ and $\sin\theta$ is said to be *even* if

$$h(-\cos\theta,-\sin\theta) = h(\cos\theta,\sin\theta);$$

if

$$h(-\cos\theta,-\sin\theta) = -h(\cos\theta,\sin\theta)$$

then h is said to be *odd*. Verify Euler's claim that, if P is an even function and Q is odd, then the perpendicularity condition is satisfied by

$$g(\cos\theta,\sin\theta) = \pm\left(\frac{\sin\lambda\theta}{1+\cos\lambda\theta}\right)^{\frac{m}{n}}\frac{P+Q}{P-Q}$$

and, in general,

$$g(\cos\theta,\sin\theta) = \pm\left(\frac{\sin\lambda\theta}{1+\cos\lambda\theta}\right)^{\frac{m}{n}}\left(\frac{P+Q}{P-Q}\right)^{\delta}$$

where $\delta > 0$ is any real number (although Euler only dealt with rational powers).

There is really no end to this.

Exercise 22.6. If P_i, $1 \leq i \leq K$ are even and Q_i, $1 \leq i \leq K$ are odd, then the condition is satisfied by

$$g(\cos\theta, \sin\theta) = \pm \left(\frac{\sin\lambda\theta}{1+\cos\lambda\theta}\right)^{\frac{m}{n}} \prod_{i=1}^{K} \frac{P_i + Q_i}{P_i - Q_i}.$$

Let us briefly consider some more solution curves. For the situation of Exercise 6, Euler takes $\lambda = 1$ and begins with $P = 1$ and $Q = m \cos\theta$, which leads to

$$\frac{dr}{r} = \pm \frac{\sin\theta}{1+\cos\theta} \frac{P-Q}{P+Q} d\theta$$

or

$$\frac{dr}{r} = \pm \frac{\sin\theta}{1+\cos\theta} \frac{1 - m\cos\theta}{1 + m\cos\theta} d\theta.$$

In order to integrate, one must employ a partial fractions technique. That is, one seeks values of A and B so that

$$\frac{\sin\theta}{1+\cos\theta} \frac{1 - m\cos\theta}{1 + m\cos\theta} = \frac{A\sin\theta}{1+\cos\theta} + \frac{B\sin\theta}{1+m\cos\theta}.$$

Exercise 22.7. Show that the correct values for A and B in the previous equation are:

$$A = \frac{m+1}{1-m},$$
$$B = \frac{-2m}{1-m}.$$

Exercise 22.8. With the assistance, perhaps, of a symbolic computation package, find and graph solutions for

$$\frac{dr}{r} = \pm \frac{\sin\theta}{1+\cos\theta} \frac{1 - 2\cos\theta}{1 + 2\cos\theta} d\theta.$$

(One obtains curves with multiple branches.)

Exercise 22.9. Similarly, investigate solutions for

$$\frac{dr}{r} = \pm \frac{\sin\theta}{1+\cos\theta} \frac{1 - \frac{1}{2}\cos\theta}{1 + \frac{1}{2}\cos\theta} d\theta.$$

(Euler notes that the positive case yields a curve "similar to a parabola", whereas the negative case yields a curve "similar to a cardioid.")

Inversion. For the $\alpha = \frac{\pi}{2}$ problem, Euler's first two classes of solutions—the two "simplest", it would seem—are parabolas and cardioids. It worth observing that these two types of curves are related in another way, through *inversion*.

Given a point P and a circle with center O and radius r, draw the ray $\overrightarrow{OP}$ and let Q be the point on the ray so that

$$OQ = \frac{r^2}{OP}.$$

Then Q is called the *inverse* of P. For a fixed circle, inversion can be thought of as a mapping of the plane to itself. Points inside the circle map to points outside, and vice versa, while each point on the circle maps to itself. The inversion mapping is also its own inverse. One interesting fact about inversions is that lines and circles map to lines and circles.

What is of interest to us is that, given any circle with center F, the image, under inversion through that circle, of any parabola with focus F turns out to be a cardioid. See Figure 22.3.

This is no accident. In fact:

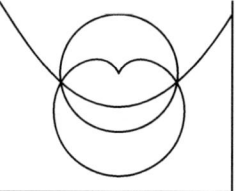

Figure 22.3. A parabola and its inversion image: a cardioid. The inversion circle is centered at the focus of the parabola.

Exercise 22.10. Show that for any $\alpha = \frac{\pi}{2}$ solution curve with center O, the inversion of the curve under any circle centered at O is also a solution to the $\alpha = \frac{\pi}{2}$ problem. Furthermore, for any $\alpha = 0$ solution curve with center O, the inversion of the curve under any circle centered at O is also a solution to the $\alpha = 0$ problem.

22.4 Conclusion

Considering the multitude of solutions for the particular angles 0 and $\frac{\pi}{2}$ along with the complete lack of solutions for other angles, Euler remarks in conclusion: "this problem merits the closest attention, and there is no doubt that consideration of it leads to a quantity of other pleasing research." [3. p. 363] Indeed, in this capsule we have only considered a few of the solution curves that Euler found; many other related questions, from Euler's article and beyond, remain to be explored.

Appendix: Remarks on Selected Exercises

Solutions to some of the Exercises, along with discussion of their use in a calculus class where polar coordinates have been introduced, are given below.

Exercise 22.2. The solution is exactly like the derivation, given in the body of this article, of the family of parabolas as solutions to the $\alpha = \frac{\pi}{2}$ problem. If you assign only one exercise, let this be the one.

Exercise 22.3.

$$\left(\frac{\sin \lambda\theta}{1 + \cos \lambda\theta}\right)' = \frac{\sin(\lambda(\theta + \pi))}{1 + \cos(\lambda(\theta + \pi))}$$
$$= \frac{\sin(\lambda\theta + \lambda\pi)}{1 + \cos(\lambda\theta + \lambda\pi)}$$
$$= \frac{-\sin(\lambda\theta)}{1 - \cos(\lambda\theta)}$$

since λ is odd. It follows that

$$\frac{\sin \lambda\theta}{1 + \cos \lambda\theta}\left(\frac{\sin \lambda\theta}{1 + \cos \lambda\theta}\right)'$$
$$= \frac{\sin \lambda\theta}{1 + \cos \lambda\theta} \frac{-\sin(\lambda\theta)}{1 - \cos(\lambda\theta)}$$
$$= \frac{\sin^2(\lambda\theta)}{1 - \cos^2(\lambda\theta)} = -1,$$

verifying the perpendicularity condition.

Exercise 22.4. Solving the differential equations is as easy as in Exercise 22.2.

Exercise 22.4 is as far as I go in a standard calculus class. A typical schedule for this material, requiring about 1.3 class periods, would be as follows:

1. Introduce the tangent property of parabolas (five minutes at the end of class.). Assign the analytic proof of this property as homework.

2. The next class day (fifty minutes) is devoted to:

    (a) demonstration that there are no solutions for $0 < \alpha < \frac{\pi}{2}$;
    (b) derivation of solutions to the $\alpha = 0$ problem
    (c) derivation of parabolas as one family of solutions to the $\alpha = \frac{\pi}{2}$ problem. Exercise 22.2 is assigned as HW.

3. At the next class meeting, go over the results of Exercise 22.2, graphing the cardioids with appropriate software. Discuss Exercise 22.3 for the case $\lambda = 3$, and assign Exercises 22.3 and 22.4 as HW. (Fifteen minutes)

4. At the next class meeting, have students report their conjectures from Exercise 22.4. (Five minutes)

22.4. Conclusion

Exercises 22.5 and beyond are not necessarily difficult. In particular, 22.5, 22.6, 22.7 and 22.8 come quickly once the students understand 4, and the problem on inversions is almost immediate when one think about polar coordinates. However, I find that there is no time for them in a regular class, since a culminating theme of these problems is the graphical exploration of complicated solution curves. I have tried a preliminary version of these exercises with one student as part of an Honors project, with enough success that I look forward to using the full set of Exercises with other Honors students in the future.

There is special value to Exercise 22.8, since the student has to realize that $\sin\theta$ and $\cos\theta$ play the role of, say, x and x^2 in the sort of partial fractions problems they have encountered previously in calculus.

Bibliography

[1] J. L. Coolidge, "The Origin of Polar Coordinates," *American Mathematical Monthly*, **59**, #2, (Feb. 1952), 78–85.

[2] L. Euler, "Reflexions sur un probleme de geometrie traite par quelques geometres et qui est neanmoins impossible," in *Leonhardi Euleri Opera Omnia*, Ser. Prima, Volumen XXVII, edited by Andreas Speiser, Birkhäuser, Switzerland, 1954, pp. 340–363.

[3] Homer S. White, "The geometry of Leonhard Euler," in *Leonhard Euler: Life, Work and Legacy*, Robert E. Bradley and C. Edward Sandifer (eds.), Elsevier, 2007, pp. 303–321.

23

Multiple Representations of Functions in the History of Mathematics

Robert Rogers
State University of New York at Fredonia

23.1 Introduction (A Funny Thing Happened on the Way to Calculus)

During the fall semester of 2005, I was slated to teach University Calculus I to a class of mostly incoming freshmen. It had been a while since I taught both the class and freshmen, so on the first day I decided to do some review and pick my students' brains. I wrote $y = f(x) = x^2$ on the board and asked if that was a function. The unanimous answer was yes. Without exploring that too much, I drew Figure 23.1 on the board and asked if that was a function.

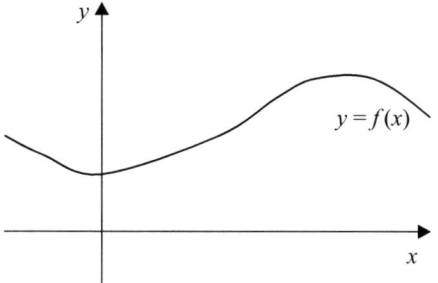

Figure 23.1. A function represented graphically

The response was overwhelmingly yes and when I asked why, the response was that it passed the vertical line test. I then wrote the following on the board.

$$c(x) = \text{ the cost (in cents) to send a first class letter weighing } x \text{ ounces through the US Postal service.}$$

When asked if that was a function, about half of the class said it was and about half said it wasn't. From the looks on people's faces, the majority of students in the class were not sure of their answers either. After having students discuss their thoughts with their neighbors, we had a class discussion about whether it was, in fact, a function or not. The gist of people who said no was that it wasn't a function because there was no formula or graph with which to determine specific values. The people who said yes said that it was still a rule that assigned a unique value to each x.

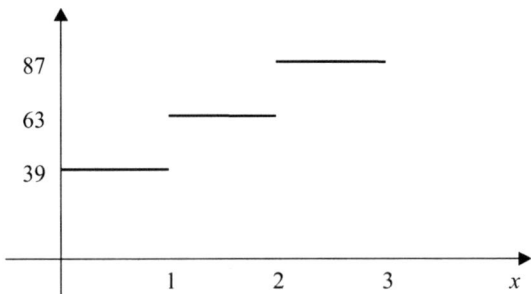

Figure 23.2. Graphic representation of 2005 postal rates.

There was still no definite consensus when I drew Figure 23.2 on the board. When asked if this was a function, the class overwhelmingly said that it was and an astute student pointed out that it was, in fact, the previous cost function, $c(x)$. By no means can this be construed as any sort of careful experiment, but I went away with the feeling that our students are comfortable with functions as formulas and graphs (possibly tables as well, though I didn't explore that), but are not as comfortable with other representations.

The NCTM standards call for the understanding of patterns, relations, and functions throughout the grade levels as part of its algebra strand. In this strand, it is expected that students represent and analyze patterns and functions using words, tables, and graphs along with the typical algebraic formulae [14]. If my episode in calculus is any indication, there seems to be a tendency to shift toward an analytic or graphical representation of functions while suppressing a verbal or tabular form. The ability to recognize a function given in verbal or tabular form is paramount in applications where a problem is given in words or data is given in a table. In the quest to make students fluent with algebraic symbolism, we should not forget that these other representations draw upon students' experiences with patterns and functions at the elementary level. Older students must be reminded that these alternate representations are examples of functions as well, so that they are comfortable working with functions in all forms: analytic, graphical, verbal, and tabular.

The evolution of the function concept contains examples of all of these representations. Many of these episodes can be utilized to provide meaningful applications illustrating the various representations of the function concept. In a number of instances, the topics involved in these episodes are already part of the curriculum. In these cases, teachers can enhance their students' understanding of both the topic and the underlying concept of function. As such, this paper will provide some of these historical perspectives on functions along with suggestions for classroom activities that will promote these various views of what is meant by a function.

23.2 Area of a Circle

23.2.1 Historical Background

An argument can be made that the early notion of a function was, in fact, verbal. Consider Problem 50 from the *Rhind Papyrus* [9, p. 20]:

> Example of a round field of diameter 9. What is the area? Take away 1/9 of [the diameter] 9; the remainder is 8. Multiply 8 times 8; it makes 64. Therefore, the area is 64.

The same area-of-circle recipe is used in the following problems from the *Rhind Papyrus* [5, p. 17–18]

> Problem 41. Find the volume of a cylindrical granary of diameter 9 and height 10.

> Problem 42. Find the volume of a cylindrical granary of diameter 10 and height 10.

Although the Egyptians used numbers instead of arbitrary variables, it appears that the recipe is meant to transcend a particular example. This recipe is meant to be a practical solution to the computation of circular area. Furthermore, this recipe does not make reference to the ratio of the circumference to the diameter (what we now know as π) as our modern definition does. This was because finding the area of a circle represented a different problem than that of finding its circumference.

The connection between the circumference of a circle and its area is made explicit in Archimedes' determination of the area of a circle in his treatise *Measurement of a Circle* [5, p. 148].

Proposition 1. *The area of any circle is equal to a right-angled triangle in which one of the sides about the right angle is equal to the radius, and the other to the circumference of the circle.*

This description is typical of Greek quadrature problems; that is, to construct (classically, with tools allowed by the axioms in Euclid's *Elements*: a straight edge and compass) a square whose area is equal to that of a given figure [3, p.12-17]. Since the quadrature of a triangle was known, then one could presumably "square the circle" if one could produce the circumference of the circle. This cannot be done with Euclidean tools, though in the same treatise Archimedes provides that the ratio of the circumference to the diameter of a circle is between $3\frac{10}{71}$ and $3\frac{1}{7}$. [3, p. 97].

23.2.2 In the Classroom

The fact that these relations were presented with their respective descriptions was of necessity, as formal equations did not exist at the time. Teachers at the elementary level are faced with a similar situation, as their students do not possess the algebraic sophistication to produce formal equations. Teachers at the middle school level are expected to bridge the gap between intuition and formalism and by doing so, set the stage for later functional understanding. The area of a circle provides an opportunity for the introduction of this functional understanding.

The Egyptian recipe can be utilized in the classroom as a way of deriving a formula for the area, A, of a circle in terms of its diameter, d. Specifically, students can be asked to compute the areas of circles of various diameters using the above recipe. From this, the general function

$$A = A(d) = \left(d - \frac{d}{9}\right)^2 = \left(\frac{8}{9}d\right)^2$$

can be derived to demonstrate the functional relation between the input d and the output A. Letting r denote the radius, this yields the formula

$$A = A(r) = \left(\frac{16}{9}\right)^2 r^2 \approx 3.16 r^2.$$

Again, the functional notation can be introduced to emphasize this input-output relationship and students need not be encumbered by the extra issue of the number π.

The determination of circular area by Archimedes also provides the opportunity to translate words into formulas. In particular, if r and C represent the radius and circumference of a circle, respectively, then its area, A, can be given by the function $A = A(r, C) = \frac{1}{2} r \cdot C$. Using Archimedes approximation, $C \approx \frac{22}{7}(2r) \approx 3.14 (2r)$, yields the familiar $A = A(r) \approx 3.14 r^2$ which can be compared to the Egyptian formula. These can also be compared to the modern formula for the area, emphasizing the fact that there is a functional dependency between A and r. A nice reversal would be to start with our modern formula, $A(r) = \pi r^2$, and have students state this in the recipe manner of the Egyptians [square the radius and multiply by π].

23.3 Ptolemy's Table

23.3.1 Historical Background

A number of mathematical tables exist from ancient civilizations, including tables of reciprocals, squares, and square roots. This input-output view of a function is recaptured in the tabling button of modern hand held calculators. Of particular note is Claudius Ptolemy's chord table from approximately 100 A.D. This table provides the lengths of the chords of arcs ranging from $\frac{1}{2}°$ to $180°$ in increments of $\frac{1}{2}°$. In general, the chord of an arc of a circle is the line segment joining the endpoints of the arc as seen in the Figure 23.3.

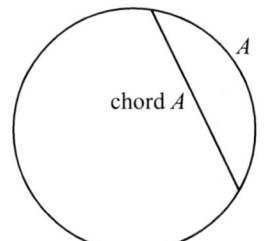

Figure 23.3. The chord of an arc.

One could make an argument that finding the length of a chord is a more natural geometric problem than finding the sine of a central angle, though as will be seen these are essentially equivalent. More specifically, in a unit circle, the sine of a central angle is half of the length of the chord of twice the angle. This relationship might help students make sense of the various trigonometric functions. With this in mind, a brief etymological history of sine might prove useful in the classroom [11, p. 200].

In dealing with spherical trigonometry used to study astronomy, Hindu mathematicians determined that it was convenient to deal with half-chords. The Hindu mathematician Aryabhata (476–550 A.D.) frequently used the abbreviation *jya* for the word *ardha-jya* (half chord). This was phonetically translated into *jiba* by subsequent Arabic mathematicians. Since Arabic is written without vowels, then this was written as *jb*. When the Arabic works were translated into Latin, *jb* posed a problem as there is no such word as *jiba* in Arabic (recall it was translated phonetically). The closest "real" Arabic word was *jaib* which means "cove" or "bay". The Latin word for this is *sinus*, which becomes our modern sine. Once it is realized that sine represents a half-chord, it is straightforward to see that cosine represents the complement's sine [Figure 23.4].

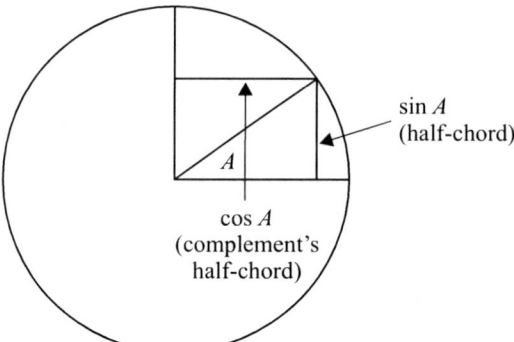

Figure 23.4. Sine and cosine of an angle

Given this relationship, it is clear that Ptolemy's chord table is a precursor to our more modern sine table. Figure 23.5 provides part of Ptolemy's chord table with a translation into English [13].

The measurements are written in base 60 and are measured in units that are 1/60 of the radius of the circle. Hence the radius of the circle is 60 parts and the chord of 60° would also be 60 parts since it is equal to the radius. Notice in the table that the chord of 180° is 120 parts which is the diameter.

23.3.2 In the Classroom

As an alternative to plotting sine and cosine curves, Ptolemy's chord table affords the opportunity to examine these trigonometric functions in a tabular form. Students can be asked to use the tabling button to produce their own version of Ptolemy's chord table. To do this, students must first understand the translation from chords to half chords. Associating an arc with its central angle, we obtain this relationship [Figure 23.6].

Letting $r = 60$, we can create any entry from Ptolemy's table (in base 10 instead of base 60). For example,

$$\text{chord } 7° = 120 \sin 3.5° \approx 7.325824744.$$

Comparing this to the value in Ptolemy's table above, we have

$$(7; 19, 33)_{60} = 7 + \frac{19}{60} + \frac{33}{60^2} \approx 7.325833333.$$

Using the tabling feature in a calculator, students can reproduce Ptolemy's chord table while reinforcing the notion that a function can be represented in table form.

Though this is a bit aside from the topic of functional representation, the etymology of sine and cosine provides an alternative way to memorize that sine is the opposite divided by the hypotenuse and cosine is the adjacent divided by the hypotenuse without a pneumonic. This also clears up the mystery of why the secant of an angle is defined to be the reciprocal of cosine and not of sine. Specifically, consider the Latin origins of tangent (*tangere to touch*) and secant (*secare to cut*). Geometrically, the relationship among these sides of a right triangle can be seen in the unit circle in Figure 23.7.

Recognizing that cotangent and cosecant are the complement's tangent and secant, respectively, completes the etymology of all of the trigonometric functions.

23.3. Ptolemy's Table

Κανόνιον τῶν ἐν κύκλῳ εὐθειῶν			Table of Chords		
περιφε-ρειῶν	εὐθειῶν	ἑξηκοστῶν	arcs	chords	sixtieths
∠′	ο λα κε	ο α β ν	½°	0;31,25	0;1,2,50
α	α β ν	ο α β ν	1°	1; 2,50	0;1,2,50
α∠′	α λδ ιε	ο α β ν	1½°	1;34,15	0;1,2,50
β	β ε μ	ο α β ν	2°	2; 5,40	0;1,2,50
β∠′	β λϛ δ	ο α β μη	2½°	2;37, 4	0;1,2,48
γ	γ η κη	ο α β μη	3°	3; 8,28	0;1,2,48
γ∠′	γ λθ νβ	ο α β μη	3½°	3;39,52	0;1,2,48
δ	δ ια ιϛ	ο α β μζ	4°	4;11,16	0;1,2,47
δ∠′	δ μβ μ	ο α β μζ	4½°	4;42,40	0;1,2,47
ε	ε ιδ δ	ο α β μϛ	5°	5;14, 4	0;1,2,46
ε∠′	ε με κζ	ο α β με	5½°	5;45,27	0;1,2,45
ϛ	ϛ ιϛ μθ	ο α β μδ	6°	6;16,49	0;1,2,44
ϛ∠′	ϛ μη ια	ο α β μγ	6½°	6;48,11	0;1,2,43
ζ	ζ ιθ λγ	ο α β μβ	7°	7;19,33	0;1,2,42
ζ∠′	ζ ν νδ	ο α β μα	7½°	7;50,54	0;1,2,41
⋮	⋮	⋮	⋮	⋮	⋮
ροδ∠′	ριθ να μγ	ο ο β νγ	174½°	119;51,43	0;0,2,53
ροε	ριθ νγ ι	ο ο β λϛ	175°	119;53,10	0;0,2,36
ροε∠′	ριθ νδ κζ	ο ο β κ	175½°	119;54,27	0;0,2,20
ροϛ	ριθ νε λη	ο ο β γ	176°	119;55,38	0;0,2,3
ροϛ∠′	ριθ νϛ λθ	ο ο α μζ	176½°	119;56,39	0;0,1,47
ροζ	ριθ νζ λβ	ο ο α λ	177°	119;57,32	0;0,1,30
ροζ∠′	ριθ νη ιη	ο ο α ιδ	177½°	119;58,18	0;0,1,14
ροη	ριθ νη νε	ο ο ο νζ	178°	119;58,55	0;0,0,57
ροη∠′	ριθ νθ κδ	ο ο ο μα	178½°	119;59,24	0;0,0,41
ροθ	ριθ νθ μδ	ο ο ο κε	179°	119;59,44	0;0,0,25
ροθ∠′	ριθ νθ νϛ	ο ο ο θ	179½°	119;59,56	0;0,0,9
ρπ	ρκ ο ο	ο ο ο ο	180°	120; 0, 0	0;0,0,0

Figure 23.5. Ptolemy's chord table.

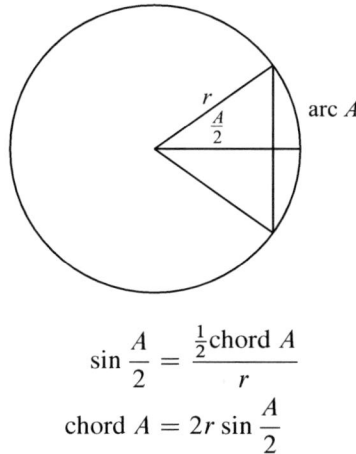

$$\sin \frac{A}{2} = \frac{\frac{1}{2}\text{chord } A}{r}$$
$$\text{chord } A = 2r \sin \frac{A}{2}$$

Figure 23.6. Conversion from sines to chords.

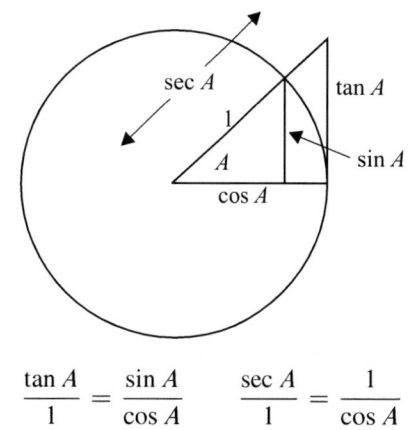

$$\frac{\tan A}{1} = \frac{\sin A}{\cos A} \qquad \frac{\sec A}{1} = \frac{1}{\cos A}$$

Figure 23.7. Defining tangent and secant.

23.4 From Geometry to Analysis to Set Theory

23.4.1 Historical Background

The fact that our current students are comfortable with the graphical representation of a function demonstrates the power of analytic geometry. The blending of algebra and geometry was very much evident in the mathematics of Medieval Islam [9, p. 238–87]. In point of fact, the word algebra evolves from the Arabic word *al-jabr* (the reunion of broken parts) [11, p. 21] and represents the restoration of a quantity subtracted on one side of an equation to the other side. This being said, equations were described verbally and the solutions are also given in recipe form. The blending comes from the fact that the verification of these recipes comes in the form of a geometric proof. The fact that algebraic techniques were seen as a powerful tool for solving mathematical problems and the access of Italian merchants to Arabic mathematics helped the rise in prominence of solving equations (and the use of the Hindu Arabic number system) in the Renaissance. However, the use of symbolic notation was still in the beginning of its development during the transfer of these techniques from the Arabic world to Europe. People such as Raphael Bombelli (1526–1572) and François Viète (1540–1603) made great strides in introducing and utilizing an increasing amount of symbolic notation to further refine and exploit these algebraic ideas.

Of particular importance is the development of analytic geometry by Pierre de Fermat (1601–1665) and René Descartes (1596–1650) [9, p. 432–42]. Both were applying algebraic techniques to re-examine the geometry of the Greeks and both were able to exploit a more modern symbolic notation to assign equations to curves. The ability to represent a curve with an equation was a driving force in the development of the Calculus by Isaac Newton (1642–1727) and Gottfried Wilhelm Leibniz (1646–1716) [1, Chapter V], and is still a powerful influence on our students' perceptions of what constitutes a function. In fact, Newton and Leibniz's calculus was a calculus of curves rather than functions. However, foundational issues caused a re-examination of the techniques of calculus and with this came a shift away from a geometrical curve toward the notion of a function as an algebraic expression.

An influential text of the eighteenth century and an example of this viewpoint is Book I of Euler's 1748 work *Introductio in Analysin Infinitorum* (*Introduction to the Analysis of the Infinite, Vol. I and II*) [4]. Book I contains no diagrams and though it does rely on intuitive appeals to infinitely large and small quantities to produce series representations, it has functions as its primary focus. In fact, Euler provides the following definition.

> A *function* of a variable quantity is an analytic expression composed in any way whatsoever of the variable quantity and numbers or constant quantities.

This viewpoint, which dominated the later half of the 18th century and the early part of the 19th century, is remarkably similar to the viewpoint that my students held in my calculus class. The fact that they gravitate to that formulation is not so surprising given the algebraic manner in which they have been asked to use functions. Furthermore, according to Euler, it was a perfectly reasonable way to prepare for understanding the Calculus. Euler was forced to refine this notion of a function in his response to Jean le Rond d'Alembert's solution to the problem of a vibrating string published in 1747 [7, p. 2–13]. This problem was one of the outstanding problems of its day and was a basis for future studies in many physical considerations such as heat flow and potential theory. The problem is to find the displacement $y = y(x,t)$ at position x and time t of an elastic string stretched between two points on the x-axis (one of which is the origin), given an initial displacement $y(x,0) = f(x)$ and an initial velocity $\frac{\partial y}{\partial t}(x,0) = g(x)$. D'Alembert showed that this displacement must satisfy the wave equation

$$\frac{\partial^2 y}{\partial x^2} = \frac{1}{c^2}\frac{\partial^2 y}{\partial t^2},$$

and showed that if $y(0,t) = 0$, then the general solution of this equation can be written as

$$y(x,t) = F(ct+x) - F(ct-x),$$

where F is an arbitrary (differentiable) function determined by the initial conditions. A disagreement that ensued between Euler and d'Alembert was in the nature of the functions f and g describing the initial conditions. D'Alembert insisted that these conditions be defined by the same formula throughout the interval between the endpoints of the string [6, p. 90]. Euler countered that this was too restrictive and contended that the initial shape of the string could be composed of a number of different curves which may or may not fit together smoothly. For example, the initial

23.4. From Geometry to Analysis to Set Theory

displacement of a plucked string could be that of an inverted V which would require two different lines coming together to a point. This function could not be described by a single (differentiable) expression. In hindsight, the use of a piecewise defined function to describe these more general displacements may seem very natural, but it required a shift in viewpoint of what, in fact, constituted a function. In the second volume of *Introductio*, Euler made a distinction between functions defined by a single expression and "mixed" functions [7, p. 6]. Since these new functions did not have to be differentiable at various points, then it seemed, to d'Alembert at least, that they should not be allowed to be part of a solution to a differential equation.

Daniel Bernoulli (1700–1782) entered this discussion by proposing that the general solution to the vibrating string problem could be written as an infinite sum of trigonometric functions such as

$$y(x,t) = \alpha \sin \frac{\pi x}{l} \cos \frac{\pi c t}{l} + \beta \sin \frac{2\pi x}{l} \cos \frac{2\pi c t}{l} + \cdots,$$

where l is the length of the string.

Bernoulli did not provide a mathematical proof of this and the idea met criticism from others (including Euler) that this representation would not be able to capture the most general functions possible [7, p. 9–10]. Indeed, how could a single expression be used to represent a function defined in a piecewise fashion? The answer came in 1807, with the submission of the manuscript, *Sur la Propagation de la chaleur* (*On the Propagation of Heat*), by Jean Baptiste Joseph Fourier (1768–1830) [2, Chapter 1]. In analyzing the flow of heat in a two-dimensional plate, Fourier was able to make Bernoulli's idea precise by determining how to represent a function as a trigonometric series. For example, Fourier determined that on the interval $(-1, 1)$, the constant function $f(x) = 1$ could be written as

$$1 = \frac{4}{\pi} \left(\cos \frac{\pi x}{2} - \frac{1}{3} \cos \frac{3\pi x}{2} + \frac{1}{5} \cos \frac{5\pi x}{2} - \cdots \right).$$

Given the fact that

$$\cos\left(\frac{(2n-1)\pi(x+2)}{2}\right) = \cos\left(\frac{(2n-1)\pi x}{2} + (2n-1)\pi\right) = -\cos\left(\frac{(2n-1)\pi x}{2}\right),$$

this says that the graph of

$$f(x) = \frac{4}{\pi} \left(\cos \frac{\pi x}{2} - \frac{1}{3} \cos \frac{3\pi x}{2} + \frac{1}{5} \cos \frac{5\pi x}{2} - \cdots \right)$$

would be given by Figure 23.8.

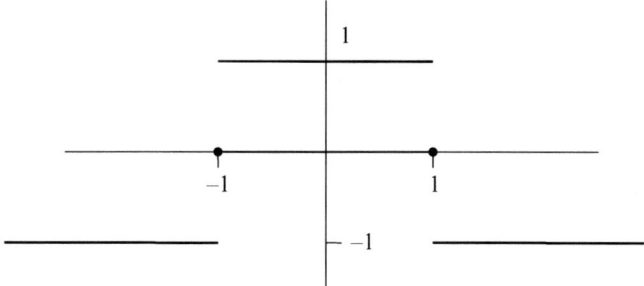

Figure 23.8. Graph of $\frac{4}{\pi} \left(\cos \frac{\pi x}{2} - \frac{1}{3} \cos \frac{3\pi x}{2} + \frac{1}{5} \cos \frac{5\pi x}{2} - \cdots \right)$.

It was very disconcerting to have a piecewise defined (even discontinuous) function represented by a single sum of continuous functions. Fourier's original manuscript was rejected, but his revisions were subsequently published by Académie des Sciences in 1822 [12]. These ideas caused mathematicians in the nineteenth century to re-examine the tried and true techniques employed in the eighteenth century and, in particular, the nature of what constituted a function. In doing so, the notion of a function became more abstract. For example, Johann Peter Gustav Lejeune Dirichlet (1805–1859) provided the following definition in 1837:

If now to any x there corresponds a unique, finite y, ...then y is called a function of x for this interval....
This definition does not require a common rule for the different parts of the curve; one can imagine the curve as being composed of the most heterogeneous components or as being drawn without following any law [8, p. 201].

As an example, consider the following function due to Dirichlet

$$f(x) = \begin{cases} 0 & \text{if } x \text{ is irrational} \\ 1 & \text{if } x \text{ is rational} \end{cases}.$$

This function defies drawing. An even more remarkable example came in 1872 from Karl Weierstrass (1815–1897). Weierstrass showed that the function given by the Fourier series

$$f(x) = \sum_{n=0}^{\infty} b^n \cos(a^n \pi x)$$

is continuous at every value x provided $0 < b < 1$, but is not differentiable at any value x when a is an odd integer with $ab > 1 + \frac{3\pi}{2}$ [2, p.260]. If one were to think of the tangent line to a path as pointing in the direction that one is traveling at a given instant, then the above would be a continuous path from one point to another which has the property that at any point, a person traversing this path is not going in any particular direction. Clearly, such a function cannot be drawn and, in fact, is fractal in nature.

The German mathematician Georg Cantor (1845–1918) utilized this more abstract notion of a function as a correspondence between sets in his study of the infinite. One particular example was how to distinguish between the set of rational numbers and the set of irrational numbers. Both sets are infinite. Furthermore, between any two rational numbers is an infinite number of irrational numbers and between any two irrational numbers is an infinite number of rational numbers. Surprisingly, Cantor showed that these sets represented two different sizes of infinity. Cantor demonstrated that there is a one to one correspondence (function) between the set of positive integers and the set of all rational numbers, but there is none between the set of positive integers and the set of irrational numbers. The set of irrational numbers represents a larger infinity than the set of rational numbers. As much as any other topic, Cantor's work ushered in the mathematics of the twentieth century and the rise of the study of set theory.

This set theoretical approach is evident in the "modern definition" of a function which is essentially the one given by Bourbaki in 1939 [10, p. 298].

Let A and B be sets. A function f from A to B is a subset of the cartesian product $A \times B$ such that if $x \in A$, then there exists a unique element $y \in B$ such that $(x, y) \in f$. In this case we denote y by $f(x)$.

23.4.2 In the Classroom

Even though d'Alembert's wave equation involves partial derivatives, it is not beyond the reach of a calculus class to show that d'Alembert's solution,

$$y(x,t) = F(ct + x) - F(ct - x),$$

satisfies it, as we have

$$\frac{\partial^2 y}{\partial x^2} = F''(ct + x) - F''(ct - x) \quad \text{and} \quad \frac{\partial^2 y}{\partial t^2} = c^2\big(F''(ct + x) - F''(ct - x)\big).$$

Further analyzing this solution, it can be seen that if F is an odd function, then d'Alembert's solution becomes

$$y(x,t) = F(x + ct) + F(x - ct)$$

which is the sum of two horizontal translates of $F(x)$. This can be animated using a computer algebra system, or at least graphed for various values of t to see the wave nature of the solution. For example, consider $F(x) = \sin x$ on $[0, 2\pi]$. Figure 23.9 represents d'Alembert's solution, $y(x,t) = \sin(x + t) + \sin(x - t)$, with $c = 1$.

Even though it is a more abstract concept than that of the wave equation, the notion of a function as a correspondence between sets can be utilized at a more elementary level. This is particularly true when looking at patterns in the

23.5 Conclusion

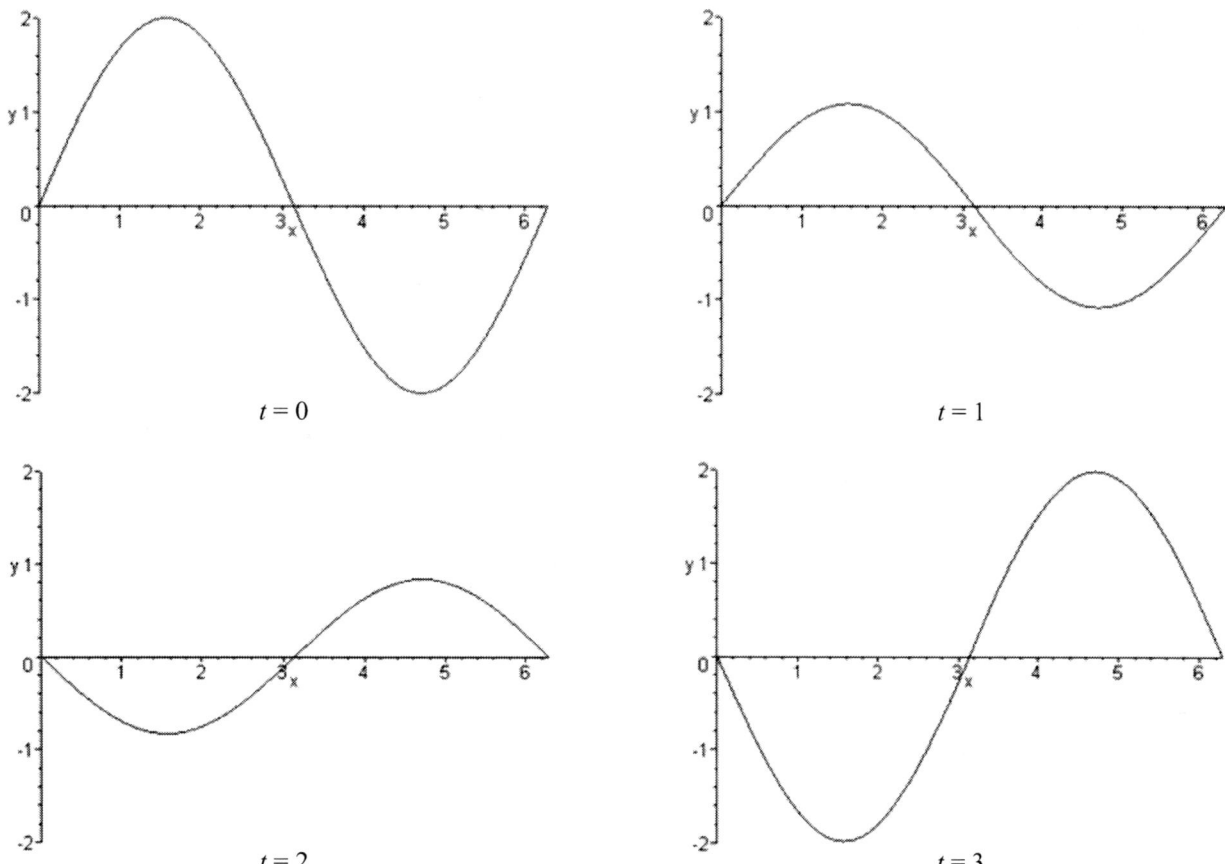

Figure 23.9. d'Alembert's solution $y(x,t) = \sin(x+t) + \sin(x-t)$ at times $t = 0, 1, 2, 3$.

elementary grades. For example, asking what the 24th even number is demonstrates the function $f(n) = 2n$. While a teacher at that level would certainly not use such a formula, this sets the stage for later grades. In those later grades, teachers should draw upon these connections so that their students have a more in-depth understanding of the notion of a function than just a graph or a formula.

23.5 Conclusion

A saying in biology is "Ontogeny recapitulates Phylogeny." This debunked theory says that the development of an individual embryo mimics that of the evolution of the entire species. While this may not be true in biology, it may have its merits in mathematics education. With an NCTM strand such as patterns, relations, and functions transcending the grades, it is paramount that teachers of mathematics at all levels be cognizant that the function concept evolved over a long period of time to meet the demands made upon it in solving various problems. Perhaps our students' understanding needs to follow this course. Teachers at the elementary level need to see where the subject is heading so that they may set the table. Teachers at the middle and high school level need to be aware of their incoming students' understanding and make connections between their students' informal concepts of function and the notions that they will be teaching. In many cases, teachers can accomplish this merely by adjusting their presentation of standard topics so as to embrace this previous knowledge. By allowing students to make connections between these various representations, teachers will empower their students to recognize and use all aspects of the concept of a function.

Bibliography

[1] C. Boyer, *The History of the Calculus and its Conceptual Development*, Dover, New York, 1959.

[2] D. Bressoud, *A Radical Approach to Real Analysis*, The Mathematical Association of America, Washington, DC, 1994.

[3] W. Dunham, *Journey Through Genius, the Great Theorems of Mathematics*, Wiley, New York, 1990.

[4] L. Euler, *Introduction to the Analysis of the Infinite, Books I and II*, translated by J. Blanton, Springer-Verlag, New York, 1988.

[5] J. Fauvel and J. Gray, *The History of Mathematics, a Reader*, The Open University and MacMillan Press, London, 1987.

[6] J. Grabiner, *The Origins of Cauchy's Rigorous Calculus*, MIT Press, Cambridge, MA, 1981.

[7] I. Grattan-Guinness, *The Development of the Foundations of Mathematical Analysis from Euler to Riemann*, MIT Press, Cambridge, MA, 1970.

[8] E. Hairer and G. Wanner, *Analysis by Its History*, Springer-Verlag, New York, 1996.

[9] V. Katz, *A History of Mathematics, an Introduction, 2nd ed.*, Addison-Wesley, Reading, Ma, 1998.

[10] I. Kleiner, "Evolution of the Function Concept: A Brief Survey," *College Mathematics Journal*, Vol. 20, No. 4 (1989), 282–300.

[11] S. Schwartzman, *The Words of Mathematics, an Etymological Dictionary of Mathematical Terms Used in English*, The Mathematical Association of America, Washington, DC, 1994.

[12] http://www-gap.dcs.st-and.ac.uk/ history/Biographies/Fourier.html

[13] press.princeton.edu/books/maor/chapter_2.pdf

[14] standards.nctm.org/document/appendix/alg.htm

24

The Unity of all Science: Karl Pearson, the Mean and the Standard Deviation

Joe Albree
Auburn University Montgomery

24.1 Introduction

In statistics, Karl Pearson's (1857–1936) *method of moments* unified the arithmetic mean, the standard deviation, and a number of further statistical calculations. It may be surprising to learn that the underlying concepts of the method of moments come from physics. For Karl Pearson, though, this development was a natural one.

After introducing the story of Karl Pearson's journey to the study of statistics, we present a set of practical data values which can be analyzed directly or grouped into classes and then analyzed. As we obtain the mean and standard deviation of this data set, we will see how the physics of the first and second moments aid in our computations, and of even more importance, give us insights into the results of the calculations.

24.2 Karl Pearson: Historical preliminaries

Even though Karl Pearson has been heralded as "the founder of the twentieth-century science of statistics" [2, p. 447], his story had a much different beginning. Carl Pearson (as he was christened) grew up in an upper middle class Victorian London home. In 1875, Carl earned a scholarship at King's College, Cambridge, where he studied the works of Charles Darwin (1809–1882) and of Benedict Spinoza (1632–1677), and German history. He graduated with honors in mathematics (1879). After graduation, Pearson traveled and studied in Germany, where he became so enamored with the works of Karl Marx that he changed the legal spelling of his name from Carl to Karl; to his friends and colleagues, he was also known as K.P. When he returned to London, he was admitted to the bar (1881), and as his father wished, he practiced law for a short time.

In his first academic position, beginning in 1884 at University College, London, Karl Pearson taught mathematics and mechanics to engineering students, and he did research in elasticity and the philosophy of science. Pearson believed that all knowledge was based on sense perception and that the task of science was

Figure 24.1. Karl Pearson (1857–1936)[22]

to summarize the routines of experience by means of laws expressed in mathematical form [11]. In the 1880s, Karl Pearson was well on his way to a respectable, if a bit eccentric, career as a traditional applied mathematician and philosopher [3, 7, 18].

The year 1889 marked an epiphany for Karl Pearson, and the catalyst was the book *Natural Inheritance* by Francis Galton (1822–1911), a half-cousin of Charles Darwin. Galton was the father of *eugenics*, a social movement holding that human mental and physical characteristics are heritable and that steps should be taken to ensure the race is constantly genetically improved. Karl Pearson described his transformation as follows:

> It was Galton who first freed me from the prejudice that sound mathematics could only be applied to natural phenomena under the category of causation. Here for the first time was a possibility, I will say a certainty, of reaching knowledge — as valid as physical knowledge was then thought to be — in the field of living forms and above all in the field of human conduct [7, pp. 18–19].

When W. F. R. Weldon (1860–1906) became Jodrell professor of zoology at University College, London, in 1890, Karl Pearson met his professional soul mate. Frank Weldon (as he was known) provided a great deal of the statistical data that inspired Pearson to develop many of his statistical techniques and theorems [12].

Pearson's first technical work in statistics was a sequence of 18 research papers, all with basically the same title:

"Contributions to the Mathematical Theory of Evolution" 1894 [9]
"Contributions to the Mathematical Theory of Evolution II" 1895 [10]
"Mathematical Contributions to the Theory of Evolution III" 1896
⋮
"Mathematical Contributions to the Theory of Evolution XIX" 1916

Number XVII was never published. Pearson's method of moments appears in several of these works.

24.3 A data set

Karl Pearson included practical sets of data in almost all of his contributions to statistics (over 400 papers and books from 1890 to 1936). In this spirit, we will work our way through a set of public health data (Figure 24.2).

With $n = 12$ data values (the county infant mortality rates), we have a sample size for which the graphical representations and the computations can be performed using the data values themselves (see Figure 24.3 and the first column of Table 24.1), or, by grouping the data values into five convenient classes (see Figure 24.4 and the first column of Table 24.2):

INFANT MORTALITY RATES BY COUNTY

Combined infant mortality rates for 2004-2006 for area counties and the state are given below. The rate represents the number of infant deaths per 1,000 live births.

Montgomery.................9.3
Autauga......................9.8
Elmore.......................6.4
Lowndes....................12.9
Macon.......................13.6
Bullock......................9.1
Crenshaw...................4.1
Pike.........................13.7
Lee..........................9.5
Dallas.......................6.5
Chilton......................9.8
Tallapoosa..................8.2
Alabama.....................9.3

Source: Alabama Department of Public Health

Figure 24.2. Infant Mortality Rates for 12 Alabama Counties, reprinted with permission [6].

x	$x - \bar{x}$	$(x - \bar{x})^2$
4.1	-5.3	28.09
6.4	-3.0	9.00
6.5	-2.9	8.41
8.2	-1.2	1.44
9.1	-.3	.09
9.3	-.1	.01
9.5	.1	.01
9.8	.4	.16
9.8	.4	.16
12.9	3.5	12.25
13.6	4.2	17.64
13.7	4.3	18.49
112.9	$.1 \approx 0$	95.75

Table 24.1. Infant Mortality Rates — Data Values

24.4. The Mean and the First Moment

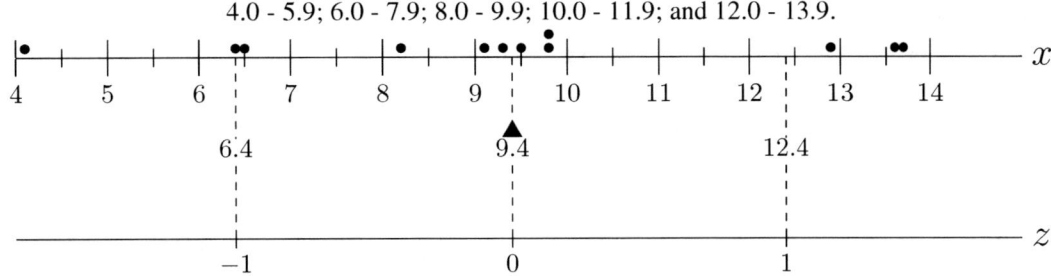

4.0 - 5.9; 6.0 - 7.9; 8.0 - 9.9; 10.0 - 11.9; and 12.0 - 13.9.

Figure 24.3. Infant Mortality Rates — Dotplot

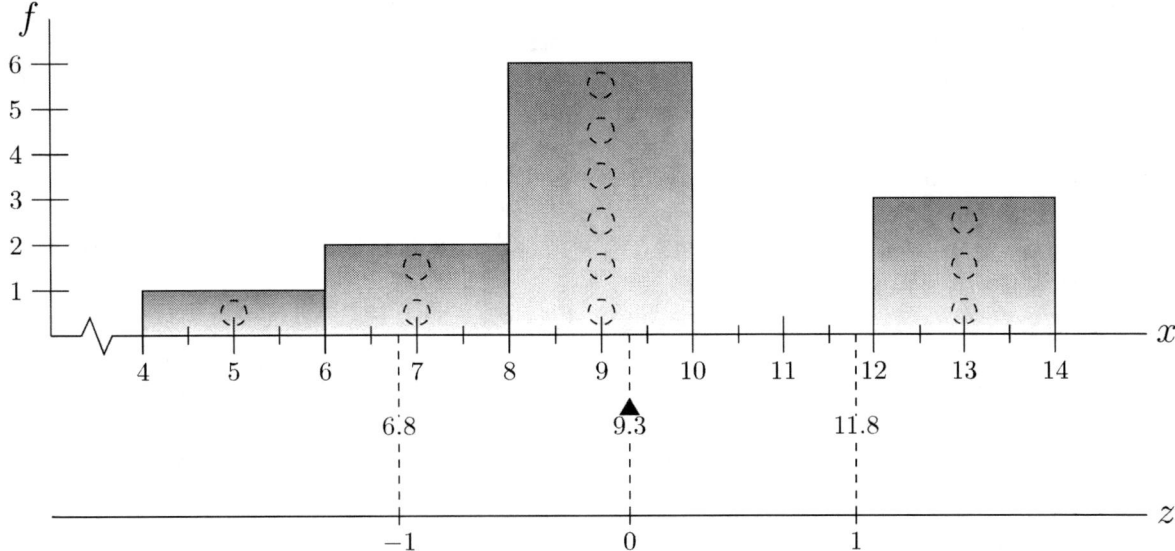

Figure 24.4. Infant Mortality Rates — Histogram

classes	$\hat{x}$	f	$\hat{x}f$	$\hat{x} - \bar{x}$	$(\hat{x} - \bar{x})f$	$(\hat{x} - \bar{x})^2$	$(\hat{x} - \bar{x})^2 f$
4.0-5.9	5	1	5	−4.3	−4.3	18.49	18.49
6.0-7.9	7	2	14	−2.3	−4.6	5.29	10.58
8.0-9.9	9	6	54	−.3	−1.8	.09	.54
10.0-11.9	11	0	0	1.7	0	2.89	0
12.0-13.9	13	3	39	3.7	11.1	13.69	41.07
−	−	12	112	−	$.4 \approx 0$	−	70.68

Table 24.2. Infant Mortality Rates — Data in Classes

24.4 The Mean and the First Moment

Certainly, anyone can calculate the "average" of a set of data values which is not too large (see Table 24.1).

Example 1A. Infant Mortality Rates, data values. The *mean* is:

$$\bar{x} = \frac{\sum x}{n} = \frac{4.1 + 6.4 + \cdots + 13.7}{12} = \frac{112.9}{12} = 9.408333\ldots \approx 9.4.$$

(This analysis will be continued in Examples 1B, 1C, and 1D.)

Example 2A. Infant Mortality Rates, data in classes. First, we count the number of data values in each of the five classes specified above; these are called the *frequencies*, f. Second, the "middle" of each class, denoted $\hat{x}$ and termed the *class mark* or the *class representative*, is determined. For our classes, the marks are 5, 7, 9, 11, and 13. Note that these values may, or may not, be data values themselves, but we *assume* that all the data values in a given class are equal to this representative (see Figure 24.4). Thus, we could also call the $\hat{x}$ values "pseudo data values." With this understanding, when we perform computations with statistical data grouped into classes, we calculate the *mean*:

$$\bar{x} = \frac{\sum(\hat{x} \cdot f)}{\sum f} = \frac{(5 \cdot 1) + (7 \cdot 2) + (9 \cdot 6) + (11 \cdot 0) + (13 \cdot 3)}{1 + 2 + 6 + 0 + 3} = \frac{112}{12} = 9.333\ldots \approx 9.3.$$

(This analysis will be continued in Examples 2B and 2C.)

In his famous works [9, 10], Karl Pearson used the physics term *centroid* for what we call the *mean*. Engineers today often use the term *centroid* of a physical body ("center-like") in place of the *center of mass* of that body (physicists make a technical distinction between the center of mass and the *center of gravity*, but this will be of no concern to us).

We think of the set of all the dots in Figure 24.3, and other sets of data values, as physical particles whose totality makes a whole physical body. Similarly, consider all of the bars in a *histogram*, like Figure 24.4, as being welded together into one polygonally shaped sheet, say of cardboard, or thin plastic, or plywood, whose particles are the atoms of the material. Each of these objects is a two-dimensional physical body. In physics, when a physical body is in any kind of motion, each particle of that body is subject to one or more forces. The *centroid* of that body is the point at which all of the body's particles could, in theory, be concentrated. For example, the centroid of a wooden baseball bat is about 60 to 65 percent of the way from the knob end of the bat. It is easy to find the centroid of a wooden baseball bat: measuring 60 to 65 percent of the bat's length from the knob end should yield the point where the bat balances horizontally and that will be its centroid.

In statistics, the *centroid* is the point where the totality of all the data values *balance*. In determining the mean in his statistical work, Karl Pearson borrowed the well-known (in physics) formula for the *first moment* (see below). In theory one may calculate the first moment about any point (see Question 4), but our main interest is in the first moment about the mean, which we shall call the *first centroidal moment*. Then, to say that a physical body, like a wooden baseball bat, balances at its centroid means that the first centroidal moment is zero [5, 10, p. 346, 15, 21, pp. 306–311].

Principle 1. *For any set of statistical data values, $x_1, x_2, \ldots x_n$, the mean is the balance point, i.e., the first moment about the mean, or the **first centroidal moment**, is* 0.

We denote the first moment about the mean by $\mathbf{M_1}$. *The Law of the Lever* of Archimedes (ca. 287–212 BCE) is the main idea behind the explanation of this principle and also provides the key to the statistical calculation of $\mathbf{M_1}$. As every child who has ever played on a see-saw knows, the play is only fun when the see-saw is balanced, for then leg power makes the see-saw go up and down. Suppose two children of different weights balance on opposite sides of a see-saw: the heavier child (of weight H) sits close (distance h) to the place where the see-saw pivots, called the *fulcrum*; and, the smaller child (of weight S) sits farther away (distance s) from the see-saw's fulcrum. The distance between any weight (for instance a child) on the see-saw and the fulcrum is termed that weight's *lever arm*. Thus, comparing the two lever arms for our children, we have $s > h$. For each child on the see-saw, the first moment about the fulcrum will be

(the child's lever arm) times (the child's weight),

where, by convention, the lever arm of the child to the left of the fulcrum is negative and the lever arm of the child to the right of the fulcrum is positive. Now, for the physical body of the whole set of the two children on the see-saw to balance, Archimedes' law of the lever specifies that the first moment about the fulcrum is zero,

$$\mathbf{M_1} = hH + sS = 0.$$

In statistics, each data value, or dot on a dotplot, can be considered a child of weight 1 (ignoring units of weight) and each lever arm will be the distance between that data value and the mean.

Example 1B. For the given infant mortality rate data values in Figures 24.2 and 24.3 and Table 24.1 (think 12 children on a see-saw), the first centroidal moment is just the sum of the lever arms,

$$\mathbf{M_1} = \sum(x - \bar{x}) = (4.1 - 9.4) + (6.4 - 9.4) + \cdots + (13.7 - 9.4) = 0.1 \approx 0.$$

We believe these calculations are most easily carried out in a table; see the first two columns of Table 24.1. [Activities 1 and 3]

Example 2B. When the data set is divided into classes, as in Example 2A, then each lever arm equals the distance between the mean and the class mark, and the frequencies correspond to the weights of the children on the see-saw.

$$\mathbf{M}_1 = \sum (x - \bar{x})f = (5 - 9.3)1 + (7 - 9.3)2 + (9 - 9.3)6 + (11 - 9.3)0 + (13 - 9.3)3 = 0.4 \approx 0.$$

These calculations can also be efficiently accomplished in a table; see the first six columns of Table 24.2. [Activities 2 and 4]

What are the rewards in statistics for calculating $\mathbf{M}_1$? First, since the mean is, arguably, the most important of all the statistics one may compute and if all the lever arms and calculations to obtain the first moment are exactly correct but $\mathbf{M}_1 \neq 0$, then there is an error in our value of the mean. Before we attempt any further statistical analysis, we had better either explain or fix this error! Second, $\mathbf{M}_1$ is just the first of the two centroidal moments that we will study. In some of the theory supporting his further statistical developments, Karl Pearson used all of the centroidal moments up to and including the *sixth*, and he even introduced some ratios of some of these higher moments (these computations are beyond our scope). We also note that K.P. employed calculus and differential equations [9, pp. 72–74]. Then he immediately applied his results to the analysis of measurements of the foreheads of certain species of crabs. Pearson and his colleague Frank Weldon intended that these researches would be steps in the grand campaign to formulate a mathematical description of Darwin's theory of evolution [2, 3, 10]. In Weldon's own words, "It cannot be too strongly urged that the problem of animal evolution is essentially a statistical problem" [12, p. 106].

24.5 The Second Moment and the Standard Deviation

What physics calls lever arms, statistics usually terms *deviations*. Thus, in Example 1B, the first centroidal moment is simply the sum of the deviations about the mean.

Example 1C. The *second centroidal moment* of the infant mortality rate data values is the sum of the *squares* of the deviations:

$$\mathbf{M}_2 = \sum (x - \bar{x})^2 = (4.1 - 9.4)^2 + (6.4 - 9.4)^2 + \cdots + (13.7 - 9.4)^2 = 95.75.$$

The third column of Table 24.1 shows an efficient way to accomplish these calculations.

In physics, the second moment is referred to as the *moment of inertia*. *Inertia* is a numerical measure of a physical body's tendency to either remain at rest or remain moving in a straight line; one can recall Isaac Newton's (1642–1727) First Law of Motion [5, 15]. For the infant mortality rate data values (considered as one physical body which is balanced at $\bar{x} \approx 9.4$ in Figure 24.3), the second moment is a measure of the *force* (we are not concerned with the units) required to rotate this body around $\bar{x}$ as an axis which is perpendicular to the plane of the data values body [Activities 7 and 8].

In statistics, in order that the standard deviation be independent of the size of the data set (i.e., the *sample size*, or n), it would be natural to divide $\mathbf{M}_2$ by n, and this is what Karl Pearson did [9, p. 78]. Unfortunately, Pearson overlooked a subtle consideration called the *degrees of freedom*, which we do not have space here to explain. For our work, a data set consisting of n real numbers has $n - 1$ degrees of freedom. The result is important in its own right:

$$s^2 = \frac{\mathbf{M}_2}{n-1} = \frac{\sum (x - \bar{x})^2}{n-1}.$$

Thanks to Sir Ronald Fisher (1890–1962), another pioneer of statistics during the first half of the 20th century, this result is called the *variance* of a set of data values [4]. Numerically, the difference in dividing the second moment by n or by $n - 1$ is not very significant, but the distinction is important in statistical theory.

The third and final step in Karl Pearson's calculation of the *standard deviation* was simply to take the square root of the variance,

$$s = \sqrt{\frac{\mathbf{M}_2}{n-1}} = \sqrt{\frac{\sum (x - \bar{x})^2}{n-1}}.$$

Pearson chose this terminology to replace several older names for the same, or close to the same calculations (see Question 6). He used the Greek lower case letter σ for the standard deviation [7, pp. 154, 156, 9, p. 75, 20]. In current practice, s denotes the standard deviation of a *sample* data set and σ is used for the standard deviation of the whole *population* of data values [17].

Principle 2. *For any sample set of data values, $x_1, x_2, \ldots, x_n$, the standard deviation is determined in the following steps:*

- *first, calculate the second centroidal moment, the moment of inertia, $\mathbf{M_2}$;*
- *second, divide $\mathbf{M_2}$ by $n - 1$ to obtain the variance s^2; and*
- *finally, the square root of the variance will be the standard deviation, s, of the given set of sample data values.*

Example 1D. For the infant mortality rates, the standard deviation is calculated in the following three steps:

$$\mathbf{M_2} = 95.75;$$

$$s^2 = \frac{95.75}{12 - 1} \doteq 8.7045455;$$

$$s = \sqrt{8.7045455} \doteq 2.9503467 \approx 3.0.$$

Physics again gives us insight into the information the standard deviation provides about not only infant mortality rates but sets of data values in general. The idea comes from trying to keep one's balance on a very narrow ledge; imagine a child walking on top of a fence, or a tightrope walker. When such a person stretches his/her arms out horizontally, which is the natural action, the effect is to *spread* the person's weight out from his/her centroid. Spreading the person's weight increases the lever arms (and also the squares of the lever arms) so that the person's moment of inertia about the narrow edge is increased and this makes it harder for his/her body to rotate around the centroid, i.e., making it harder for this person to fall off of the fence or tightrope.

Conclusion: the *standard deviation* of a set of data values is a measure of how *spread out* these data values are.

When the data is in classes, we use the same three steps as summarized above to calculate the standard deviation of the data set. However, note, referring to Example 2B, that we are actually going to compute the second moment of the *class marks*. Also see the last two columns of Table 24.2.

Example 2C. For the infant mortality rates grouped into the five classes specified above, we have:

$$\mathbf{M_2} = \sum (\hat{x} - \bar{x})^2 f$$
$$= (5 - 9.3)^2 1 + (7 - 9.3)^2 2 + (9 - 9.3)^2 6 + (11 - 9.3)^2 0 + (13 - 9.3)^2 3$$
$$= 70.68;$$

$$s^2 = \frac{\mathbf{M_2}}{n - 1} = \frac{70.68}{12 - 1} \doteq 6.4254546;$$

$$s = \sqrt{6.4254546} \doteq 2.534848 \approx 2.5.$$

Reviewing Figures 24.3 and 24.4 in the light of our numerical results, the x-axis is our data axis, and here its units are *infant mortalities per 1,000 live births*. The mean in each case is marked with the symbol ▲ (suggestive of a see-saw's fulcrum) to signify the centroid. Then, in Figure 24.3,

$$1 \text{ standard deviation } below \text{ the mean is } \bar{x} - s = 9.4 - 3.0 = 6.4$$
$$1 \text{ standard deviation } above \text{ the mean is } \bar{x} + s = 9.4 + 3.0 = 12.4;$$

and, in Figure 24.4,

$$1 \text{ standard deviation } below \text{ the mean is } \bar{x} - s = 9.3 - 2.5 = 6.8$$
$$1 \text{ standard deviation } above \text{ the mean is } \bar{x} + s = 9.3 + 2.5 = 11.8.$$

By this means, whatever the units of the data values are, we have established the scale on the z-axis, the "standard" axis for statistics whose units are always *standard deviations*.

24.6 In the Classroom

For an introductory statistics course, at the college or secondary level, the underlying idea is to imitate the unity that Karl Pearson brought to this material by merging, as closely as possible, the mean and standard deviation (see also Question 4). To work through all of the cases and details of the mean and "balance" ideas and the calculation of the second moment may require most of a regular 50 minute class period. In the next class meeting, I reinforce the physics of the first moment and then devote most of the class time to having the students experience the inertia described by the second moment and to working through the various cases to obtain the standard deviation. I end the second class meeting by explaining the 2/3 Rule and/or Chebyshev's Inequality by coloring in appropriate areas of previously prepared histograms. On tests, I give the students all of the formulas for the mean and standard deviation, but only applied problems appear on my tests because I strongly discourage simple plugging into formulas by emphasizing the ideas, processes, and applications of these computations.

24.7 Conclusions

Karl Pearson made extensive use of moments in his pioneering contributions to statistics. For a set of data values, $x_1, x_2, \ldots, x_n$, the next step would be the *third centroidal moment*, $\mathbf{M_3} = \sum (x - \bar{x})^3$. The third moment about the mean is a key step in measuring the asymmetry of the distribution of the data values about the mean, called the *skewness* of the data. There is a brief intuitive introduction to skewness in [17, pp. 67–68]. Drawing upon physics to accomplish statistical analysis and then generalizing the mathematics illustrates one aspect of Karl Pearson's scientific philosophy, "The unity of science consists alone in its method, not its materials" [2, p. 449, 11]. Karl Pearson's legacies are immense, in some instances they are controversial, and they are certainly many faceted. "Although Karl Pearson made contributions to statistical technique which now appear to be of enduring importance, those techniques are not the principal reason we should remember him ... What he did in moving the scientific world from a state of sheer disinterest in statistical studies over to a situation in which a large number of well trained persons were eagerly at work developing new theory, gathering and analyzing statistical data from every content field, computing new tables, and reexamining the foundations of statistical philosophy, is an achievement of fantastic proportions" [19, p. 22].

24.8 Activities and Questions

Question 1. Explain why we obtained two different results for $\bar{x}$ in Examples 1A and 2A.

Question 2. Referring to Table 24.2, what is the value, if any, of calculating either $\frac{\sum \hat{x}}{5}$ or $\frac{\sum \hat{x}}{12}$?

Activity 1. Make what we shall call a *statistical beam*. Suggested materials: one 3/4 inch by 1 and 1/2 inch piece of clean (no knots) wood stock; 50 inches of masking or athletic tape; 50 to 100 one inch cuphooks; and 12 to 15 heavy straight steel brackets. Suggested construction: apply the tape the full length of one side of the wood stock and mark the scale on the tape: $4.1, 4.2, 4.3, \ldots, 13.9$. On the underside of the wood stock, carefully screw in cuphooks at the data values. Finally, hang one of the heavy straight metal brackets on each cuphook. See Figure 24.5.

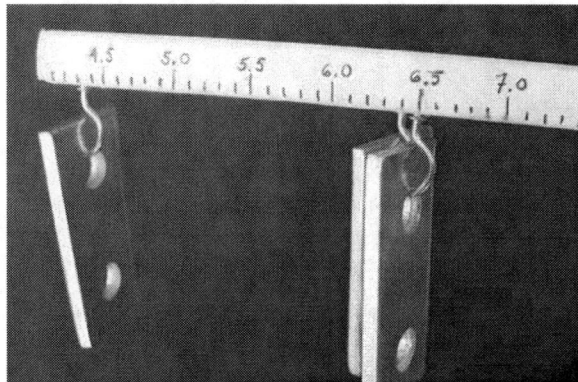

Figure 24.5. A Portion of a Statistical Beam.

Activity 2. Make a two-dimensional *histogram polygon* in the exact shape of the histogram in Figure 24.4. Suggestions: Photocopy the histogram, enlarging it as much as possible so that it will fit onto the available cardboard or plywood. Cut the photocopy along the perimeter of the bars and just below the x-axis so that you have a narrow strip of paper from 10.0 to 12.0 connecting the last bar on the right to the rest of the histogram. Attach your paper histogram polygon to the cardboard or plywood (glue stick works well) and then carefully cut the cardboard or plywood around the perimeter.

Question 3. Explain why M_1 in Example 1B is not exactly zero.

Question 4. When *we* are presented with a set of data values, our first inclination is to find the "average," i.e., we calculate the *mean*. On the other hand, if he had been given the same set of data vales in 1894 or 1895, Karl Pearson probably would have begun by calculating the first six *moments* about an axis, call it x_a, of his choosing. From these moments, Pearson then calculated the mean, the standard deviation, and other statistics descriptive of the given data set [9, 10]. This seems unnecessarily complicated as a method for obtaining the mean, but note that our calculation of the standard deviation starts with the second moment. Complicated, even backward, as a way to get the mean, this procedure does always work! To demonstrate this fact in as easy a way as we can and to capture the main idea, suppose we are given just $n = 4$ data values: x_1, x_2, x_3, x_4. Then the first moment about the arbitrary axis, x_a is,

$$\mathbf{M_{1a}} = \sum (x - x_a) = (x_1 - x_a) + (x_2 - x_a) + (x_3 - x_a) + (x_4 - x_a).$$

Show, algebraically, that if the first moment about x_a is zero, then x_a is the mean of the given set of data values.

Activity 3. Consider the statistical beam constructed in Activity 1. With the weights attached to this beam at the infant mortality rate data values, carefully balance the beam at approximately 9.4. Explain why this balance may not be exactly at 9.4.

Activity 4. Carefully balance the histogram polygon of Activity 2 at approximately 9.3. Explain why this balance point may not be exactly 9.3.

Question 5 Assume that you start this semester with a 3.0 GPA, on a 4.0 system, and you are now taking three courses: Sociology (3 hours credit); Chemistry (4 hours credit); and Statistics (3 hours credit). You have a rock-solid B in your Chemistry course, but you are most likely going to make a C in Sociology. Explain using the word "balance" what you are going to have to achieve in Statistics to retain your 3.0 GPA. Every student should be able to calculate his/her GPA!

Activity 5. Consider the following set of made-up data values: 8.3; 8.4; 8.5; 9.0; 9.1; 9.3; 9.5; 9.7; 9.7; 9.9; 10.4; and 11.1. Make a dotplot, then calculate the mean and first moment. Attach weights to the statistical beam at these data values and note where the beam balances now. (To be continued in Activity 8.)

Activity 6. One can find many exercises in introductory statistics textbooks to practice calculating the mean and first moment, and balancing the statistical beam and the histogram polygon. For instance, in [17], problems 2, 3, 5, 12, 18, 19, and 20 on pages 69–71 are all appropriate. One should replace the tape on the statistical beam to record the scale of data values in each of these problems. In general, to reuse the statistical beam, one should choose a data set whose range is slightly less than a power of 5 or a power of 10. (To be continued in Activity 9.)

Activity 7. In an open space, grasp the statistical beam containing the weights attached at the infant mortality rates of Activity 1 at its balance point, approximately 9.4, with your arm straight and holding the beam away from your body. It is intended that the beam be able to swing CW and CCW about 90 degrees in a plane parallel to the floor. Hold the beam perfectly still, pointing it straight forward, for at least five seconds. Then, rotate the beam CW and CCW, as indicated, four or five times and carefully note how much force you have to exert to accomplish these rotations. This force is directly related to the second centroidal moment of the statistical beam.

Question 6. One of the older names for what Karl Pearson called the *standard deviation* was the "root-mean-square." Explain how this older terminology describes our calculations of s.

24.8. Activities and Questions

Activity 8. Continued from Activity 5. For the data values given in Activity 5, calculate the second moment and the standard deviation. With the weights attached to the statistical beam at these data values, repeat Activity 7. Be sensitive to the force that is now required to rotate the beam in a horizontal plane, and in particular compare this force to that required in Activity 7. Explain the difference in these two forces.

Activity 9. Continued from Activity 6. For additional practice with the standard deviation, problems 2, 3, 5, 12, 18, 19, and 20 in [17, pp. 87–90] are convenient because they employ the same data sets as the corresponding exercises notes in Activity 6. As in Activities 7 and 8, students should be sensitive to the physical forces, the moments of inertia, of horizontal rotation of the statistical beams.

Activity 10. It is always valuable to be able to perform these computations and Activities using data values obtained by students from their biology labs, their political science surveys and textbooks, their nursing trials, etc.

Bibliography

[1] R.L. Finney, M.D. Weir and F.R. Giordano, *Thomas' Calculus*, 10th edition, Addison Wesley, Boston, 2001.

[2] C.C. Gillespie, editor, *Dictionary of Scientific Biography*, 16 volumes, Charles Scribners' Sons, New York, 1970–1980.

[3] J.B.S. Haldane, "Karl Pearson, 1857-1936. A Centenary Lecture Delivered at University College," in *Studies in the History of Statistics and Probability*," E.S. Pearson and M. G. Kendall, editors, Hafner Press, New York, 1970.

[4] M.G. Kendall, "Ronald Aylmer Fisher, 1890–1962," in *Studies in the History of Statistics and Probability*," E.S. Pearson and M.G. Kendall, editors, Hafner Press, New York, 1970.

[5] D.J. McGill and W.W. King, *Engineering Mechanics: Statics*, PWS Engineering, Boston, 1985.

[6] *Montgomery* (AL) *Advertiser*, September 19, 2007, unsigned editorial, p. 2B.

[7] E.S. Pearson, *Karl Pearson: An Appreciation of Some Aspects of His Life and Work*, Cambridge University Press, Cambridge, 1938.

[8] ——, and M.G. Kendall, editors, *Studies in the History of Statistics and Probability*, Hafner Press, New York, 1970.

[9] K. Pearson, "Contributions to the Mathematical Theory of Evolution," *Philosophical Transactions of the Royal Society*, 185 (1894), 71–110.

[10] ——, "Contributions to the Mathematical Theory of Evolution II," *Philosophical Transactions of the Royal Society*, 186 (1895), 343–414.

[11] ——, *The Grammar of Science*, J.M. Dent and Sons Ltd., London, 1937. (There were previous editions in 1892, 1900, and 1911.)

[12] ——, "Walter Frank Raphael Weldon, 1860–1906," in *Studies in the History of Statistics and Probability*, E.S. Pearson and M. G. Kendall, editors, Hafner Press, New York, 1970.

[13] T. Porter, *The Rise and Fall of Statistical Thinking, 1820–1900*, Princeton University Press, Princeton, NJ, 1986.

[14] ——, *Karl Pearson: The Scientific Life in a Statistical Age*, Princeton University Press, Princeton, NJ, 2004.

[15] F.W. Sears, M.W. Zemansky, and H.D. Young, *College Physics*, 6th edition, Addison-Wesley, Reading, MA, 1987.

[16] S.M. Stigler, *The History of Statistics: The Measurement of Uncertainty Before 1900*, The Belknap Press of Harvard University Press, Cambridge, MA, 1986.

[17] M.F. Triola, *Elementary Statistics*, 9th edition, Pearson/Addison Wesley, Reading, MA, 2004.

[18] G. Udney Yule and L.N.G. Filon, "Karl Pearson, 1857–1936," *Obituary Notices of Fellows of the Royal Society*, 3 (1936), 72–110.

[19] H. Walker, "The Contributions of Karl Pearson," *Journal of the American Statistical Association*, 53 (1958), 11–22.

[20] ——, *Studies in the History of Statistical Method*, Arno Press, New York, 1975. (Reprint of the 1929 edition).

[21] J. Walker, *Halliday/Resnick Fundamentals of Physics*, 8th edition extended, John Wiley and Sons, New York, 2008.

[22] www-history.mcs.st-and.ac.uk/PictDisplay/Pearson.html

25

Finding the Greatest Common Divisor

J.J. Tattersall
Providence College

25.1 Introduction

One of the more important mathematical concepts students encounter is that of the greatest common divisor (gcd), the greatest positive integer that divides two integers. It can be used to solve indeterminate equations, compare ratios, construct continued fraction expansions, and in Sturm's method to determine the number of real roots of a polynomial. For a development of these applications, see [1]. Most of the significant applications of the gcd require that it be expressed as a linear combination of the two given integers. The gcd and its associated linear equation provide an efficient way to find inverses of elements in cyclic groups, to compute continued fractions, to solve linear Diophantine equations, and to decrypt and encrypt exponential ciphers. In order to calculate the gcd and determine the required linear combination, most textbooks present the ancient but effective Euclidean approach putting an algebraic strain on many students. A more innovative technique, Saunderson's algorithm, offers a much more efficient approach. The algorithm can be introduced in number theory, modern algebra, computer science, cryptology, and other courses that require a method to find the greatest common divisor of two integers and its associated linear combination.

25.2 Historical Background

Euclid's *Elements*, written around 300 B.C., consists of a deductive chain of 465 propositions in thirteen "books" or what we would refer to as chapters. The *Elements* serves as a synthesis of Greek mathematical knowledge and has made Euclid the most successful textbook author of all time. It is one of the most important texts in intellectual history. He was a great synthesizer for it is believed that relatively few of the geometric theorems in the book were his invention. Books VII, VIII, and IX are an exception for they deal with number theory where positive integers are represented as lengths of line segments and appear to be original mathematical contributions. The division algorithm, introduced in Book VII, states that given two line segments the segment with the smaller magnitude can be marked off on the segment of the larger magnitude until the length that remains has length less than or equal to the smaller magnitude. Rewriting the statement in algebraic notation, for two positive integers a and b with $a > b$ there exist unique integers q and r with the property that $a = bq + r$ where $0 \leq r < b$. The Euclidean algorithm, the method commonly taught to determine the gcd of two integers, follows from repeated applications of the division algorithm with the gcd appearing as the last nonzero remainder in the process. For example, to determine the gcd of 12378 and

3054, we apply the Euclidean method to obtain

$$12378 = 3054 \cdot 4 + 162$$
$$3054 = 162 \cdot 18 + 138$$
$$162 = 138 \cdot 1 + 24$$
$$138 = 24 \cdot 5 + 18$$
$$24 = 18 \cdot 1 + 6$$
$$18 = 6 \cdot 3 + 0$$

Hence 6, the last non-zero remainder, is the gcd of 12378 and 3054. Using substitution and working backwards from the penultimate step through each previous step, we obtain the gcd as a linear combination of the two numbers:

$$6 = 24 - 1 \cdot 18 = 24 - 1 \cdot (138 - 24 \cdot 5)$$
$$= 6 \cdot 24 - 138 = 6 \cdot (162 - 138 \cdot 1) - 138$$
$$= 6 \cdot 162 - 7 \cdot 138 = 6 \cdot 162 - 7 \cdot (3054 - 162 \cdot 18)$$
$$= 132 \cdot 162 - 7 \cdot 3054 = 132 \cdot (12378 - 3054 \cdot 4) - 7 \cdot 3054$$
$$= 132 \cdot 12378 - 535 \cdot 3054$$

Thus, 6, the gcd of 12378 and 3054, is expressed as a linear combination of 12378 and 3054. It is straightforward to see that using the Euclidean approach in determining the gcd, in particular with two integers that require a large number of iterations of the division algorithm, will lead to a lengthy algebraic process in order to represent the gcd as a linear combination of the two given numbers. Saunderson's algorithm provides a more economical way to determine the gcd of two integers and in the process the gcd is represented as a linear combination of the two numbers. In order to see how it works, let us begin again with $a = 12378$ and $b = 3054$. Construct the following two fundamental equations:

$$1 \cdot a - 0 \cdot b = 12378$$
$$0 \cdot a - 1 \cdot b = -3054$$

Since 3054 goes into 12378 four times leaving a remainder of 162, we multiply the second equation by 4 and add it to the first equation to obtain the third equation below. Since 162 goes into 3054 eighteen times leaving a remainder of 138, we multiply the third equation by 18 and add the result to the second equation to obtain the fourth equation. Continuing this process until we reach zero, we find that 6, the last nonzero remainder, is the gcd of 12378 and 3054 and $6 = 132 \cdot 12378 - 535 \cdot 3054$.

$$
\begin{array}{rll}
1 \cdot a - 0 \cdot b &= 12378 & \\
0 \cdot a - 1 \cdot b &= -3054 & [4] \\
a - 4b &= 162 & [18] \\
18a - 73b &= -138 & [1] \\
19a - 77b &= 24 & [5] \\
113a - 458b &= -18 & [1] \\
132a - 535b &= 6 & [3] \\
509a - 2063b &= 0 &
\end{array}
$$

Note that the multipliers listed in the penultimate column are the same as the quotients that appeared in the Euclidean approach. In addition and more importantly, there is no necessity to use substitution to get the desired linear combination. To see what makes Sanderson's algorithm work apply the division algorithm to 12378 and 3054 to get $12378 = 4 \cdot 3054 + 162$. Rearranging the terms gives us step (3) above. The division algorithm applied to 3054 and 162 yields $3054 = 162 \cdot 18 + 138$. Multiplying the first equation by 18 and substituting yields $18 \cdot 12378 = 72 \cdot 3054 + (18 \cdot 162) = 72 \cdot 3054 + (3054 - 138) = 73 \cdot 3054 - 138$ or equivalently $18 \cdot 12378 - 73 \cdot 3054 = -138$

which is the equation in step (4). Continuing in this manner will eventually establish step (8) and justify the process. A formal inductive proof of Saunderson's algorithm can be found in [4, p. 73]. In some instances it may be necessary to multiply the penultimate equation by -1 in order to obtain a positive gcd. For example, if $a = 51329$ and $b = 2437$, we have:

$$
\begin{aligned}
1 \cdot a - 0 \cdot b &= 51392 \\
0 \cdot a - 1 \cdot b &= -2437 & [21] \\
a - 21b &= 152 & [16] \\
16a - 337b &= -5 & [30] \\
481a - 10131b &= 2 & [2] \\
978a - 20599b &= -1 & [2] \\
2437a - 51329b &= 0
\end{aligned}
$$

Hence, the gcd of 2437 and 51329 is unity and $(-978)51329 + (20599)2437 = 1$

25.3 More Historical Background

Nicholas Saunderson (1682–1739) overcame a great handicap to become an extremely diligent mathematician and professor at Cambridge University. He was blinded by smallpox when he was just a year old. The virus damaged his cornea and a secondary staphylococcus infection completely destroyed his eyes. As a youth, he taught himself to read by tracing out letters on gravestones with his fingers. He was taught the rudiments of arithmetic by his father, a tax collector for the Penistone district in Yorkshire, England. At age twenty five, with the encouragement of his friends and teachers, he decided to go to Cambridge University not as a student but as a teacher of mathematics. He felt that Cambridge offered him the best opportunity to pursue his favorite studies.

At Cambridge, with the approval and encouragement of the Lucasian Professor of Mathematics, William Whiston, he was permitted to give lectures and tutor mathematics. When Whiston was dismissed from the Lucasian Chair in 1711 for his Arian beliefs, Saunderson was chosen as his successor. As Lucasian Professor, he found it difficult to divide the day amongst all who applied for his instructions. He set high standards for himself and instilled a high regard for truth in his students. Prime Minister Horace Walpole recalled that when his son was at Cambridge he was tutored briefly in mathematics by Saunderson. After a few sessions, Saunderson told the lad that it would be cheating to take his money, for he could never learn what he was trying to teach him. The British statesman Lord Chesterfield considered Saunderson to be an excellent lecturer and a professor who did not have the use of his eyes, but taught others to use theirs.

In 1733, Saunderson, at the urging of his students, friends, and several of the senior fellows at Cambridge, decided to spare time from his lectures and tutoring to write up his algebra notes. The project occupied the greater part of the last six years of his life. Through the efforts of his wife, his son, and his Lucasian successor John Colson the *Elements of Algebra* containing his algorithm was published posthumously in 1740.[3] To read more about Saunderson's life and works see [4].

25.4 In The classroom

A natural way to introduce Saunderson's method would be with an example related to solving an applied problem akin to one of those mentioned in the introduction. Using a historical approach and sufficient classroom time, one could introduce the concept of a greatest common divisor by means of the Euclidean technique. At that time or later in the course, Saunderson's algorithm could be introduced. One well-constructed example will be enough to convince most students of the efficiency of Saunderson's algorithm over the Euclidean method. In upper-level courses that require applications of the gcd as the linear combination of the two given numbers it would be more expedient to introduce only Saunderson's algorithm.

25.4.1 Taking it further

As an extension of the concept of a greatest common divisor, one might introduce methods a few of the methods that have been devised to calculate an upper bound for the number of steps required to calculate the gcd of two integers using either Saunderson's or the Euclidean algorithm. In 1845, Gabriel Lamé, a civil engineer and graduate of the École Polytechnique in Paris, used properties of the Fibonacci sequence, $1, 1, 2, 3, 5, 8, 13, \ldots$, to show that the number of iterations required to determine the gcd of the two integers is less than five times the number of digits in the smaller of the two integers. In 1970, John Dixon, a mathematician at Carleton University in Ottawa, improved the bound by showing that the number of iterations is less than or equal to $(2.078)[\log(a) + 1]$, where a is the larger of the two positive integers. For our earlier Euclidean example using 12378 and 3054, Lamé's result implies that it will take less than twenty steps to determine the gcd. Dixon's result implies that it will take fewer than or equal to ten steps, which is a much more accurate measure, for in our example it took six steps. For more details on Lamé's and Dixon's results see [5, p. 72]. The life and works of other blind mathematicians or mathematician who have gone blind such as Lawrence Baggett, William Gee Bickley, Evgenii P. Dolzhenko, Leonhard Euler, Bernard Morin, Joseph Antoine Ferdinand Plateau, Lev Pontryargin, Anatoli Georgievich Vitushkin, Vladimir Ivanovich Zubov and other mathematicians with sever handicaps might also be introduced as a topic of discussion.[1]

25.5 Conclusion

Even though the Euclidean technique has been around and taught for many years, Saunderson's method provides a much more effective way to calculate the gcd of two integers and to express it as a linear combination of the two numbers. With a large number of steps the Euclidean method, found in most textbooks that deal with applications of the gcd, is laborious and rife with arithmetic and algebraic pitfalls. Most students who master Saunderson's algorithm will undoubtedly be able calculate the gcd and its associated linear combination more efficently than those who use the Euclidean algorithmic approach.

Bibliography

[1] J.-L. Chabert, *A History of Algorithms: From the Pebble to the Microchip*, Springer-Verlag, Berlin, 1998.

[2] A. Jackson, The World of Blind Mathematicians, *Notices of the AMS*, **49** (2002) 1246–1251.

[3] N. Saunderson, *The Elements of Algebra in Ten Books*, Cambridge University Press, 1740.

[4] J. J. Tattersall, Nicholas Saunderson: The Blind Lucasian, *Historia Mathematica* **19**(1992) 356–370.

[5] ——, *Number Theory in Nine Chapters*, 2ed ed., Cambridge University Press, Cambridge, 2005.

26

Two-Way Numbers and an Alternate Technique for Multiplying Two Numbers

J.J. Tattersall
Providence College

26.1 Introduction

In 1726, John Colson (1680–1759), a British mathematician and member of the Royal Society of London, devised an ingenious way to represent positive integers using what he called negativo-affirmative figures.[2] With his scheme positive and negative digits are intermingled and the basic arithmetic operations of addition, subtraction, and multiplication are as straightforward as in decimal arithmetic. The figures can be used to encrypt integers and have been rediscovered on several occasions. One version makes unnecessary the use of the digits 6, 7, 8, and 9, another rotates the digits 180°. Colson referred to his method as a "promiscuous scheme" to simplify the basic operations of arithmetic. In the process, he discovered a more compact and efficient way to multiply two numbers. This article is appropriate for an advanced elementary or secondary school mathematics class and represents a block of mathematical-historical material.

26.2 Historical Background

There are several ways to represent positive integers other than using the standard decimal system. For example, the internal operations of computers are executed using the binary system which is translated into the hexadecimal system making it easier for humans to understand it. Colson's negativo-affirmative figures offer students an introduction to ciphering and a different perspective on the basic arithmetic operations. For example, consider the negativo-affirmative expression $3\bar{5}7\bar{8}4$ which represents the positive integer 25624. To understand why this is true, replace every digit in $3\bar{5}7\bar{8}4$ with a bar over it with a zero to obtain 30704. Then replace every digit without a bar over it with a zero to obtain 05080. Subtracting the latter from the former we obtain the desired result 25624.

$$\begin{array}{r} 3\ 0\ 7\ 0\ 4 \\ -\ 0\ 5\ 0\ 8\ 0 \\ \hline 2\ 5\ 6\ 2\ 4 \end{array}$$

Colson referred to digits without a bar as affirmative and those with a bar as negative. A negativo-affirmative figure with a bar over the first digit represents a negative number. For example, $\bar{6}5\bar{3}2$ corresponds to -6472 since $0502 - 6030 = -6472$. Similarly, $\bar{2}3°$ represents $-17°$ since $\bar{2}3 = 03 - 20 = -17$.

The basic operations of addition, subtraction, and multiplication using negativo-affirmative figures are straightforward and in the case of multiplication more efficient. For example, to sum two or more negativo-affirmative figures add column-wise carrying appropriately. The sum of $7\bar{8}2\bar{3}5$ and $2\bar{3}\bar{8}4\bar{7}$ is obtained as follows:

$$\begin{array}{r}7\;\bar{8}\;2\;\bar{3}\;5\\ +\;2\;\bar{3}\;\bar{8}\;4\;\bar{7}\\ \hline 8\;\bar{1}\;\bar{6}\;1\;\bar{2}\end{array}$$

Note that $5 + \bar{7} = 5 + (-7) = -2 = \bar{2}$. In addition, $\bar{8} + \bar{3} = (-8) + (-3) = -11 = \bar{1}\bar{1}$.

Subtracting two negativo-affirmative figures is just as straightforward. First negate the subtrahend (put a bar over each digit that does not have a bar over it and remove the bars from the digits that have them). Then add the two numbers. For example, to subtract $2\bar{5}9\bar{2}6$ from $8\bar{4}256$, first convert the $2\bar{5}9\bar{2}6$ to $\bar{2}5\bar{9}2\bar{6}$, then add it to $8\bar{4}256$.

$$\begin{array}{r}8\;\bar{4}\;2\;5\;6\\ +\;\bar{2}\;5\;\bar{9}\;2\;\bar{6}\\ \hline 6\;1\;\bar{7}\;\bar{3}\;0\end{array}$$

Colson used an efficient moving multiplier technique to calculate the product of two negativo-affirmative figures. The process is a variant of the "junction of doors" method which has Hindu origins. The algorithm is less wasteful of space as the "zig-zag" method that is normally taught in elementary school. For example, to multiply $3\bar{7}6$ by $14\bar{8}$, write the digits of the multiplier in reverse order to obtain $\bar{8}41$. Place it above $3\bar{7}6$ with the $\bar{8}$ in $\bar{8}41$ directly over the 6 in $3\bar{7}6$ as shown below.

$$\begin{array}{ccc} & & \bar{8}\;4\;1\\ 3 & \bar{7} & 6 \end{array}$$

Multiply the 6 and the $\bar{8}$ to obtain $\bar{4}\bar{8}$ and write it below $3\bar{7}6$ in the following manner:

$$\begin{array}{ccc} & & \bar{8}\;4\;1\\ 3 & \bar{7} & 6\\ \hline & & \bar{8}\\ & \bar{4} & \end{array}$$

Treat the $\bar{8}41$ as if it were a sliding door and shift it over one space to the left and evaluate the inner product of $[\bar{7}, 6]$ and $[\bar{8}, 4]$ to obtain $\bar{7}\cdot\bar{8} + 6\cdot 4 = 56 + 24 = 80$ which is inserted with the 8 below the 3 in $3\bar{7}6$ and the 0 below the $\bar{7}$ in the following manner:

$$\begin{array}{ccc} & & \bar{8}\;4\;1\\ 3 & \bar{7} & 6\\ \hline & 0 & 8\\ 8 & \bar{4} & \end{array}$$

Shift the $\bar{8}41$ over one slot to the left and evaluate the inner product of $[3, \bar{7}, 6]$ and $[\bar{8}, 4, 1]$ to obtain $3 \cdot \bar{8} + \bar{7} \cdot 4 + 6 \cdot 1 = \bar{32} + \bar{28} + 6 = \bar{46}$. Continue the shift and evaluation process until the 1 in $\bar{8}41$ is over the 3 in $3\bar{7}6$, then add the third and fourth rows to determine that the product of $3\bar{7}6$ and $14\bar{8}$ is $031\bar{2}4\bar{8}$.

$$\begin{array}{rrrrrr} & & & \bar{8} & 4 & 1\\ & & & 3 & \bar{7} & 6\\ \hline & & 3 & 5 & \bar{6} & 0 & \bar{8}\\ & 0 & 0 & \bar{4} & 8 & \bar{4} &\\ \hline & 0 & 3 & 1 & \bar{2} & \bar{4} & \bar{8} \end{array}$$

Colson preferred reducing negativo-affirmative figures to what he called "small figures" where the digit 6 is replaced by $1\bar{4}$, 7 by $1\bar{3}$, 8 by $1\bar{2}$, and 9 by $1\bar{1}$. The result of this transformation is that these "small figures" contain neither $6, 7, 8, 9$ or $\bar{6}, \bar{7}, \bar{8}, \bar{9}$. For example, to encrypt $2\bar{7}3\bar{9}4$ into "small figures" make the appropriate substitutions to obtain $2(1\bar{3})3(\bar{1}\,1)4$. The leading 1 in $1\bar{3}$ is added to the digit that precedes the insertion, in this case 2 to yield $3\bar{3}$. We perform the same operation with the $\bar{1}\,1$ to obtain $(2 + 1)\bar{3}(3 + \bar{1})\,\bar{1}\,4$ or $3\bar{3}4\bar{1}4$.

26.3 In The Classroom

Negativo-affirmative arithmetic represents a different way to think and visualize the basic arithmetic operations. Students who master addition, subtraction, and multiplication with negativo-affirmative figures will sharpen their basic arithmetic skills. Constructing arithmetic problems for classroom exercises is straightforward. Cryptograms such as the following encrypted message: "Meet me tonight at $1\bar{3}$ o'clock at $2\bar{4}3\bar{7}$ Main Street." can be devised for students. Besides standard arithmetic exercises, students can determine the missing digits in problems such as:

$$
\begin{array}{rccccc}
 & \bar{3} & Y & \bar{2} & W & \bar{4} \\
+ & X & 8 & Z & \bar{1} & U \\
\hline
 & 3 & 5 & 1 & 0 & 5
\end{array}
$$

The "junction of doors" method of multiplication can be taught as an alternative and more efficient way to multiply decimal numbers. For example, to determine the product of 785 and 243, reverse the digits of 243 and place the 3 in 243 above the 5 in 785 and multiply the 5 and 3 to obtain 15,

$$
\begin{array}{ccc}
3 & 4 & 2 \\
7 & 8 & 5 \\
\hline
 & & 5 \\
\hline
 & 1 &
\end{array}
$$

Now shift the 342 over one space to the left and evaluate the inner product of [8, 5] and [3, 4] to obtain $8 \cdot 3 + 5 \cdot 4 = 44$.

$$
\begin{array}{cccc}
 & 3 & 4 & 2 \\
 & 7 & 8 & 5 \\
\hline
 & & 4 & 5 \\
 & 4 & 1 &
\end{array}
$$

Continuing the process, we deduce that the product of 785 and 234 is 190 755

$$
\begin{array}{ccccc}
3 & 4 & 2 & & \\
 & 7 & 8 & 5 & \\
\hline
 & 4 & 4 & 3 & 4 & 5 \\
1 & 4 & 6 & 4 & 1 & \\
\hline
1 & 9 & 0 & 7 & 5 & 5
\end{array}
$$

26.4 Taking It Further

Colson's negativo-affirmative figures were rediscovered in 1840 by the French mathematician Augustin-Loius Cauchy who used the associative and distributive laws to establish the equivalence of the "junction of doors" and "zig-zag" methods by varying the way in which numbers are distributed and associated.[1] For example, in multiplying 243 and 785 using the "junction of doors" method we would associate and distribute as follows:

$$
\begin{aligned}
243 \cdot 785 &= (200 + 40 + 3)(700 + 80 + 5) \\
&= (200 \cdot 700) + (200 \cdot 80 + 40 \cdot 700) + (200 \cdot 5 + 40 \cdot 80 + 3 \cdot 700) + (3 \cdot 80 + 40 \cdot 5) + (3 \cdot 5) \\
&= 140\,000 + 44\,000 + 6\,300 + 440 + 15 \\
&= 190\,755
\end{aligned}
$$

Whereas, the standard "zig-zag" method shown below

$$
\begin{array}{cccccc}
 & & & 7 & 8 & 5 \\
 & & & 2 & 4 & 3 \\
\hline
 & & 2 & 3 & 5 & 5 \\
 & 3 & 1 & 4 & 0 & \\
1 & 5 & 7 & 0 & & \\
\hline
1 & 9 & 0 & 7 & 5 & 5
\end{array}
$$

can be justified by associating and distributing in the following manner:

$$243 \cdot 785 = (200 + 40 + 3)(700 + 80 + 5)$$
$$= (200 \cdot 700 + 200 \cdot 80 + 200 \cdot 5) + (40 \cdot 700 + 40 \cdot 80 + 40 \cdot 5) + (3 \cdot 700 + 3 \cdot 80 + 3 \cdot 5)$$
$$= 157\,000 + 31\,400 + 2\,355$$
$$= 190\,755.$$

For over forty years, the Trinity College Cambridge Mathematical Club kept its account ledgers using only negativo-affirmative figures. In the 1980s, in order to disseminate the advantages of using Colson's figures, a group of Cambridge University students formed the Colson Society. For several years they published the *Colson News*, a journal that championed the use of Colson's representation of the integers. The group refined his technique replacing $\bar{1}, \bar{2}, \bar{3}, \bar{4}$ with 1,ᄅ,Ɛ,ㆆ,ϛ respectively, referring to them as 'neg', 'doub', 'trip', 'quad', and 'quin'. Since each of the five digits was doing double duty as a normal digit and as a digit rotated 180 degrees, they referred to these figures as 'two-way' numbers. Members of the Society mastered the use of 'two-way'-tables for addition and multiplication. The secretary of the Society kept records for twenty years using only 'two-way' numbers. One of the more unidentified literary scholars in the group Society composed a poem in tribute to their hero:

> There was a professor named Colson,
> Who devised a method most wholesome.
> He invented an easy new system,
> That defied conventional wisdom.
> He got rid of all digits from six to nines,
> And replaced them by neg, doub, trip, quad, and quin signs.
> So addition, subtraction become simpler at least,
> And long division no longer was quite such a beast.[5]

Most of his life Colson was employed by booksellers. In 1736, the first complete English translation of Newton's *Method of Fluxions* helped build his reputation as a scholar. Colson was brought to Cambridge through the interests of Robert Smith, Master of Trinity College, in order to lecture at the University succeeding Nicholas Saunderson the blind Lucasian Professor of Mathematics. However, during Colson's tenure in the Lucasian chair, there is no evidence that he lectured, tutored, or did any original mathematical research.

Later in life, Colson became enchanted with Maria Gaetana Agnesi's *Instituzioni analitiche ad uso della gioventu italiana*, which had been published privately in 1748. Agnesi was the oldest of twenty one children. She wrote the book for her own private amusement and as a study guide for her younger brothers. In the preface of his translation, Colson stated that the chief rationale for his work was to give "British youth the benefit of the work as well as the youth of Italy and to give the women of England an example of what the female mind can accomplish when applied in the right direction" and to "implore the women of England to forget games of whist, quadrille, back-gammon, and all other games of chance and to take up analytics." In his zeal in translating the work with a newly acquired knowledge of the Italian language, he mistranslated the word *versiera* as "witch", in reference to the description of a curve, designating it for posterity as the "Witch of Agnesi." The word, however, was in reference to the versed sine curve. Colson's translation of Agnesi's work was published posthumously in 1801.[6] For more information on Agnesi's life and work, see [4].

Negativo-affirmative arithmetic is especially apropos for students looking for a challenge. The "junction of doors" method of multiplication can be introduced to show students that there are other ways that one way to multiply two numbers and can led to the discussion of other ways. An interesting and beneficial classroom project would be presenting a number of ways to multiply two numbers. For example, the Egyptian or Russian peasant method of halfing and doubling, the Chinese stick method, or the gelosia or lattice method (with or without Napier's bones) can be introduced and their merits discussed.

26.5 Conclusion

Working with negativo-affirmative figures or "two-way" numbers and their associated operations will help students realize that there is not just one way to represent numbers and a single way to perform the basic arithmetic operations.

Even though Colson is better known for his translation errors than his translating ability, undoubtedly, he is best remembered for his efforts to promote important mathematical works and for his clever invention of negativo-affirmative figures.

Bibliography

[1] A. Cauchy, Sur les moyens d'eviter les erreurs dans les calculus numérique. *Comptes Rendus de l'Académie des Sciences, Paris* **11** (1840), 789–798.

[2] J. Colson,, A short account of negativo-affirmative arithmetic. *Philosophical Transactions of the Royal Society of London* **34** (1726), 161–173.

[3] ——, *Donna Maria Gaetana Agnesi's Analytical Institutions*, John Hellins, ed. London: Taylor and Wilks, 1801.

[4] A. Cupillari, *A Biography of Maria Gaetana Agnesi, an Eighteenth Century Woman Mathematican*, Edwin Mellen Press, Lewiston, New York, 2007.

[5] E. Hillman, ed., *The Colson News* **1**, (1983), 44.

[6] H.C. Kennedy, The witch of Agnesi exorcised. *Mathematics Teacher* **62** (1969), 480–482.

27

The Origins of Integrating Factors

Dick Jardine
Keene State College

27.1 Introduction

In a differential equations course, students learn to use integrating factors to solve first order linear differential equations, and in the process reinforce learning of key concepts from their calculus courses. This capsule offers some differential equations solved by the originators of the technique of using an integrating factor, though they did not use that expression. Solving differential equations via integrating factors is difficult for some students, particularly those who try to memorize a formula. We advocate that students learn to derive the method and solve differential equations using the product rule and the fundamental theorem of calculus, as advocated in a number of modern texts [2, 3]. Memorizing the formula would not be in the spirit of the originators of the method, Johann Bernoulli (1667–1748) and Leonhard Euler (1707–1783), nor does formula memorization lead to deep learning of fundamental mathematical processes. Understanding why integrating factors work, as offered in this historical perspective, can deepen student understanding of calculus topics such as the product rule, the fundamental theorem of calculus, and basic integration techniques. This capsule provides some historical information about the work of Bernoulli and Euler, and we offer student activities that will connect that history to enable more thorough learning of the method of integrating factors.

27.2 Historical preliminaries

Johann Bernoulli was a colleague of Gottfried Leibniz (1646–1716) and is acknowledged as one of the foundational figures in the development of the calculus. In the early 1690's he prepared lectures in the nascent calculus for Guillaume de l'Hôpital (1661–1704), who is credited with writing the first text on the calculus. As noted by V. Frederick Rickey [7], Bernoulli published his work on the integral calculus in his *Opera omnia* in 1742. A German translation was published by Gerhard Kowalewski in 1914 [1], and Figure 27.1 is a page from that work. There we see how Bernoulli used an integrating factor to solve a differential equation. In the example, the integrating factor is x/y^2. Bernoulli used the technique to solve a variety of homogeneous first order equations.

In 1720 Leonhard Euler was admitted to the University at Basel, and shortly thereafter began to study under Johann Bernoulli. Euler wrote the following about his teacher:

> ...I soon found the opportunity to become acquainted with the famous professor Johann Bernoulli, who made it a special pleasure for himself to help me along in the mathematical sciences. Private lessons, however, he categorically ruled out because of his busy schedule: However, he gave me a far more beneficial

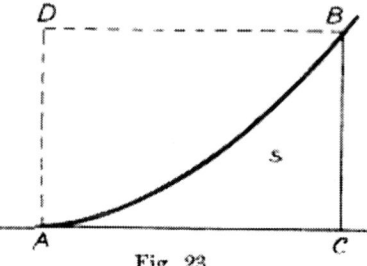

Figure 27.1. Bernoulli's use of an integrating factor

advice, which consisted in myself taking a look at some of the more difficult mathematical books and work through them with great diligence, and should I encounter some objections or difficulties, he offered me free access to him every Saturday afternoon, and he was gracious enough to comment on the collected difficulties, ...which certainly is the best method of making auspicious progress in the mathematical sciences [5].

Euler credits his mentor with inspiring him to make progress in mathematics, and Euler's advances led to his becoming the most prolific mathematician of not only his time, but arguably of all time. His mathematical innovations significantly influence the way we do mathematics today, not only because of his brilliant discoveries but also due to his useful notation and the clarity of his exposition. I believe it is important for us to help our students understand the connection between successful mathematicians and their predecessors, and this connection between Euler and Johann Bernoulli provides just such an opportunity.

Euler's innovations built on the work of his predecessors, including Johann Bernoulli. In his *Nova methodus innumerabiles aequationes differentialis secundi gradus reducendi ad aequationes differentialis primi gradus (A new method of reducing innumerable differential equations of the second degree to equations of first degree)* [8], Euler solved what we would classify as a linear first order differential equation that was the result of the reduction of a second order equation. In doing so, Euler extended the method of Johann Bernoulli to nonhomogeneous equations. In the worksheet accompanying this capsule, students will solve Bernoulli's homogeneous problems and Euler's nonhomogeneous example as well.

27.3 Mathematical preliminaries: Integrating factors

The method of employing an integrating factor for first order linear equations employs key ideas from the calculus, providing an opportunity for students to revisit important calculus skills. For the sake of brevity, we choose not to reproduce the derivation of the integrating factor here, referring the reader to a number of differential equations texts, some that emphasize the use of the fundamental theorem of calculus and the product rule [2], and others that start with the notion of an exact equation [3]. For a given equation, there are an infinite number of integrating factors, which is explained in other texts [6]. We focus on the method commonly taught in an undergraduate differential equations course, and which is seen in Euler's work [8].

Given a linear differential equation in standard form

$$\frac{dy}{dx} + p(x)\,y = q(x) \tag{27.1}$$

the integrating factor $\mu(x)$ (our notation, not Euler's) can be found by evaluating

$$\mu(x) = e^{\int p(x)dx}.$$

Many of our students have little trouble differentiating the exponential function since $\frac{d(e^u)}{dx} = e^u \frac{du}{dx}$. Once they get over the notation issues, students can differentiate the right side of the expression for $\mu(x)$ as Euler would, applying the fundamental theorem of calculus $\frac{d}{dx}\left(\int p(x)dx\right) = p(x)$ to find that $\frac{d\mu}{dx} = p(x)\mu$. Using that result and our knowledge of the product rule, we observe that,

$$\frac{d(\mu\,y)}{dx} = \mu\frac{dy}{dx} + y\frac{d\mu}{dx} = \mu\frac{dy}{dx} + y\,p(x)\mu. \tag{27.2}$$

Multiplying Equation 27.1 by the integrating factor $\mu = \mu(x)$ produces,

$$\mu\frac{dy}{dx} + \mu\,p(x)y = \mu\,q(x). \tag{27.3}$$

The reason for multiplying the left side of the original differential equation by the integrating factor, as Bernoulli and Euler would, is to put that side of the differential equation into the form that we recognize from Equation 27.2, so that we can write

$$\frac{d(\mu\,y)}{dx} = \mu\,q(x). \tag{27.4}$$

Students initially are bewildered at how anyone "observed" or "noted" such relationships. My best explanation is that Leibniz, Bernoulli, and Euler spent many hours determining those and many other useful results with the calculus. Because of their effort, they developed useful mathematical intuition about such relationships. With similar effort, our students can obtain similar intuition.

We would like students to recognize that the transition from Equation 27.3 to Equation 27.4 is nothing more than "undoing" the product rule. To ensure they have found a correct integrating factor for a specific differential equation, students should check by applying the product rule and observing that the result is the same as Equation 27.3. If it isn't, they have made an error that should be corrected before proceeding further.

The final steps in the process are to integrate both sides of Equation 27.4 and do a little algebra to obtain the general solution,

$$y(x) = \frac{c}{\mu} + \frac{1}{\mu}\int \mu\,q(x)dx.$$

The resulting equation is not to be memorized by students, but students should practice the *process* of arriving at the solution. That process involves writing the differential equation in standard form, finding the integrating factor, multiplying the equation by the integrating factor, "undoing" the product rule (and checking!), and then doing the final integration and algebra. Students should work enough exercises to become very comfortable with the process, and the activities proposed later in this capsule are designed toward that end.

27.4 Bernoulli's and Euler's use of integrating factors

Both Johann Bernoulli and Euler used integrating factors to solve differential equations. In this section we review the integrating factors they used before we move on to student activities.

In finding curves via integration, Bernoulli derived and solved first order homogeneous differential equations of various forms. For example, the differential equation Bernoulli solved in Figure 27.1 is

$$2y\,dx - x\,dy = 0.$$

He multiplied the equation by the factor $\frac{x}{y^2}$ so that he could subsequently manipulate the equation in essentially the following way:

$$\frac{x}{y^2}(2y\,dx - x\,dy) = 0$$

$$\frac{2x}{y}dx - \frac{x^2}{y^2}dy = 0$$

$$\frac{d}{dx}\left(\frac{x^2}{y}\right) = 0.$$

Integrating both sides of the last equation Bernoulli obtained the solution $x^2 = by$, which we would write as $y = cx^2$. By working through this example, which is one of the activities in the Appendix, students learn what mathematical problems motivated Bernoulli to use integrating factors. In this case, he was interested in obtaining an unknown curve which had specified characteristics that could be described using differentials. The curve turned out to be a parabola, found by solving a differential equation using an integrating factor.

As noted by Sandifer [8], Euler reduced the order of a second order differential equation to obtain the following first order equation:

$$dt + 2tz\,dt - t\,dz + tt\,dz = 0.$$

Euler [4] rewrote the equation in the form

$$dz + \frac{2z\,dt}{t-1} + \frac{dt}{tt-1} = 0,$$

which we would write in standard form as the linear first order equation

$$\frac{dz}{dt} + \frac{2}{t-1}z = \frac{1}{t^2-1}.$$

As we mentioned earlier, Euler knew that the differential of $c^{\int \frac{2}{t-1}dt} = c^{2\int \frac{dt}{t-1}}$ was $\frac{2\,dt}{t-1}c^{\int \frac{2}{t-1}dt}$, which he left in the integral form[1]. He multiplied the entire differential equation by $c^{2\int \frac{dt}{t-1}}$ to obtain

$$c^{2\int \frac{dt}{t-1}}dz + \frac{2c^{2\int \frac{dt}{t-1}}z\,dt}{t-1} + \frac{c^{2\int \frac{dt}{t-1}}dt}{tt-1} = 0$$

Euler recognized the first two terms are the result of applying the product rule to $z\,c^{2\int \frac{dt}{t-1}}$. He integrated the entire equation with result

$$c^{2\int \frac{dt}{t-1}}z + \int \frac{c^{2\int \frac{dt}{t-1}}dt}{tt-1} = a.$$

He then completed the integrations, and using l to symbolize the natural logarithm, he wrote the solution of the reduced equation as

$$(t-1)^2 z + t - lt = a.$$

---

[1] In [4], Euler used c rather than e as the base of the natural logarithm.

27.5 Student activities

The more practice students get in using the method of integrating factors, the better. In the Appendix is a worksheet that I have used in my Differential Equations and Applied Mathematics courses. Students complete the worksheet in class with a partner. This is done after they have completed the reading from the course text and attempted assigned exercises in preparation for class. Prior to the worksheet, they will have seen me derive the method and work through at least one example problem. The activity in the Appendix is offered for consideration and modification at the instructor's discretion. From my experience, students have greater difficulty with the algebraic manipulations required in this activity than they do with the calculus.

The purpose of the worksheet is to give students opportunities to practice solving differential equations using integrating factors. The differential equations they solve are the actual equations solved by Bernoulli and Euler. Additional questions are posed to give students practice solving the differential equations in alternative ways, and to encourage students to reach the conclusion that integrating factors are not unique. Most modern differential equations texts identify only one method for finding an integrating factor. Asking additional questions helps expand student understanding, to include coming to the realization that there are infinitely many integrating factors for a differential equation, not just the one obtained in the manner presented by their textbook author.

In addition to the worksheet, brief (five minute) student presentations on Euler and Bernoulli are also part of that day's classroom activities. Those students presenting will also have completed one page historical essays on the mathematicians. The essays are submitted at the beginning of class.

27.6 Summary and conclusion

Filling in gaps that may exist after the students' historical presentations, I elaborate as necessary to provide as complete a historical perspective as possible. From my course evaluations, I have learned that many students are interested in knowing where the mathematics they are studying came from, and relating how Euler extended the work of his teacher Johann Bernoulli is important for students to understand. It helps them realize that mathematics is not just discovered by geniuses but developed by real people working with one another, building on one another's knowledge. Learning to solve differential equations via the integrating factor is tough for students, but seeing the integrating factor through a historical lens can facilitate student understanding. Additionally, the student essays, oral presentations, and completed worksheets are useful artifacts for certification and for program assessment purposes, to validate that students are learning the history of the mathematics while learning the mathematics.

Bibliography

[1] Johann Bernoulli as translated by Gerhard Kowalewski in *Die erste integralrechnung eine auswahl aus Johann Bernoullis Mathematischen vorlesungen uber die methode der integrale und anderes, aufgeschrieben zum gebrauch des herrn marquis de l'Hospital in den jahren 1691 und 1692 als der verfasser sich in Paris aufhielt*, available from the The Cornell Library Historical Mathematics Monographs, http://mathbooks.library.cornell.edu:8085/Dienst/UIMATH/1.0/Display/cul.math/Bern002.

[2] Paul Blanchard, Robert L. Devaney, Glen R. Hall, *Differential Equations*, 3rd edition, Brooks/Cole, Pacific Grove, CA, 2006.

[3] William Boyce and Richard DiPrima *Elementary Differential Equations and Boundary Value Problems* 8th edition, Wiley, New York, 2005.

[4] Leonhard Euler, "Nova methodus innumerabiles aequationes differentiales secundi gradus reducendi ad aequationes differentiales primi gradus," *Comm. Acad Sci, Imp. Petropol.* 3 (1728) 1732, 124–137, E10 found at www.eulerarchive.org.

[5] E. Fellman, *Leonhard Euler*, translated by E. Gautschi and W. Gautschi, Birkhäuser, Basel, 2007, p. 5.

[6] E. L. Ince, *Ordinary Differential Equations*, Dover, New York, 1956.

[7] V. Frederick Rickey, *Integrating Factors,* unpublished manuscript, April 3, 2007.

[8] C. Edward Sandifer, *The Early Mathematics of Leonhard Euler*, Mathematical Association of America, Washington, DC, 2007.

[9] George F. Simmons and Steven G. Krantz, *Differential Equations: Theory, Technique, and Practice*, 1st Edition, McGraw-Hill, New York, 2007.

[10] George Gabriel Stokes, "On the proof of the Proposition that $(Mx + Ny)^{-1}$ is an Integrating Factor of the Homogeneous Differential Equation $M + N\frac{dy}{dx} = 0$," *Cambridge Mathematical Journal*, **IV**, p. 241, May 1845, *Mathematical and physical papers* Cambridge University Press, London, 1880 as found in the University of Michigan Historical Math Collection, historical.library.cornell.edu/math/.

Appendix: Student activities

Listed below are some activities I have used in my courses acquainting students with the early use of the integrating factor, and giving them practice in solving differential equations.

1. Leonhard Euler (1707–1783) is responsible for the way that we have come to know the calculus and the study of differential equations. One example of his work is the publication of the concept of the integrating factor in *Nova methodus innumerabiles aequationes differentials secundi gradus reducendi ad aequationes differentials primi gradus (A new method of reducing innumerable differential equations of the second degree to equations of the first degree)*. In that 1728 paper he arrived at the equation $dt + 2tz\, dt - t\, dz + tt\, dz = 0$, written in terms of the differentials dt and dz. He solved the differential equation using an integrating factor, though not exactly in the way that you will in completing this activity. The notion of an integrating factor was known earlier by the Bernoulli's, Euler's mentors and colleagues.

    (a) Algebraically manipulate the differential equation from the form that Euler wrote it to the standard form for a first order linear differential equation (DE).

    (b) Find the integrating factor using the coefficient of the linear term of the DE.

    (c) Multiply each term of the linear DE by the integrating factor, then combine the terms on the left side by recognizing the product rule.

    (d) Integrate both sides and then algebraically solve for z to obtain the solution.

2. Johann Bernoulli (1667–1748) used what we now call an integrating factor to solve the differential equation $2y\, dx - x\, dy = 0$.

    (a) Write the differential equation in the form of a first order linear DE.

    (b) Solve the DE using the method of integrating factors.

    (c) Bernoulli used the integrating factor $\frac{x}{y^2}$. Is that the same integrating factor you obtained? Does Bernoulli's integrating factor work as well as yours? Is there just one unique integrating factor for a given DE?

    (d) Solve the DE using another method that you have learned in our course.

3. Another differential equation that Bernoulli solved by his method of multiplying by a factor was

$$3y\, dx = x\, dy + y\, dx.$$

Convert the equation to the standard form for a first order linear DE and solve. Be sure to check your solution.

28

Euler's Method in Euler's Words

Dick Jardine
Keene State College

28.1 Introduction

Euler's method is a technique for finding approximate solutions to differential equations addressed in a number of undergraduate mathematics courses. Various current texts include Euler's method for calculus [4], differential equations [1], mathematical modeling [9], and numerical methods [2] students. Each of those courses are opportunities to give students an opportunity to read Euler's own brief description of the algorithm, and in the process come to understand the technique and its limitations from Euler himself. This capsule includes historical information about Euler and his development of the approximation method. Additionally, I describe Student Assignments (Appendix A) I use to connect that history to the mathematics the students are learning. The activities are designed to deepen student understanding of Euler's method specifically and reinforce learning of calculus skills in general. I also include a translation of Euler's writing on the topic (Appendix B).

28.2 Historical preliminaries

Leonhard Euler (1707–1783) was one of the most gifted of all mathematicians. Excellent biographies of Euler, some identifying the voluminous quantity of his mathematical writing, are available [6], which interested readers are encouraged to explore. One of Euler's many gifts was his ability to write mathematics clearly and understandably. The great French mathematician Pierre-Simon Laplace (1749–1827) had this to say of Euler's writing: "Read Euler, read Euler. He is the master of us all"[5]. From my experience our students find Euler readable, particularly Euler's textbooks on the calculus, with a little help from the instructor as described in the sections that follow.

While in the service of the Russian Empress Catherine the Great at St. Petersburg, Euler published a text on the integral calculus, *Institutionum calculi integralis* [8], a portion of which appears on the next page. Euler wrote at least some of this volume while he was at the Berlin Academy and employed by Frederick the Great of Prussia, prior to his return to the St. Petersburg Academy in 1766. Previously, Euler had written precursors to this three volume integral calculus text. In 1755, he published *Institutiones Calculi Differentialis*, his text on the differential calculus. His "precalculus" book, *Introductio in Analysin Infinitorum*, was published in 1748.

As the calculus had only been discovered within a hundred years of these publications by Euler, his texts were among the first on this relatively new mathematics. Euler read the works of the inventors of the calculus, Sir Isaac Newton (1643–1727) and Gottfried Leibniz (1646–1714), as well as those of their respective disciples, to include

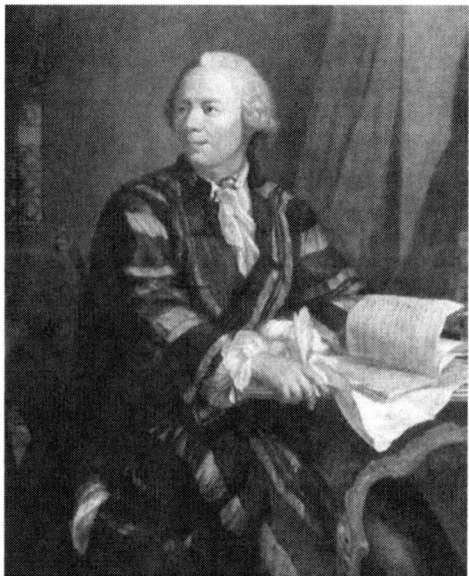

Figure 28.1. Leonhard Euler (1707-1783)

Brook Taylor (1685–1731) and Johann Bernoulli (1667–1748). Euler adopted the best of their notation, overlooked the worst, and included many of his own innovations. Euler's texts were widely read by his peers and successors, and the notation and terminology we use today in our undergraduate calculus and differential equations texts are largely due to Euler.

In the *Institutiones Calculi Differentialis*, Euler stated that "...the main concern of integral calculus is the solution of differential equations..." [7]. Largely because of his extensive coverage of solutions to differential equations, it took three volumes for Euler to address the integral calculus, while the differential calculus was covered in just one.

Our attention is on just one topic in the first of those three volumes on the integral calculus. The first section of that

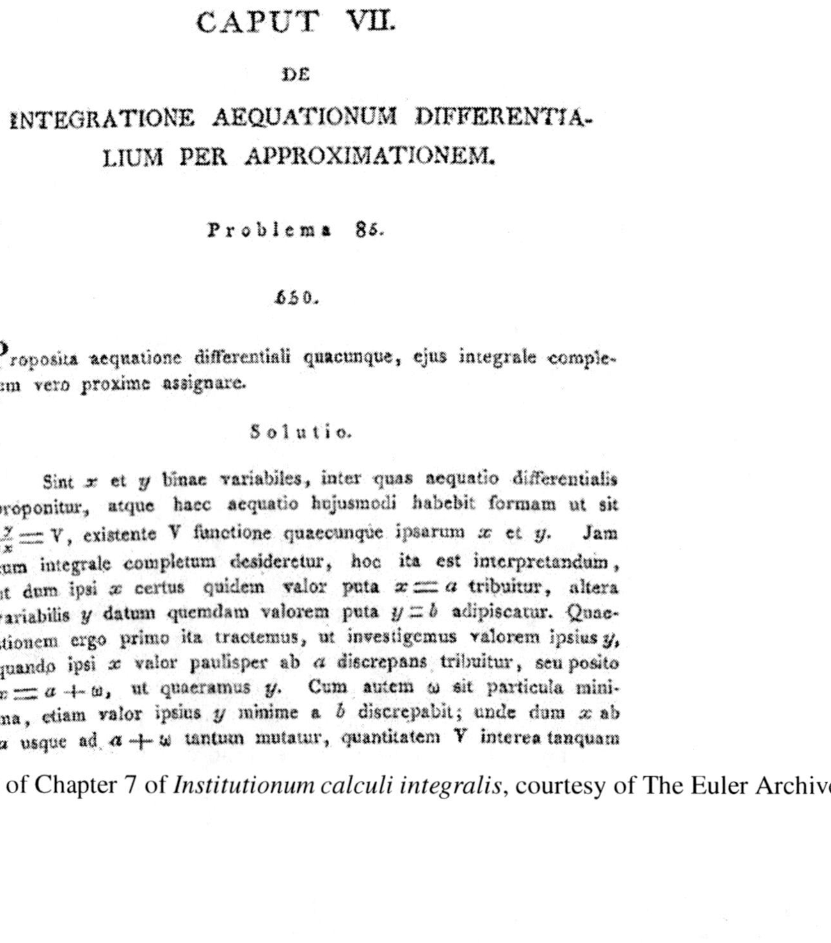

CAPUT VII.

DE

INTEGRATIONE AEQUATIONUM DIFFERENTIALIUM PER APPROXIMATIONEM.

Problema 85.

650.

Proposita aequatione differentiali quacunque, ejus integrale completum vero proxime assignare.

Solutio.

Sint x et y binae variabiles, inter quas aequatio differentialis proponitur, atque haec aequatio hujusmodi habebit formam ut sit $\frac{\partial y}{\partial x} = V$, existente V functione quaecunque ipsarum x et y. Jam cum integrale completum desideretur, hoc ita est interpretandum, ut dum ipsi x certus quidem valor puta $x = a$ tribuitur, altera variabilis y datum quemdam valorem puta $y = b$ adipiscatur. Quaestionem ergo primo ita tractemus, ut investigemus valorem ipsius y, quando ipsi x valor paulisper ab a discrepans tribuitur, seu posito $x = a + \omega$, ut quaeramus y. Cum autem ω sit particula minima, etiam valor ipsius y minime a b discrepabit; unde dum x ab a usque ad $a + \omega$ tantum mutatur, quantitatem V interea tanquam

Figure 28.2. The start of Chapter 7 of *Institutionum calculi integralis*, courtesy of The Euler Archive.

first volume is on integral formulas; the remaining two sections of the book are on the solution of differential equations. Depicted above is the beginning of the seventh chapter of the second section, translated in Appendix B. After some mathematical preliminaries, we describe these opening paragraphs of chapter 7, as they contain the algorithm that we call Euler's method. Euler obtained exact solutions for many differential equations in his calculus text, but he acknowledged that there were many differential equations for which the best he could do was obtain an approximation to the exact solution. Chapter 7 begins with Euler's description of his algorithm to approximate solutions, and continues with improvements, to include the use of power series to solve differential equations.

28.3 Euler's description of the method

Euler's method is a crude method for approximating solutions to differential equations. It is crude for reasons that Euler explained in the corollaries contained in the first couple of pages of chapter 7 of *Institutionum calculi integralis*. We discuss those later. In this section we only briefly review Euler's method, since details can be found in your course text, and others [1, 2, 9].

We start with a first order differential equation, $\frac{dy}{dt} = f(t, y)$, with initial values $y(t_0) = y_0$. The differential equation is converted to a difference equation, $y_{k+1} = y_k + \Delta t f(t_k, y_k)$, where $t_{k+1} = t_k + \Delta t$ with step size Δt. Approximate solutions are computed recursively starting with the known initial value $y_0 = y(t_0)$. Euler provides one method for deriving the recursive relationship; other methods can be found in the references [1, 2, 4]. Euler's derivation is different than we find in modern texts, and students who read and understand his approach reinforce their understanding of key ideas of the integral calculus and differential equations.

In sections 650 to 653 of chapter 7 (see Appendix B), Euler described the algorithm for obtaining an approximation to the solution of a first order differential equation. In section 650 he derived and discussed the implementation; in the remaining three sections he provided a summary and offered warnings about the error associated with the method. In my courses, I provide students with a copy of the translation, which fits on one sheet of paper printed front and back. We review Euler's derivation together interactively in class, as I describe in the Student Activities section.

The differential equation that Euler solved has the form $\frac{dy}{dx} = V(x, y)$, with initial values $x = a$ and $y = b$. His goal was to incrementally find the value of y when x changed just a little, or when $x = a + \omega$ in his notation. In our notation, we write $x_{k+1} = x_k + \Delta x$, with $x_0 = a$, and $\omega = \Delta x$. Euler made the assumption that $V(x, y)$ was constant in the small interval, or $A = V(a, b)$. He then integrated the resulting differential equation $\frac{dy}{dx} = A$ and found the value of the integration constant so that the solution satisfied the initial data $x = a$ and $y = b$. When he evaluated the resulting solution $y = b + A(x - a)$ at $x = a + \omega$, Euler obtained

$$y = b + A\omega.$$

That last equation is equivalent to the recursion formula, in our modern notation, $y_{k+1} = y_k + \Delta x f(x_k, y_k)$. Euler obtained the next x using $x = a + \omega$, or $a' = a + \omega$. He found the next y using $y = b + A\omega$, or $b' = b + A\omega$. The value b' was Euler's approximate solution to the differential equation at $x = a'$. Euler computed the next approximation by first evaluating $V(a', b')$ to obtain A', and then substituted the new values into $y = b + A\omega$ to get $b'' = b' + A'\omega$, where b'' is the numerical solution at $x = a' + \omega = a''$. Euler then repeated this process iteratively, obtaining approximate values for the solution as far from the initial values as he desired.

In the corollaries, Euler explained to his readers the caveats of using this approximation method. I have found that students readily understand Euler's descriptions, and as a result come away from the reading with a solid understanding of the limitations of this numerical method. Section 651 is the first corollary, in which Euler reiterated that successive values of x and y are obtained by repeated calculations. In corollary two, Euler pointed out that the error can be reduced by making the incremental steps small, but even with that, the error accumulates. In the third corollary Euler stated that not only does the error depend on the step-size, but also on the variability of the function $V(x, y)$ in the interval. He specified that if $V(x, y)$ varies greatly in the interval, the error of the approximation is large. In the corollaries, then, Euler articulated key ideas concerning numerical methods, which an instructor today can use to focus student learning on these important concepts.

28.4 Student Activities

Instructors can engage students in a variety of learning activities using Euler's description of the algorithm used to approximate solutions to differential equations. Some of those activities are described here, which are offered for consideration and modification at the instructor's discretion.

Since the translation (see Appendix B) is brief, assigning the translation as a student reading assignment as part of the preparation for class is a place to start. This assignment can supplement students' reading of the corresponding section of the course text, and the following questions can be provided to guide students' reading of the translation. The questions can also be used for student homework, in-class activities, or writing assignments. I have done all three, which provided my department with documented evidence for program assessment purposes of student work in the history of the mathematics that students are learning. The questions help students become "active readers" of Euler's text. As we all know, reading a mathematics text is different than reading a text in other disciplines, and the study questions help our students learn to read mathematics.

- What words would we use to describe what Euler meant by the expression "a complete integral"?
- Euler uses the notation $\frac{\partial y}{\partial x} = V$ for the differential equation. What would we use? Why might the partial derivative notation be appropriate?
- How does Euler justify transforming the differential equation $\frac{\partial y}{\partial x} = V$ to $\frac{\partial y}{\partial x} = A$?
- If we know $V(x, y)$, how do we determine the value of the constant that Euler labels A?
- How does Euler arrive at $y = b + A(x - a)$? Show the steps necessary to arrive at the solution $y = b + A(x - a)$ to the differential equation $\frac{\partial y}{\partial x} = A$ and the given initial data.
- Euler uses the Greek letter ω to represent a small quantity. What is the corresponding parameter in the version of Euler's method described in our course text?
- Which equation in the translation is Euler's version of the difference equation $y_{k+1} = y_k + \Delta t f(t_k, y_k)$?
- What is the point Euler is trying to make in Corollary 1?
- In Corollary 2, what are the two significant points about the error made in implementing this algorithm?
- What does Corollary 3 state about the relationship of the function V and the error of the algorithm?

One way for students to process the translation is to have them read the translation in class as an entire-class activity. Going around the room, I have each student read one sentence of Euler, and explain what she just read. Other members of the class can comment as they'd like on the interpretation. Then the next student reads the next sentence of Euler, and he explains what he just read, with others adding to the discussion as appropriate. The process is repeated until the reading is completed. Historians of mathematics read original sources in this manner, as exemplified by the Arithmos [11] reading group in the Northeast and Oresme [12] in the Midwest. Another approach I have used is to project the translation on a screen at the front of the room and work through the derivation on an adjacent chalkboard interactively with students.

In completing the reading and answering these questions, students will obtain a deeper understanding of Euler's method than they would by simply reading the course text or passively listening to a lecture on the topic. There are more student activities in Appendix A, and some of those activities may be used as a classroom activity or as out-of-class student projects, with students working individually or in groups as the instructor prefers.

28.5 Summary and conclusion

Reading Euler's introduction to methods for approximating the solution of differential equations can be a meaningful activity for students learning Euler's method. By learning from the "master of us all," students will gain an understanding of the origins of the method and an understanding of why this mathematical method was invented. Additionally, they will gain an appreciation for our modern notation and its origins. Most importantly, Euler clearly describes some of the important practices and cautions to be observed in implementing the method, which should deepen student understanding of the algorithm if they actively read Euler's work.

Bibliography

[1] Paul Blanchard, Robert L. Devaney, Glen R. Hall, *Differential Equations,* 3rd edition, Brooks/Cole, Pacific Grove, CA, 2006.

[2] Richard Burden and J. D. Faires, *Numerical Analysis*, 7th ed., Brooks/Cole, Pacific Grove, CA, 2001.

[3] Jean-Luc Chabert, et al., *A History of Algorithms: From the Pebble to the Microchip,* Springer, Berlin, 1999.

[4] David W. Cohen and James M. Henle, *Calculus: The Language of Change,* Jones and Bartlett, Sudbury, MA, 2005.

[5] William Dunham, *Euler: The Master of Us All*, The Mathematical Association of America, Washington, 1999, p. xiii.

[6] William Dunham, ed., *The Genius of Euler: Reflections on his Life and Work*, The Mathematical Association of America, Washington, 2007.

[7] Leonhard Euler, *Foundations of Differential Calculus*, translated by John D. Blanton, Springer, New York, 2000, p. 167.

[8] Leonhard Euler, *Institutionum Calculi integralis,* vol. I, St. Petersburg, 1768. Available from The Euler Archive (www.eulerarchive.org).

[9] Frank R. Giordano, Maurice D. Weir, William P. Fox, *A First Course in Mathematical Modeling*, Brooks/Cole, Pacific Grove, CA, 2003.

[10] Herman Goldstine, *A History of Numerical Analysis From the 16th Through the 19th Century,* Springer, New York, 1977.

[11] Arithmos, http://www.arithmos.org, accessed May 29, 2009.

[12] Oresme, http://www.nku.edu/~curtin/oresme.html. accessed May 29, 2009.

Appendix A: Student Assignments

Listed here are general descriptions of some additional activities for students, which instructors can consider adapting for their courses. In every case, depending on how the instructor would like to implement these activities, an appropriate level of detail would have to be added. For example, in the first bulleted assignment, the instructor may specify the step-size, the differential equations, and the technology students are to use in completing the assignment.

The goal of the assignments is that in completing these activities, students will have a deeper understanding of Euler's method and the associated mathematics. An additional assignment, more appropriate for a senior seminar or an academic conference, would be for students to continue the translation of subsequent sections of Chapter 7. Such a translation could be submitted for publication at The Euler Archive (www.eulerarchive.org).

1. Euler does not provide a specific example of the method in this chapter of his text. Choose an appropriate differential equation, approximate the solution as Euler describes, and create a table similar to that found in the translation but displaying actual numerical values.

2. The translator tried to retain the punctuation, capitalization and vocabulary used by Euler. Rewrite the translation using the notation and language that you find in our course text.

3. Implement Euler's method with differential equations for which you can determine the exact solution. Use your examples to demonstrate each of the points that Euler makes about error in the last two corollaries.

4. Euler derived the relationship used for iteration, $y = b + A\omega$, by solving the general form for the differential equation. Our course text derives the relationship using a graphical method. Explain each of the steps in the following alternative methods for deriving Euler's method:

    (a) Using the definition of the derivative and given the differential equation $\frac{dy}{dt} = f(t, y)$:

    $$\frac{dy}{dt} = \lim_{\Delta t \to 0} \frac{y(t + \Delta t) - y(t)}{\Delta t}$$

    $$\frac{dy}{dt} \approx \frac{y(t + \Delta t) - y(t)}{\Delta t}$$

    $$\frac{dy}{dt} = f(t, y) \approx \frac{y(t + \Delta t) - y(t)}{\Delta t}$$

    $$y(t + \Delta t) \approx y(t) + \Delta t f(t, y)$$

    $$y_{k+1} = y_k + \Delta t f(t_k, y_k)$$

    (b) Using Taylor series and given the same differential equation:

    $$y(t) = y(t_0) + y'(t_0)(t - t_0) + \frac{y''(t_0)(t - t_0)^2}{2} + \cdots$$

    $$y(t_0 + \Delta t) = y(t_0) + y'(t_0)(\Delta t) + \frac{y''(t_0)(\Delta t)^2}{2} + \cdots$$

    $$y(t_0 + \Delta t) = y(t_0) + \frac{dy}{dt}(t_0)(\Delta t) + \frac{y''(t_0)(\Delta t)^2}{2} + \cdots$$

    $$y(t_0 + \Delta t) = y(t_0) + \Delta t f(t_0, y_0) + \frac{y''(t_0)(\Delta t)^2}{2} + \cdots$$

    $$y_{k+1} = y_k + \Delta t f(t_k, y_k)$$

    (c) Explain how the derivation using Taylor series may be used to obtain an estimate for the error involved in implementing Euler's method.

Appendix B: Original source translation

This is the author's translation of the initial sections of Leonhard Euler's, *Institutionum calculi integralis*, vol. I, St. Petersburg, 1768, posted in The Euler Archive. The punctuation and notation of the original were retained in this translation, which sometimes makes the translation seem awkward.

CHAPTER VII
ON
THE INTEGRATION OF DIFFERENTIAL EQUATIONS BY APPROXIMATION

Problem 85.
650.

Whenever presented a differential equation, find its complete integral very approximately.

Solution

The pair of variables x and y appear in a differential equation, and moreover this equation has the form $\frac{\partial y}{\partial x} = V$, the function V itself a function of x and y. We desire the complete integral, which is interpreted that as long as x is assigned a certain value $x = a$, the other variable y takes on a given value $y = b$. Therefore our primary goal is to find the value of y so that when x takes on a value that differs little from a, or we assume $x = a + \omega$, then we can find y. Since ω is a very small quantity, then the value of y itself differs minimally from b; so while x varies a little from a to $a + \omega$, one may consider the quantity V as a constant. When we specify $x = a$ and $y = b$ then $V = A$, and by virtue of the small change we have $\frac{\partial y}{\partial x} = A$, for that reason when integrating $y = b + A(x - a)$, a constant being added of course, so that when $x = a$ we have $y = b$. Therefore given the initial values $x = a$ and $y = b$, we obtain the approximate next values $x = a + \omega$ and $y = b + A\omega$, so that proceeding further in a similar way over the small interval, in the end arriving at values as distant as we would like from the earlier values. These operations can be placed for ease of viewing, displayed successively in the following manner.

Variable	successive values
x	$a, a', a'', a''', a^{IV}, \ldots\ 'x, x$
y	$b, b', b'', b''', b^{IV}, \ldots\ 'y, y$
V	$A, A', A'', A''', A^{IV}, \ldots\ 'V, V$

Certainly from the given initial values $x = a$ and $y = b$, we have $V = A$, then for the second we have $b' = b + A(a' - a)$, the difference $a' - a$ as small as one pleases. From here in putting $x = a'$ and $y = b'$, we obtain $V = A'$, and from this we will obtain the third $b'' = b' + A'(a'' - a')$, when we put $x = a''$ and $y = b''$, we obtain $V = A''$. Now for the fourth, we have $b''' = b'' + A''(a''' - a'')$, from this, placing $x = a'''$ and $y = b'''$, we shall obtain $V = A'''$, thus we can progress to values as distant from the initial values as we wish. The first sequence of x values can be produced successively as desired, provided it is ascending or descending over very small intervals.

Corollary 1.

651. Therefore one at a time over very small intervals calculations are made in the same way, so the values, on which the next depend, are obtained. As values of x are done iteratively in this way one at a time, the corresponding values of y are obtained.

Corollary 2.

652. Where smaller intervals are taken, through which the values of x progress iteratively, so much the more accurate values are obtained one at a time. However the errors committed one at a time, even if they may be very small, accumulate because of the multitude.

Corollary 3.

653. Moreover errors in the calculations arise, because in the individual intervals the quantities x and y are seen to be constant, so we consider the function V as a constant. Therefore the more the value of V changes on the next interval, so much the more we are to fear larger errors.

29

Newton's Differential Equation
$$\frac{\dot{y}}{\dot{x}} = 1 - 3x + y + xx + xy$$

Hüseyin Koçak
University of Miami

"But this will appear plainer by an Example or two."
— Newton (1671)
After outlining his general method for
finding solutions of differential equations.

29.1 Introduction

In this note we redress Newton's solution to his differential equation in the title above in a contemporary setting. We resurrect Newton's algorithmic series method for developing solutions of differential equations term-by-term. We provide computer simulations of his solution and suggest further explorations.

The only requisite mathematical apparatus herein is the knowledge of integration of polynomials. Therefore, this note can be used in a calculus course or a first course on differential equations. Indeed, the author used the content of this paper while covering the method of series solutions in an elementary course in differential equations. Additional specific examples studied by the luminaries in the early history of differential equations are available in [1]. This work was supported by the National Science Foundation's Course, Curriculum, and Laboratory Improvement Program under grant DUE-0230612

29.2 Newton's differential equation

Newton's book [6], *ANALYSIS Per Quantitatum, SERIES, FLUXIONES, AC DIFFERENTIAS: cum Enumeratione Linearum TERTII ORDINIS* consists of one dozen problems. The second problem

PROB. II An Equation is being proposed, including the Fluxions of Quantities, to find the Relations of those Quantities to one another

is devoted to a general method of finding the solution of an initial-value problem for a scalar ordinary differential equation in terms of series. The equation in the title of the present paper (see also Figure 29.1) is the first significant example in the section on PROB. II.

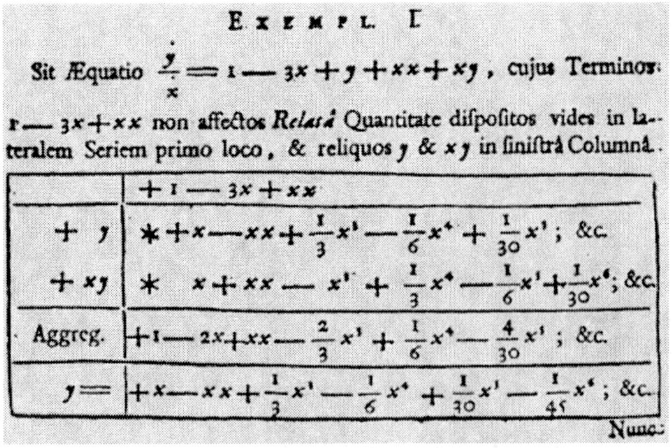
EXEMPL. I

Sit Æquatio $\frac{\dot{y}}{\dot{x}} = 1 - 3x + y + xx + xy$, cujus Terminos $1 - 3x + xx$ non affectos *Relatâ* Quantitate dispositos vides in lateralem Seriem primo loco, & reliquos y & xy in sinistrâ Columnâ.

	$+1 - 3x + xx$
$+y$	$* + x - xx + \frac{1}{3}x^3 - \frac{1}{6}x^4 + \frac{1}{30}x^5$; &c.
$+xy$	$* \quad x + xx - x^3 + \frac{1}{3}x^4 - \frac{1}{6}x^5 + \frac{1}{30}x^6$; &c.
Aggreg.	$+1 - 2x + xx - \frac{2}{3}x^3 + \frac{1}{6}x^4 - \frac{4}{30}x^5$; &c.
$y =$	$+x - xx + \frac{1}{3}x^3 - \frac{1}{6}x^4 + \frac{1}{30}x^5 - \frac{1}{45}x^6$; &c.

Nunc:

Figure 29.1. Original text of Newton's differential equation.

Quantities	x	Correlate Quantity
	y	Relate Quantity
	$\dot{x}$	Fluxion of x
	$\dot{y}$	Fluxion of y
Equation	$\frac{\dot{y}}{\dot{x}} = 1 - 3x + y + xx + xy$	

Table 29.1. Newton's quantities in his differential equation.

Newton thought of mathematical quantities as being generated by a continuous motion. He called such a flowing quantity a *fluent* (variable), and referred to its rate of change as the *fluxion* of the quantity and denoted it by a dot over the quantity. He denoted the change of *Relate Quantity* (dependent variable) with respect to the *Correlate Quantity* (independent variable) with the ratio of their fluxions (see Table 29.1).

Let us interpret Newton in our current calculus jargon. If we consider the relate quantity $y(t)$ and the correlate quantity $x(t)$ to be generated by continuous motions in time t then their fluxions $\dot{y}$ and $\dot{x}$ are

$$\dot{y} = \frac{dy}{dt}, \qquad \dot{x} = \frac{dx}{dt}$$

and the ratio of their fluxions becomes

$$\frac{\dot{y}}{\dot{x}} = \frac{dy}{dx}.$$

Thus, Newton's proposed equation, "including the Fluxions of Quantities," can be written as

$$\frac{dy}{dx} = 1 - 3x + y + x^2 + xy$$

whose solution $y(x)$ will yield "the Relations of those Quantities to one another."

29.3 Newton's solution

Newton obtained the solution of a differential equation satisfying a given initial condition in terms of series. He computed the terms of his series recursively starting with the constant term dictated by the initial condition. At each stage of his series solution, he kept only the terms up to a fixed order and he inserted the series into his differential equation. Then integrated the resulting polynomial to obtain the next order term in his series.

Now, we will paraphrase Newton's steps (see also [3]) and obtain several terms of his series solution $y(x)$ of his differential equation $\frac{dy}{dx} = 1 - 3x + y + x^2 + xy$ satisfying the initial condition $y(0) = 0$. You can read the original calculation of this series in Newton's words in the Suggested Explorations below.

To satisfy the initial condition $y(0) = 0$, the constant term of the solution must be 0; so, start with the zeroth-order term

$$y = 0 + \cdots.$$

Now, insert this for y in the differential equation to obtain

$$\frac{dy}{dx} = 1 + \cdots$$

for the lowest order term for the derivative (ignoring the higher-order terms). Next, integrate this with respect to x to obtain the first-order term in the series:

$$y = x + \cdots.$$

Inserting this form for y into the differential equation yields

$$\frac{dy}{dx} = 1 - 2x + \cdots$$

for the first-order terms, ignoring the higher-order terms. Integration of these terms gives

$$y = x - x^2 + \cdots.$$

The next iteration of this process gives

$$\frac{dy}{dx} = 1 - 2x + x^2 + \cdots \quad \text{and} \quad y = x - x^2 + \frac{1}{3}x^3 + \cdots.$$

Newton performed several more iterations and arrived at the solution

$$y = x - x^2 + \frac{1}{3}x^3 - \frac{1}{6}x^4 + \frac{1}{30}x^5 - \frac{1}{45}x^6 + \cdots.$$

It is prudent to verify that a proposed solution of a differential equation indeed satisfies the differential equation. Here is how Newton demonstrates the validity of his solution:

DEMONSTRATION

56. And thus we have solved the Problem, but the demonstration is still behind. And in so great a variety of matters, that we may not derive it synthetically, and with too great perplexity, from its genuine foundations, it may be sufficient to point it out thus in short, by way of Analysis. That is, when any Equation is propos'd, after you have finish'd the work, you may try whether from the derived Equation you can return back to the Equation propos'd ... And thus from $\dot{y} = 1 - 3x + y + xx + xy$ is derived $y = x - x^2 + \frac{1}{3}x^3 - \frac{1}{6}x^4 + \frac{1}{30}x^5 - \frac{1}{45}x^6$, &c. And thence by Prob. I. $\dot{y} = 1 - 2x + x^2 - \frac{2}{3}x^3 + \frac{1}{6}x^4 - \frac{2}{15}x^5$, &c. Which two values of $\dot{y}$ agree with each other, as appears by substituting $x - xx + \frac{1}{3}x^3 - \frac{1}{6}x^4 + \frac{1}{30}x^5$, &c. instead of y in the first Value.

29.4 Phaser simulations

A series solution of an initial-value problem, in principle, should yield better approximations to the solution as more terms of the series are included. In Figure 29.2, third through sixth-order series approximations of the solution of Newton's differential equation satisfying the initial condition $y(0) = 0$ are plotted.

A carefully computed actual solution of the differential equation satisfying the initial condition $y(0) = 0$ is plotted as the lower curve in Figure 29.3. It was indicated above that one can expect better approximations as more terms of the series are included. However, this expectation holds only locally near the initial condition, but not globally. Indeed, the fourth-order approximation appears to resemble the actual solution more than the fifth-order approximation.

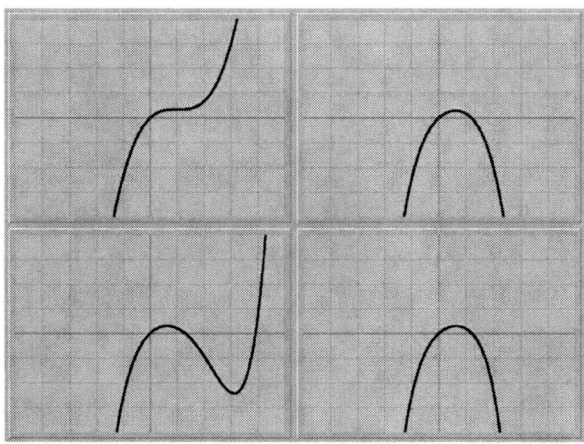

Figure 29.2. Third through sixth-order (upper-left, upper-right, lower-left, lower-right) polynomial approximations of Newton's series solution $y = x - x^2 + \frac{1}{3}x^3 - \frac{1}{6}x^4 + \frac{1}{30}x^5 - \frac{1}{45}x^6 + \cdots$ are plotted.

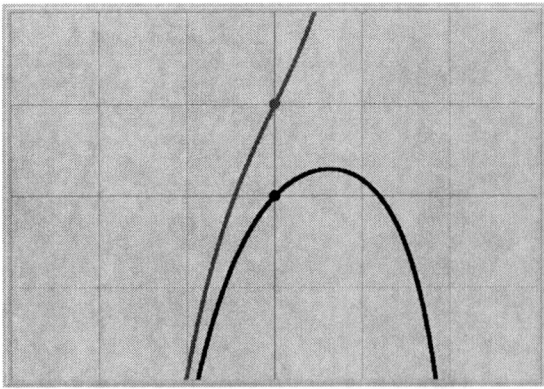

Figure 29.3. A carefully computed solution of Newton's differential equation $\frac{dy}{dx} = 1 - 3x + y + x^2 + xy$ satisfying the initial condition $y(0) = 0$ is plotted as the lower curve. The additional solution as the upper curve satisfies the initial condition $y(0) = 1$.

Newton also computed a series solution of his differential equation satisfying the initial condition $y(0) = 1$. A carefully computed graph of this solution is plotted as the upper curve in Figure 29.3. More generally, Newton computed an "infinite variety" of solutions of his differential equation satisfying the initial condition $y(0) = a$ for any real number a. More information about these solutions are contained in the Suggested Explorations below.

At www.phaser.com/modules/history/newton/index.html an interactive version of this paper is available. With simple mouse clicks on Figure 29.3 at this Phaser Web site [1], you can generate accurate solutions of Newton's differential equation satisfying any initial condition.

29.5 Remarks: Newton, Leibniz, and Euler

Newton's differential equation is a scalar *linear* differential equation. The solution of a linear differential equation of the form
$$\frac{dy}{dx} = a(x)y + b(x)$$
satisfying the initial condition $y(x_0) = y_0$ is given by the formula [4]:
$$y(x) = e^{\int_{x_0}^{x} a(u)\,du} \left[y_0 + \int_{x_0}^{x} e^{-\int_{x_0}^{x} a(u)\,du} b(s)\,ds \right].$$

Indeed, using this formula, with $a(x) = 1 + x$, $b(x) = 1 - 3x + x^2$, $x_0 = 0$, and $y_0 = 0$, one obtains the following closed-form solution of Newton's differential equation satisfying the initial condition $y(0) = 0$:
$$y(x) = 4 - x + e^{(x+1)^2/2} \left(3\sqrt{2\pi} \left[\mathrm{erf}((x+1)/\sqrt{2}) - \mathrm{erf}(1/\sqrt{2}) \right] - 4e^{-1/2} \right).$$

Notice, however, that the solution above involves the *error function*
$$\mathrm{erf}(x) = \frac{2}{\sqrt{\pi}} \int_0^x e^{-t^2/2} dt$$

which is not an elementary function. Full details of the calculations leading to this solution is available at the Phaser Web site [1].

Like Newton, Leibniz also devoted a great deal of his attention to solving differential equations. His approach, however, was quite different from that of Newton's. Leibniz sought mostly closed-form solutions in terms of known functions. In fact, he is often credited with the discovery of the method of separation of variables. This method, however, is of limited utility as told by Ince [5] in his brief history of early differential equations:

> One of the earliest discoveries in the integral calculus was that the integral of a given function could only in very special cases be finitely expressed in terms of known functions. So it is also in the theory of differential equations. That any particular equation should be integrable in a finite form is to be regarded as a happy accident; in the general case the investigator has to fall back, as in the example just quoted, upon solutions expressed in infinite series whose coefficients are determined by recurrence formulae [5].

29.6. Suggested Explorations

Indeed, Newton could "solve" any differential equation (see the Suggested Explorations below) using his series method, including the ones that Leibniz could not integrate. It is interesting to speculate whether Newton suspected that his differential equation could not be integrated in terms of elementary functions.

Newton's series method can generate approximate solutions of any desired accuracy; however, the series solution is valid only near a given initial condition. Another method of generating approximate solutions of differential equations is the method of Euler [2] which is commonly presented as the simplest algorithm in numerical analysis of differential equations. It is likely that Euler might have been trying to rectify the shortcoming of the locality of the series method by devising a new approximation method capable of generating solutions away from the initial condition. Indeed, Euler writes [2]:

> ...thus we can progress to values as distant from the initial values as we wish.

Unlike Newton, Euler does not present a specific differential equation to demonstrate the effectiveness of his method. However, he does point out a new kind of difficulty with his method in the following Corollary:

> *Corollary 2. 652. Where smaller intervals are taken, through which the values of x progress iteratively, so much the more accurate values are obtained one at a time. However the errors committed one at a time, even if they may be very small, accumulate because of the multitude.*

29.6 Suggested Explorations

Newton [6] developed detailed computational rules, and tricks, for manipulating series. He then applied these developments to many interesting examples of differential equations. In this closing section we present several additional examples studied by Newton and suggest further explorations with PHASER. If you need help with some of the lengthy calculations for these examples, consult the master.

1. Newton's original calculations for his series solution for his differential equation we have studied here are depicted in Figure 1. Here is the original text accompanying his table of calculations:

    > *Let the equation $\frac{\dot{y}}{\dot{x}} = 1 - 3x + y + x^2 + xy$ be proposed, whose terms $1 - 3x + x^2$(which are not affected by the Relate Quantity y) you see disposed collaterally in the uppermost partition; and the rest, y, and xy, in the left hand column. This done, first I multiply the initial term, 1, into the Correlate Quantity, x, and it makes x; which being divided by the number of dimensions 1, I place in the quote above written; then substituting these terms instead of y in the marginal terms y and $+xy$, I have $+x$ and $+xx$, which I write overagainst them to the right hand. After which from the rest I take the least terms $-3x$ and $+x$, whose aggregate $-2x$ multiplied into x becomes $-2xx$, this divided by the number of its dimensions 2, gives $-xx$ for the second term of the value of y in the quote. In proceeding this term being likewise assumed to complete the value of the marginals $+y$ and $+xy$, there will arise $-xx$ and $-x^3$ to be added to the terms $+x$ and $+xx$, that were before inserted: which being done, I again assume the next least terms, $+xx$, $-xx$, and $+xx$, which I collect into one sum $+xx$, and thence I derive (as before) the third term $+\frac{1}{3}x^3$ to be put into the value of y. Again, taking this term $+\frac{1}{3}x^3$ into the place of marginals; from the next least terms, $+\frac{1}{3}x^3$ and $-x^3$ added together, I obtain $+\frac{1}{6}x^4$ for the fourth term of the value of y. And so on in infinitum.*

    Try to decipher Newton's calculations and compare them with our rewording of them.

2. Newton solved his equation for the initial value $y(0) = 1$ as well. His answer, in this case, is $y = 1 + 2x + x^3 + \frac{1}{4}x^4 + \frac{1}{4}x^5 + \cdots$.

    Demonstrate the validity of Newton's solution a la Newton. This solution is plotted as the upper curve in Figure 29.3 above.

3. It is very interesting to observe that Newton calculates up to sixth-order (even) terms for the lower solution while he stops at the fifth-order (odd) terms for the upper solution. Series solutions should become more accurate with additional terms; this may be true locally but not necessarily globally. Why do you think Newton stopped at the fifth-order terms for the upper solution while continued to the sixth-order terms for the lower solution?

4. Visit `www.phaser.com/modules/history/newton/index.html`
and load Figure 29.3 into your local copy of Phaser by simply clicking on the picture. Now, click the left mouse button at several locations along the vertical axis to mark additional initial conditions. Press the Go button of Phaser to see the additional solutions.

5. Newton also computed the solution of his differential equation for the initial condition $y(0) = a$:

> *I said before, that these Solutions may be perform'd by an infinite variety of ways. This may be done if you assume at pleasure not only the initial quantity of the upper Series, but any other given quantity for the first Term of the Quote, and then you may proceed as before. ... Or if you make use of any Symbol, say a, to represent the first Term indefinitely, by the same method of Operation, (which I shall here set down,) you will find $y = a + x + ax - xx + axx + \frac{1}{3}x^3 + \frac{2}{3}ax^3$, &c. which being found, for a you may substitute 1, 2, 0, $\frac{1}{2}$, or any other Number, and thereby obtain the Relation between x and y an infinite variety of ways.*

Verify his answer.

6. Find the solution satisfying the general initial condition $y(x_0) = y_0$. Hint: Find the power series expansion in powers of $(x - x_0)$.

7. Newton also studied differential equations containing terms that are more complicated than polynomials in x and y. In this case, he first expanded the differential equation itself into series and proceeded as before. Here is such an example.

> *32. And after the same manner the Equation $\frac{\dot{y}}{\dot{x}} = 3y - 2x + \frac{x}{y} - \frac{2y}{xx}$ being proposed; if, by reason of the Terms $\frac{x}{y}$ and $\frac{2y}{xx}$, I write $1 - y$ for y, $1 - x$ for x, there will arise $\frac{\dot{y}}{\dot{x}} = 1 - 3y + 2x + \frac{1-x}{1-y} + \frac{2y-2}{1-2x+x^2}$. But the Term $\frac{1-x}{1-y}$ by infinite Division gives $1 - x + y - xy + y^2 - xy^2 + y^3 - xy^3$, &c. and the Term $\frac{2y-2}{1-2x+xx}$ by a like Division gives $2y - 2 + 4xy - 4x + 6x^2y - 6x^2 + 8x^3y - 8x^3 + 10x^4y - 10x^4$, &c. Therefore $\frac{\dot{y}}{\dot{x}} = -3x + 3xy + y^2 - xy^2 + y^3 - xy^3$, &c. $+ 6x^2y - 6x^2 + 8x^3y - 8x^3 + 10x^4y - 10x^4$, &c.*

Perform the "infinite Divisions" and verify Newton's calculations.

Bibliography

[1] *PHASER : A Universal Simulator for Dynamical Systems, Based on Java Technology*, at `www.phaser.com`, [2007].

[2] L. Euleri, "Caput VII. de integratione Aequationum Differentialium per Approximationem" in *Calculi Integralis, Liber Prior. Sectio Secunda, de Integratione Aequationum Differentialium.* Impensis Academiae Imperialis Scientiarum, Petropoli, 1824. [English translation of this work is available in the article by D. Jardine in this volume.]

[3] E. Hairer, S. Norsett, and G. Wanner, *Solving Ordinary Differential Equations*, Springer-Verlag, New York, 1987. (page 4)

[4] J. Hale and H. Koçak, *Dynamics and Bifurcations*, Springer-Verlag, New York, 1991. (page 111)

[5] E. Ince, *Ordinary Differential Equations*, Dover Publications, New York, 1944. (page 530)

[6] I. Newton, *The Method of Fluxions and Infinite Series; with its Applications to the Geometry of Curve-lines* by the Inventor Sir Isaac Newton, $K^{t\cdot}$, Late President of the Royal Society. Translated from the Author's Latin Original not yet made publick. To which is subjoin'd, A perpetual Comment upon the whole Work, Consisting of Annotations, Illustrations, and Supplements, In Order to make this Treatise A Compleat Institution for the use of Learners. By John Colson. London: Printed by Henry Woodfall; And Sold by John Nourse, at the Lamb without Temple-Bar. M.DCCXXVI.

[7] J. O'Connor and E. Robertson, *Sir Isaac Newton*, at
`www-gap.dcs.st-and.ac.uk/history/Mathematicians/Newton.html`.

30

Roots, Rocks, and Newton-Raphson Algorithms for Approximating $\sqrt{2}$ 3000 Years Apart

Clemency Montelle
University of Canterbury
New Zealand

30.1 Introduction

One of the classic examples to demonstrate the so-called Newton-Raphson method in undergraduate calculus is to apply it to the second-degree polynomial equation $x^2 - 2 = 0$ to find an approximation to the square-root of two. After several iterations the solution converges quite quickly. Indeed, $\sqrt{2}$ has fascinated mathematicians since ancient times, and one of its earliest expressions is found on a cuneiform tablet written, it is supposed, some time in the first third of the second millennium B.C.E by a trainee scribe in southern Mesopotamia. While keeping this mathematical artifact firmly in its original archaeological and mathematical context, we look at the similarities and differences it shares with modern mathematical techniques, 3000 years distant.

Observing that mathematical knowledge is, to a certain extent, culturally dependent can be revelatory to students. Modern mathematical pedagogy is generally based around a cumulative approach which allows little room for lateral breadth, as it focuses on the acquisition of skills, often with scant regard for the actual manifestation or circumstances of mathematical knowledge itself. Students may have never been exposed to other contexts in which mathematics flourished, nor encountered different mathematical traditions thus far in their studies, much less non-western ones. Yet, such exposure can give them a vital and nuanced perspective on their own mathematical circumstances. Though many a mathematical problem posed may be universal, the ways in which various mathematically literate cultures attacked them and solved them are diverse, depending on many other factors related to the broader intellectual environment. This is an important observation to bequeath to future mathematically-literate generations. However, at the same time, the universality of mathematics should not be forgotten. That humans contemplated problems of an ultimately similar nature reminds us that there is a way in which mathematical problems and their solutions are, in a sense, transcendent.

This capsule is ideally suited for undergraduate Calculus courses as part of one's coverage of the derivative and its applications, particularly before or after the introduction of the Newton-Raphson Algorithm. It will introduce students to some of the key differences and indeed similarities in solutions focused on the same problem at opposite ends of mathematical history. It is intended to take 50 minutes (one lecture) and contains exercises for students at appropriate places, which are designed to help them reflect and interact with the content.

30.2 The Problem

Imagine you needed to compute the square-root of some number on a calculator, only you were not allowed to use the square-root function. Extracting the square-root of this number $c > 0$ is equivalent to determining the positive solution of the non-linear equation:
$$x^2 - c = 0$$

30.2.1 What are our options? Taking a step back …

There exist a number of options to solve equations. We have learnt the formulae needed to solve polynomials of specific degrees directly. Indeed, the solution to polynomials of the first degree:

$$ax + b = 0 \quad \text{is determined via} \quad x = -\frac{b}{a}$$

and second degree:

$$ax^2 + bx + c = 0 \quad \text{via} \quad x = \frac{-b \pm \sqrt{b^2 - 4ac}}{2a}.$$

In fact, formulae also exist for retrieving the solutions of polynomials of the third and fourth degrees—but they are fairly complicated so that they are of little practical use, and therefore not necessary to have at one's mathematical fingertips, so to speak!

However, that is where we run out. There exist no formulae, complicated or not, for the solutions of a general polynomial of fifth degree or higher—they can not be solved by a regular analytical process. This remarkable fact was proved by the Norwegian mathematician Niels Abel in 1824.

Thus, for these cases, and for non-polynomial equations, solutions are generally retrieved by means of some sort of approximation technique. One efficient and relatively straight-forward technique would be to apply the so-called Newton-Raphson algorithm. The Newton-Raphson method (sometimes called simply Newton's method) is a method for solving non-linear equations, particularly those that have no simple closed-form solution. This method was first set out by Newton in 1669 in a work called *De analysi per aequationes numero terminorum infinitas* which was not formally published until 1711, almost fifty years later. It did appear in the public domain earlier however as it was published by John Wallis in 1685. Newton's method as articulated by him only applied to polynomials however. Joseph Raphson published an expanded version, still restricted to polynomials, but in terms of successive approximations rather than the intricate sequence of polynomials which Newton gave. It was not until 1740, when mathematician Thomas Simpson expanded this method further still to apply to general nonlinear equations.[1]

30.3 Solving $\sqrt{2}$ — Second Millennium C.E. Style

30.3.1 The Newton-Raphson Method

The Newton-Raphson method reduces the problem of root-finding to the much simpler problem of determining the root of a linear equation, in this case a line which is tangent to the function at some point close to where we expect the actual root to be. In particular, in order to find a root x^* of the equation $f(x) = 0$, one selects an initial rough guess, x_n, of the actual solution (x^*), that is, somewhere near where the function intercepts the x−axis. The function can be then approximated by the tangent line at the point $(x_n, f(x_n))$. It is fair to expect that the resulting x-intercept of this tangent, say x_{n+1}, will be a better approximation to x^*.

Again, we can repeat this process with x_{n+1}, constructing the tangent line and take the 'zero' of this new tangent line to be the basis of a further approximation. Continuing in this manner, a succession of values can be generated which will, under the right circumstances, approach x^* rapidly, improving in accuracy as the number of iterations increases.

The method then is based on the computation of tangents. For the tangent to a function passing through the point $(x_n, f(x_n))$ with slope $f'(x_n)$, an equation can be generated from the point-slope form of the line, i.e., $y = mx + b$:

$$y = f(x_n) + f'(x_n)(x - x_n).$$

---

[1] See for example, Boyer [2] p. 411f.

30.3. Solving $\sqrt{2}$ — Second Millennium C.E. Style

As long as the derivative evaluated at the point x_n, that is $f'(x_n)$, is non-zero (i.e., the line can't be parallel to the x-axis), then we can find the 'zeros', i.e., find that x for which $y = 0$:

$$x - x_n = -\frac{f(x_n)}{f'(x_n)}$$

or, if x_n is the n'th approximation, the next term, x_{n+1}, can be generated from:

$$x_{n+1} = x_n - \frac{f(x_n)}{f'(x_n)} \quad n = 1, 2, 3, \ldots$$

30.3.2 Applying this to $\sqrt{2}$

There exist several adaptations of this method for calculating particular functions, such as reciprocals and square-roots. In the latter case, such an adaption has been given a name. The so-called *Mechanic's Rule* is used to approximate square-roots and is derived from the Newton-Raphson procedure. Suppose that x_n is a close approximation to the square-root of some positive number $\sqrt{c}$, then x_{n+1} is an even closer approximation, with:

$$x_{n+1} = \frac{1}{2}\left(x_n + \frac{c}{x_n}\right).$$

Exercise Can you derive this from applying the Newton-Raphson method to the appropriate equation?

Indeed, we can use this to find the square-root of 2. We set $c = 2$ and select a suitable starting point, a guess which is moderately close to what we expect the answer to be; $x_1 = 1$ would be a suitable selection. Then:

$$x_2 = \frac{1}{2}\left(1 + \frac{2}{1}\right)$$
$$= 1.5$$

and successive approximations converge very rapidly:

$$x_3 = 1.41666667\ldots$$
$$x_4 = 1.41421568\ldots$$
$$x_5 = 1.414213562\ldots$$
$$x_6 = 1.414213562\ldots$$

so that beyond the fifth iteration, your standard pocket calculator can no longer register successive improvements. In fact,

$$\sqrt{2} = 1.4142135623731\ldots$$

Exercise What about other square-roots? By analogy, apply the mechanic's rule to find $\sqrt{3}$.

The above algorithm is in fact quadratically convergent. This means the error is essentially squared at each iteration, or in other words, the algorithm roughly doubles the number of correct digits after each iteration.

Exercise By inspection, confirm this rate of convergence in the above sequence of approximations to $\sqrt{2}$.

Another notable feature of the Newton-Raphson method is that, except for possibly the initial approximation, all iterations *overestimate* the square-root.

Exercise Let the error at any iteration be given by:

$$x_n - \sqrt{c}$$

Using the definition of quadratic convergence, that is, the square of the error at one iteration being proportional the error in the subsequent iteration, show why all iterations *overestimate* the square-root.

30.4 Time Warp: Solving $\sqrt{2}$ — Second Millennium B.C.E. Style

30.4.1 Mathematics in Mesopotamia

We have seen how to compute the square-root of any number without the square-root button on the calculator, however, we did need techniques such as those from calculus to help us. So what did mathematicians do before calculus, let alone calculators? Historical texts reveal that even the earliest mathematically literate societies were concerned with computing square-roots, and judging by the accuracy of some of their approximations, they had some pretty powerful methods to do so.

As far back as the second millennium BCE, we see that scribes in the Ancient Near East (or more specifically 'Mesopotamia', as it is commonly referred to) were interested in square-roots. They were useful for the many practical problems they needed to solve, and it may be that they were also interested in them for their own sake.[2] Fascinatingly enough, judging from the evidence that remains, these Mesopotamian scribes too used techniques comparable to those used in the Newton-Raphson method, despite being 3000 years before the development of calculus. In order to appreciate this incredible link, we must first become a little acquainted with the way in which mathematics was practiced in the Ancient Near East.

Mathematical activity from Ancient Mesopotamia can be found on clay tablets that were fashioned in particular shapes and inscribed upon with a stylus. This style of writing is commonly referred to as 'cuneiform' to reflect the wedge-shape (from the Latin *cuneus*) configurations. In many cases, tablets that were deemed important to keep were fired at high temperatures, so that they would become rock hard; a fortunate practice for the historian! Indeed, their near-indestructible state has meant that these artifacts have survived many millennia, unlike other scholarly documents written on paper and other organic surfaces which have been forever lost through natural decay.

In addition to the medium, Mesopotamian mathematics has many distinct features.[3] Perhaps the most notable of these is the way in which the number system was structured. Mathematics was carried out using sexagesimal-based arithmetic, or base-60 (think hours and minutes), so that, unlike our base-10 system which uses numerals arranged according to multiples of tens, a number written in sexagesimal notation operates in multiples of 60s.

However, embedded within their use of base-60 is a surprising orientation towards base-10. Perhaps because of the overwhelming burden of devising and working with 59 (or even 60) different symbols for each element necessitated by arithmetic in base-60, they instead adopted only two different symbols. One of these was a vertical wedge representing the unit, and the other, ten. Any sexagesimal number would be represented as a combination of (multiples) of these:[4]

Figure 30.1. The Vertical Wedge (1) and the Winkelhaken (10)

To represent numbers sexagesimally, scholars have developed conventional notation, which is now near-universal amongst assyriologists. Numbers expressed in cuneiform are transliterated as follows: a simple space is left between successive sexagesimal places (although some scholars prefer a comma), units and/or tens that are absent are indicated by a place-holding zero, and numbers are generally represented, faithful to their original articulation, without indication of their absolute value (that it is the place a number is put that gives it its value). However, as has been adopted here, a scholar may choose to indicate (their reckoning of) a number's absolute value by the strategic placement of a semi-colon, the sexagesimal equivalent of a decimal point; all numbers to the left of the semi-colon are multiples of successive powers of 60, and all numbers to the right are multiples of successive reciprocal powers of 60. Thus the number

---

[2] For evidence of their practical use see, for example, their inclusion in scribal coefficient lists in Robson [7].
[3] For an excellent and thorough overview see Robson's account in Katz [7] pp. 138–196.
[4] For a fuller description of how the system works see Neugebauer and Sachs [8] p. 2 for example. Of much use and great fun is the online sexagesimal calculator created by Benno Van Dalen.

30.4. Time Warp: Solving $\sqrt{2}$ — Second Millennium B.C.E. Style

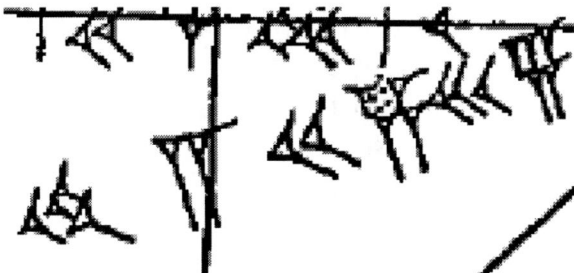

Figure 30.2. A number in cuneiform (from a close-up of Figure 30.3)

can be translated as:[5]

$$42\ \ 25\ \ 35$$

As you can immediately appreciate from the above example, the lack of additional mathematical notation entailed that mathematicians operated without the concept of "place-value". Place-value means that it is the *place* a number is put that gives it its absolute value, and 'unoccupied' places are held by a place-holding zero. The sexagesimal system had no such feature, that is, the absolute value of a number can only be determined (if at all) by context. Therefore, when one sees a simple vertical wedge, one can not know for certain whether this means "1", "60", or even "1/60", or, for that matter, any multiple of powers of sixty thereof. Thus, when modern scholars add the semicolon and commas, they are doing so after the fact—these are not present in the originals.

30.4.2 YBC 7289

The Old Babylonian mathematical corpus[6] is as yet the best documented period of Mesopotamian mathematics in scribal schools. One of the most famous mathematical documents from this period is the tablet known simply by its museum accession number YBC7289,[7] well-known because of its mathematical contents: it contains a remarkably accurate approximation to $\sqrt{2}$.

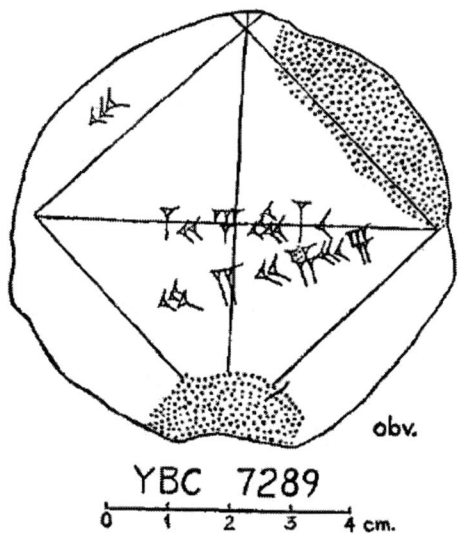

Figure 30.3. YBC 7289 from Neugebauer and Sachs

---

[5] Start in the bottom left-hand corner where you see the four 'winkelhakens'. This gives you 40. Then continue towards the right to encounter two vertical wedges. This gives you 2. Taking these together gives the first sexagesimal place '42'. Continue towards the right for the rest. You will notice that the five units that are part of the '25' are slightly imperfect—this has been represented in the transliteration by tiny dots. This indicates damage to the surface of the actual cuneiform tablet, but context allows us to reconstruct the five with certainty (compare, for example with the shape and size of the five vertical wedges in the next sexagesimal place).

[6] Old Babylonian refers to the historical period 2000–1600 B.C.E.

[7] The tablet is part of the Yale Babylonian Corpus, hence its designation.

Exercise You have already seen one of the numbers on this tablet-can you figure out the other two? Can you determine how they are related to one another?

The tablet contains three numbers which are carefully arranged on a diagram: a square with its two diagonals. The numbers are:

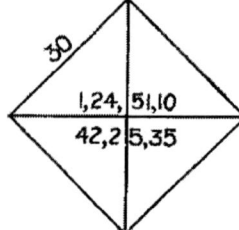

Figure 30.4. Details from YBC 7289 from Neugebauer and Sachs

To the left on the diagram there is the number 30, and roughly centered on the intersection of the two diagonals are the numbers 1 24 51 10 and 42 25 35. The relationship among these three numbers becomes clear when we examine their mathematical connection, consistent with their placement on the diagram. If we multiply 1 24 51 10 and 30 together, there results 42 25 35. Furthermore, 1 24 51 10 could mean:

$$1 + \frac{24}{60} + \frac{51}{60^2} + \frac{10}{60^3}.$$

This amount, rendered in base-10 (more recognizable to the modern eye), is 1.414212963, which is very good approximation to $\sqrt{2}$. Now the numbers must be reconciled with their position on the diagram. Suppose it is the side of a square that is 30 units, then its diagonal, namely 42;25,35, can be found by multiplying 30 with $\sqrt{2}$, or 1;24,51,10. So, 1;24,51,10 is the amount by which one can multiply any given side of a square to produce the length of its diagonal.

Further analysis will reveal that 1;24,51,10 is the best three-sexagesimal-place approximation to $\sqrt{2}$:

$$(1; 24, 51, 09)^2 = 1; 59, 59, 56, 48, 19, 21 \ldots$$
$$(1; 24, 51, 10)^2 = 1; 59, 59, 59, 38, 01, 40 \ldots$$
$$(1; 24, 51, 11)^2 = 2; 00, 00, 02, 27, 44, 01 \ldots$$

A further interesting relation between the numbers that one can uncover is:

$$\frac{\sqrt{2}}{2} = 0; 42, 25, 35 = \frac{1}{\sqrt{2}}$$

so that 42 25 35 and 1 24 51 10 are indeed reciprocal pairs (that is, their product is 1).[8]

30.4.3 Possible Mathematical Reconstructions

Let us consider now how would you compute the square-root of two, without a calculator *at all* (let alone one that has a broken square-root button!) and without calculus. In order to investigate some plausible approaches as to how this accurate number was found, not only must you devise an appropriate strategy, but you must bear in mind the additional constraints of only performing operations that are part of the repertoire of an ancient Mesopotamian mathematician.

Now, the precision with which this number was recorded (the three sexagesimal places) provide our biggest clue as to how it may have been generated. Combining this with what historians know about the circumstances and practices of Mesopotamian mathematicians, they can offer some reasonable ways (but indeed these are only hypothetical!) as to how this was computed. In fact, this task requires a great amount of background which is more than we can cover here, but for our purposes the more important points to keep in mind are as follows:

1. Other instances: There does exist at least one other less precise approximation to $\sqrt{2}$, namely 1;25.

---
[8]See the comments in Fowler and Robson [3, p. 368] for further reflections on this.

30.4. Time Warp: Solving $\sqrt{2}$ — Second Millennium B.C.E. Style

2. Geometry: We know most mathematical problems were predominantly geometrically orientated. Can we make sense of our procedure geometrically?

3. Division: Algorithms for division in base 60 are pretty tricky. To avoid division, scholars would instead multiply by the *reciprocal* of the original divisor. For 'irregular' numbers, that is those that have no exact reciprocal, approximations or alternative, less detectable, techniques may have been used.

4. Tolerance: Scholars were comfortable computing to a high degree of precision, commonly keeping track of at least several sexagesimal places.

Keeping all these factors in mind, let us consider the following geometrical reconstruction.[9] Imagine that $\sqrt{2}$ is in fact the side of a square, whose area will thus be 2. We are then looking to establish the length of the side of the square,

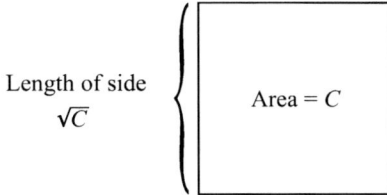

Figure 30.5. Geometrical interpretation of relationship between a number and its square-root

which will give us the square-root we were looking for. Suppose we make a guess at the length, the area of whose square will either *under-approximate* or *over-approximate* the area. Then the resulting square, in the former case, will be enclosed by the original square area 2, with an additional portion left over. We can imagine this backwards L-shaped portion left over as two equal rectangles (with long side of length equal to our initial approximation, henceforth a) and a small square (with length equal to the short side of the rectangle, henceforth b).

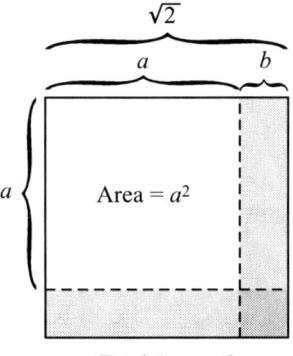

Figure 30.6. (Under-) Approximating the Area

Our intention is to find a value of a which brings the area of its square as close as possible to the square area 2. Therefore, we are looking to determine b, the short side of one of the rectangles, and add it onto a. To find b we can consider R, the area of the shaded remainder.

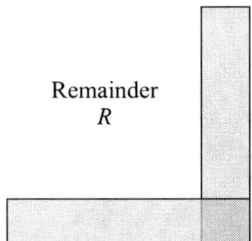

Figure 30.7. The Remainder

---
[9]This is based on the ingenious solution proposed in Fowler and Robson [3, p. 370–376].

Rather than find the area of one of the rectangles directly, we can simplify the problem and approximate its area. The area of one of these rectangles can be approximated by dividing the area of the remainder (R) in half, i.e., $\frac{1}{2}R$. We know the area of a rectangle is given by:

$$A = \text{base} \times \text{height}$$

so that using $A \approx \frac{1}{2}R$, and given we know one of the sides, a, we can conclude:

$$\frac{1}{2}R = \text{base} \times a$$

so that the base or b can be determined as follows:

$$\frac{1}{2}\frac{R}{a} = b$$

$$A \approx \frac{1}{2}R = \text{base} \times \text{height}$$
$$= b \times a$$
$$\Rightarrow b = \frac{1}{2}R \div a$$

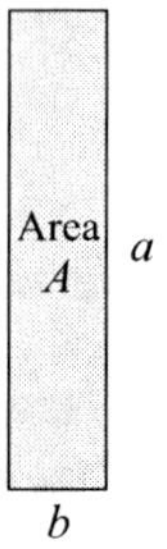

Figure 30.8. Area of the rectangle and length of the short side

The approximated base of the rectangle b can then be added (or subtracted) to our initial guess to give us a closer value, i.e., $a \pm b$. We can represent this geometrical process arithmetically as follows:

$$\sqrt{2} = \sqrt{a^2 \pm R} \approx a \pm b$$
$$= a \pm \frac{1}{2}\frac{R}{a}$$

Indeed, this is only approximate as the process disregards the area of the small square (of side length equal to the small side of the rectangle).[10] Thus we have geometrical precedence for a procedure to generate good approximations to the square-root of two.

Exercise Try this procedure sketching the process geometrically as you go. Beginning with a guess of $a = 1$. What is the length of b? What will be your next guess for a?

At this point it will be useful to examine this procedure arithmetically to see if it in fact matches the numbers that are on YBC7289. Let our guess a be such that $a < \sqrt{2}$. Suppose we begin with a_1 (our initial guess) as 1. Then:

$$\sqrt{2} = \sqrt{1^2 + 1} \approx 1 + \frac{1}{2} \times \frac{1}{1}$$
$$= 1;30$$

which is indeed a better approximation to $\sqrt{2}$. A square with side $1;30$ will over-estimate the original square with area 2. We can refine $1;30$ by repeating the procedure again (paying attention to our signs!). Our new a will be $1;30$ and thus $R = 0;15$ (extracting R from $\sqrt{2} = \sqrt{a^2 \pm R} \Rightarrow R = 1;30^2 - 2 = 0;15$):

$$\sqrt{(1;30)^2 - 0;15} \approx 1;30 - \frac{1}{2} \times \frac{0;15}{1;30}$$
$$= 1;25$$

---

[10] As Robson notes, this method is attested more explicitly elsewhere in the extant literature. See Fowler and Robson [3, p. 371].

You will recall that 1;25 is a value attested in the sources. We can further refine, by repeating the procedure yet again. A square with side 1;25 will still over-approximate our original square, but by much less:

$$\sqrt{(1;25)^2 - 0;0,25} \approx 1;25 - \frac{1}{2} \times \frac{0;0,25}{1;25}$$
$$= 1;24,51,10,35,\ldots$$

In just three steps we have reached our recognizable value (provided we use truncated rounding for which there is plenty of precedence in other texts). Thus we have a geometrically based procedure for determining square-roots, with techniques that are attested and entirely within the scope of the mesopotamian mathematical environment.

It will strike you and your students that this technique, when written up using modern notation as above (but something that is certainly *not* attested in ancient mesopotamia) is entirely equivalent to the modern day mechanic's rule for finding square-roots. Thus, if this reconstruction is right, we have a rule for determining square-roots, mathematically equivalent, yet 3000 years apart!

Exercise   Show that these two expressions are mathematically equivalent.

Exercise   What might be some of the mathematical reasons the scribes stopped where they did?

30.5 Taking it Further: Final Reflections

Our reconstruction is only a guess at how the attested value might have been determined. It is arguably the best one we have. It has to its advantage all techniques attested to in other mathematical contexts, it embodies an inherently geometrically approach, and the numbers are all in conformity with what actually remains. However, it does have its deficiencies. The main problem surrounds the issue of the divisions carried out in the algorithm—more precisely—the existence of a reciprocal for one of the numbers in the process.

Our description passed over a vital computational detail, which seems entirely trivial to us, but was an issue for the scribes. It is one that is easy to miss when one follows through the reasoning, but glaringly obvious when one undergoes the actual computation. The first 'iteration' requires multiplication (by one half), addition, and the division (by one). The second iteration requires multiplication and addition as above, but this time division by 1;30. This division would have been carried out by the scribes by multiplying by the reciprocal of 1;30. 1;30 is a 'regular' number in base-60[11] and its reciprocal is 40. (See Table 30.1.) However, the third iteration requires us to divide by 1;25, a seemingly innocuous number, but indeed as you can quickly verify, one that has no direct reciprocal.[12]

We are faced with a difficulty. It is hard to imagine how the Mesopotamian mathematicians carried out this division. 1;25 has no tabulated reciprocal and their algorithms for computing non-tabulated reciprocals also fail.[13] We could suppose that there existed some approximation for its reciprocal, although this is nowhere attested.

One possible solution to this problem could be the following. We observe that:

$$1\ 25 = 5 \times 17.$$

5 is a regular number whose reciprocal was known and 17 an irregular number whose reciprocal was approximated.[14]

$$\overline{5} = 12 \quad \overline{17} = 3\ 31\ 42 \approx 3\ 32.$$

---

[11] Finite reciprocals exist for numbers in base-60 if and only if its only prime divisors are 2, 3, and 5.
[12] For more on reciprocals see Melville [5] and Sachs [8].
[13] See Meville [5] tables 2 and 3.
[14] See Sachs [8, p. 152–153] and his discussion of M10 from the John F. Lewis Collection of Cuneiform Tablets in The Free Library of Philadelphia. The attested value is obviously in error. The tablet gives 35 17 which when multiplied by 17 gives 9 59 49, far from the required 1!. There are several other such errors on this tablet which may simply be indicative of poor execution. However, that it was computed is the main point.

x	$\bar{x}$	x	$\bar{x}$
2	30	27	2,13,20
3	20	30	2
4	15	32	1,52,30
5	12	36	1,40
6	10	40	1,30
8	7,30	45	1,20
9	6,40	48	1,15
10	6	50	1,12
12	5	54	1,6,40
15	4	1	1
16	3,45	1,4	56,15
18	3,20	1,12	50
20	3	1,15	48
24	2,30	1,20	45
25	2,24	1,21	44,26,40

Table 30.1. Reciprocals of Regular Numbers[15]

Now, let us reconsider the last iteration again, making very clear the computations we need to make to determine the result:

$$\sqrt{(1;25)^2 - 0;0,25} \approx 1;25 - \frac{1}{2} \times \frac{0;0,25}{1;25}$$

$$= 1;25 - \frac{1}{2} \frac{0;0,25}{0;05 \times 17}$$

$$= 1;25 - \frac{1}{2}(0;0,25)(12)(0;3,32)$$

$$= 1;25 - \frac{1}{2}(0;0,17,40)$$

$$= 1;25 - 0;0,8,50$$

$$= 1;24,51,10$$

1 24 51 10 is found exactly, without need to resort to truncated rounding. Such a conjecture is tempting and determines the value with attention to the nitty-gritty of sexagesimal computation procedures. However, it is only a conjecture. Others have compelling reasons to support alternative hypotheses.[16]

We wrap up our investigation into square-roots with a compelling artifactual detail, which gives us a flash of a glimpse into the individuals responsible for these computations. YBC7289 is not only famous because of its approximation to one of our favorite irrationals, but also because contained on the actual cuneiform tablet remains the fingerprint of its executor, impressed before the clay was dried, and preserved in the firing process!

30.6 Conclusion

Episodes from the history of mathematics remind us that how mathematically literate societies articulate, attack, and solve mathematical problems is dependent on a variety of factors—mathematics isn't culturally neutral. In this regard, the examination of two approaches to approximating $\sqrt{2}$, 3000 years apart, can reveal some fascinating features about mathematical practice. In order to penetrate these early cuneiform tablets we must also consider the ways in which the mathematics was presented and recorded, the number system it used, a seeming 'geometrical' orientation, the

---

[15] See Melville [5, p. 2].

[16] See Høyrup [4, p. 263] and his reconstruction based on side-and-diagonal numbers. See also an early account by Neugebauer and Sachs [6, p. 43].

30.6. Conclusion

Figure 30.9. Photo reprint of YBC7289 with fingerprint

handling of arithmetical operations, and the like. However, while observing these striking differences, we notice that the resulting reconstructed procedures are similar in essence to those developed some 3000 years later.

Some preliminary ground work may be necessary before covering this capsule. It would be ideal to undertake it after having established the necessary details of the Newton-Raphson method and the students are comfortable with the iterative procedure. Having introduced one of its applications, namely of computing square roots (mechanic's method), use the method to compute the square-root of two. Now, present the cuneiform tablet to the class, noting the various relevant details to your students as discussed above. Introduce them to the system of numeration, and the idea of sexagesimal arithmetic and get them to "read" off the numbers themselves. Emphasize the idea that even though we are missing any explanation from these ancient mathematicians as to how they obtained these numbers, the precision of the numbers on the tablet can give us some clue as to how they were generated. Then follow through the geometrical interpretation with them and show how it is equivalent to the Newton-Raphson method, while also emphasizing the obvious differences.

Bibliography

[1] Howard Anton, Irl Bivens, and Stephen Davis, *Calculus* 8th ed., John Wiley and Sons, 2005.

[2] Carl B. Boyer, *A History of Mathematics*, 2nd Ed., John Wiley and Sons, 1991, 391–414.

[3] David Fowler and Eleanor Robson, "Square Root Approximations in Old Babylonian Mathematics: YBC 7289 in Context," *Historia Mathematica* 25 (1998) 366–378.

[4] Jens Høyrup, *Lengths, Widths, Surfaces: A Portrait of Old Babylonian Algebra and Its Kin*, Sources and Studies in the History of Mathematics and Physical Sciences, Springer-Verlag, New York, Inc., 2002.

[5] Duncan J. Melville, "Reciprocals and Reciprocal Algorithms in Mesopotamian Mathematics" (electronically circulated), September 6 2005.

[6] O. Neugebauer, and A. Sachs *Mathematical Cuneiform Texts*, American Oriental Society, New Haven, Connecticut, vol. 29, 1986.

[7] Eleanor Robson, "Mesopotamian Mathematics", from Victor J. Katz, (ed.), *The Mathematic of Egypt, Mesopotamia, China, India, and Islam A Sourcebook*, pp. 138–196, Princeton University Press, Princeton, New Jersey, 2007.

[8] A. Sachs, "Babylonian mathematical texts II: approximations of reciprocals of irregular numbers in an Old Babylonian text. III: the problem of finding the cube root of a number" *Journal of Cuneiform Studies* 6, 151–156 (1952).

31

Plimpton 322: The Pythagorean Theorem, More than a Thousand Years before Pythagoras

Daniel E. Otero
Xavier University

31.1 Introduction

An amazingly sophisticated example of some of the oldest written mathematics known to humanity is the clay tablet Plimpton 322 (Figure 31.1), so called because it is item number 322 in a collection assembled by G. A. Plimpton in the 1930s and now housed at Columbia University in New York City. The tablet dates to the 19th century BCE, and can be traced to the Old Babylonian civilization that flourished in Mesopotamia, the fertile valley of the Tigris and Euphrates rivers (present-day Iraq). This exotic artifact is an ideal touchstone that can be used to spark interest in the study of representations of number and of arithmetical computational algorithms, say, by future computer scientists or prospective school teachers. It can also serve to deepen an understanding of the solution of quadratic equations by students of algebra at all levels.

31.2 Historical Background

Evidence of mathematical thinking is at least as old as the species *homo sapiens*. Older, in fact, provided we agree to classify certain animal behaviors, like the ability to differentiate quantities, to count, or even to employ geometric design in the building of shelters, as evidence of mathematical activity.

Once humans moved from hunting and foraging to farming and later to forming cities, new challenges of life required new forms of thought, including those that looked much more like what we would today identify as mathematics. Certainly, the first literate civilizations known to us also provided written evidence of their mathematics as well. (See [4] for a glimpse at various forms of mathematics across a wide array of ancient cultures.) The first urban civilizations had sprung up in the Mesopotamian River Valley some 1000 years prior to the authorship of Plimpton 322 in the wake of improvements in agriculture, trading, organization of labor, and political structures, developments in human societies that emerged over centuries.

The unknown author of the tablet, hereafter called the Scribe, lived within a century or so of two very famous Babylonians. King Hammurapi (1792?-1750? BCE) wrote a code of law well-known for its "eye for an eye, tooth for a tooth" directives in deciding personal injury disputes among his subjects [7, p. 27]. The Hebrew patriarch Abraham of Ur, another rough contemporary of the Scribe, led his clan from the bank of the Euphrates west into Canaan where

Figure 31.1. Plimpton 322. (Image used with permission from the Plimpton Collection, Rare Book and Manuscript Library, Columbia University.)

generations later they formed the kingdom of Israel.

Perhaps the most important development to come out of the old civilizations of Mesopotamia was cuneiform script (literally, "wedge-shaped," from the Latin *cuneus*, for *wedge*), an improvement over earlier, more pictographic styles of writing. This innovation led to more efficient means of recording ideas and information, and allowed thinkers to communicate their thoughts more easily. Plimpton 322 is an example of the use of cuneiform writing. Let us examine what the tablet tells us. (See [2] for a wonderfully engaging account of how one might approach this text. This paper offers an attempt to reproduce some of the adventure of that account in school classrooms.)

31.3 Reading the Tablet

Cuneiform writing is clearly recognizable on this image (Figure 31.1) of Plimpton 322. Although the tablet is damaged in spots (especially in the upper left corner where much of the writing is lost, and on the right side where a good-sized chip has come off), we can still identify a table of five columns, organized in fifteen rows, with additional script along the top edge in symbols of a different style than those in the body of the tablet.

A study of the main body of text, underneath the column headings, reveals that all the writing here is built up from just two kinds of symbol, a thin vertical wedge (𒁹), and a wide horizontal wedge (𒌋).

The fourth of the columns contains a repeated pattern of symbols, but the rightmost fifth column can be seen to contain the "words" 𒁹, 𒈫, 𒐍, 𒐎, in order from the top down. The next two rows are lost, but in row seven is the combination 𒐐, followed by the groupings 𒐏 and 𒐘. Curiously, the tenth row contains a single horizontal wedge 𒌋, under which we find the groupings 𒌋𒁹, 𒌋𒈫, 𒌋𒐍, with the rest of the column lost.

Clearly, what we are seeing is Babylonian numeration: the symbols for the numbers 1, 2, 3, etc. That is, the fifth column in this table is enumerating the rows of the table. Thin vertical wedges are used to represent units of one (clustered vertically in groups), and fat horizontal wedges denote units of ten. So the numeration appears to be decimal in form. Indeed, the entries in the fourth column are simply repetitions of the Akkadian word for *number*, so that together, the fourth and fifth columns are giving line numbers for the table.

31.4. Sexagesimal numeration

Having noted this, we observe that the first three columns contain lists of numbers as well. For instance, consider column 3: in the first row, we make out the sequence 𒐖𒐏𒐎. This denotes the pair of numbers 2 and 49. The next entry in this column contains three vertical wedges, then one horizontal wedge, two vertical wedges, a short gap, and one vertical wedge; we interpret this as the group of numbers 3, 12, 1. The next entry represents the sequence 1, 50, 49.

In this way, we can now translate the entire tablet, save some strange text at the very top for which we must consult Akkadian linguists to obtain a translation. What we obtain is presented in Table 31.1.

	width	diagonal	name
, 15	1, 59	2, 49	1
, 58, 14, 50, 6, 15	56, 7	3, 12, 1	2
1, 15, 33, 45	1, 16, 41	1, 50, 49	3
, 29, 32, 52, 16	3, 31, 49	5, 9, 1	4
48, 54, 1, 40	1, 5	1, 37	
47, 6, 41, 40	5, 19	8, 1	
43, 11, 56, 28, 26, 40	38, 11	59, 1	7
41, 33, 59, 3, 45	13, 19	20, 49	8
38, 33, 36, 36	9, 1	12, 49	9
35, 10, 2, 28, 27, 24, 26, 40	1, 22, 41	2, 16, 1	10
33, 45	45	1, 15	11
29, 21, 54, 2, 15	27, 59	48, 49	12
27, 3, 45	7, 12, 1	4, 49	13
25, 48, 51, 35, 6, 40	29, 31	53, 49	14
23, 13, 46, 40	56	53	

Table 31.1. Plimpton 322 transliterated

Notice the blanks where there is missing text from the damaged portions of the tablet. However, the missing entries in the final column are no mystery; they can be easily reconstructed. What seems much harder to do is rebuild the missing numbers at the top of the first column. Surprisingly, we will be able to accomplish even this task.

Still, some interesting observations can already be made even after only a cursory study of the tablet:

- the sequences of numbers in the first column are much longer on average than those in the other columns;
- the first column is ordered by the size of the first number in its sequence;
- there are curious gaps between numbers in the sequences in the first column (see rows 5, 6, 12 and 13, in which numbers less than 10 are preceded by wide spaces);
- and we never see numbers higher than 59, that is, we never see groups of more than five horizontal wedges at a time.

This last observation is critical to understanding how to interpret the tablet, for it indicates that Babylonian numeration is not a decimal numeration at all, but rather a *sexagesimal* numeration, namely one based on units of 60.

31.4 Sexagesimal numeration

In our familiar decimal numeration, numbers are represented by means of just ten symbols, the digits 0, 1, 2, 3, 4, 5, 6, 7, 8, 9. It is their relative position in a numeral that determines its value. For instance, we distinguish between 247 and 724 because

$$247 = 2 \cdot 10^2 + 4 \cdot 10^1 + 7 \cdot 10^0 = 200 + 40 + 7$$
$$724 = 7 \cdot 10^2 + 2 \cdot 10^1 + 4 \cdot 10^0 = 700 + 20 + 4$$

This is why we call this numeration decimal *positional* notation: the position of the digits represent successive powers of 10, with the 10^0, or ones, place at the rightmost position.

We can also extend this to encompass non-integer numbers by placing the decimal point after the ones digit and setting additional digits to the right of the point to represent negative powers of 10. So

$$2.47 = 2 \cdot 10^0 + 4 \cdot 10^{-1} + 7 \cdot 10^{-2} = 2 + (4/10) + (7/100)$$

Sexagesimal numeration behaves similarly. Consider the sequences of numbers at the top of the second and third columns on Plimpton 322. These sequences, 1, 59 and 2, 49, are really two-digit sexagesimal numbers. Hereafter, we will adopt a notation that expresses sexagesimal digits with the standard numerals $0, 1, \ldots, 59$ and separates consecutive digits with the pipe symbol '|':

$$1|59 = 1 \cdot 60^1 + 59 \cdot 60^0 = 60 + 59 = 119$$
$$2|49 = 2 \cdot 60^1 + 49 \cdot 60^0 = 120 + 49 = 169$$

These computations illustrate how to convert numbers from sexagesimal into decimal notation. As another example, consider the numbers in row 10 of the second and third columns:

$$1|22|41 = 1 \cdot 60^2 + 22 \cdot 60^1 + 41 \cdot 60^0 = 3600 + 1320 + 41 = 4961$$
$$2|16|1 = 2 \cdot 60^2 + 16 \cdot 60^1 + 1 \cdot 60^0 = 7200 + 960 + 1 = 8161$$

By the way, conversion from decimal to sexagesimal is equally straightforward. We reverse the last computation above to find the sexagesimal equivalent of 8161 by dividing 8161 by 60, marking the quotient and remainder: $8161 \div 60 = 136$ rem 1. So the remainder is our ones (the rightmost) digit. The quotient, 136, allows us to determine the remaining digits by repeating this process. Since $136 \div 60 = 2$ rem 16, the sixties digit is 16, and the 60^2-digit is 2. Therefore, $8161 = 2|16|1$.

This still leaves the question of how we are to understand the numbers of the first column. For if we treat them in the same way as the others, we are forced to deal with some *stupendously* large numbers. The number in row 10 of the first column has 8 sexagesimal digits, so its decimal equivalent would be an integer on the order of 100,000,000,000,000! Clearly, something else is afoot here.

Recall that the numbers in the first column seem to be ordered according to a pattern, namely, by the size of their *leftmost* digit. This suggests that they are not sexagesimal integers, but instead *fractions*, and that a "sexagesimal point", or more properly what is called a **radix point**, should precede each sequence of digits. Consequently, all the numbers in this first column of the table are *less* than 1, decreasing in size from the top down, in the same way that the decimal numbers 0.836, 0.776, 0.603, 0.495 are seen to be ordered from highest to lowest by virtue of the decrease of their first significant digits. Now, the Scribe had no special way to signify a radix point, as there was at this point no symbol for zero. Surprisingly, the use of zero is a much more modern invention, dating to around 500 CE in Asia. The Scribe's numeration was a type of "floating point" representation scheme, in which the relative size values of successive digits in a number was to be inferred *from context*. But we will use double pipes '||' to notate the radix point, so that the entries of the first column in rows 5, 10, and 13 should be interpreted as the following numbers:

$$0||48|54|1|40 = 48 \cdot 60^{-1} + 54 \cdot 60^{-2} + 1 \cdot 60^{-3} + 40 \cdot 60^{-4}$$
$$= 0.8150\ldots$$
$$0||35|10|2|28|27|24|26|40 = 35 \cdot 60^{-1} + 10 \cdot 60^{-2} + 2 \cdot 60^{-3} + 28 \cdot 60^{-4}$$
$$+ 27 \cdot 60^{-5} + 24 \cdot 60^{-6} + 26 \cdot 60^{-7} + 40 \cdot 60^{-8}$$
$$= 0.5861\ldots$$
$$0||27|0|3|13 = 27 \cdot 60^{-1} + 0 \cdot 60^{-2} + 3 \cdot 60^{-3} + 13 \cdot 60^{-4}$$

The decimal equivalents produced here are only 4-place approximations. However, we can give them exact values as fractions (in lowest terms), as for instance,

$$0||48|54|1|40 = 48 \cdot 60^{-1} + 54 \cdot 60^{-2} + 1 \cdot 60^{-3} + 40 \cdot 60^{-4}$$
$$= 60^{-4}(48 \cdot 60^3 + 54 \cdot 60^2 + 1 \cdot 60^1 + 40 \cdot 60^0)$$
$$= \frac{10562500}{12960000} = \frac{4225}{5184}$$

Notice that we have inserted a 0 as the second digit after the radix point in the number 0||27|0|3|13 from row 13. How did we know to do this, especially since no symbol for zero appears in the tablet at this location? The presence of the 0 is inferred on account of the interpretation of what the numbers in the tablet must signify; this interpretation is the subject of the next section below. Our Scribe, and other Babylonians, would know to read a 0 for that digit only by context! We find some tablets from later centuries that will use a separator symbol for a zero, but only to represent zeros that appear *between* other nonzero digits [1, p. 9].

31.5 So What Does It All Mean?

It was originally thought that Plimpton 322 was one of dozens of tablets that simply recorded inventories of foodstuffs and other merchandise. But in 1946, historian of mathematics Otto Neugebauer and A. J. Sachs pointed out that the Babylonian Scribe was producing some rather sophisticated mathematics in this tablet [5, p. 36ff]: the numbers in the table were the results of purposeful computations and not simply inventory records. A clue to support this claim lies in the text at the top of the tablet. These column headings include the words *width*, *diagonal*, and *name*, respectively. That the fifth column is enumerating the rows of the table has already been ascertained, but the other labels indicate that the numbers represent dimensions of a right triangle as in the figure below.

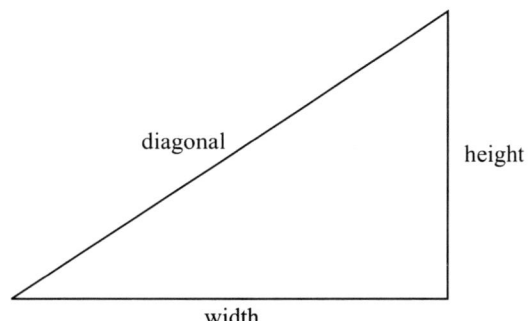

Figure 31.2. Dimensions of a right triangle

They confirmed this conjecture by checking that the numbers in columns 2 and 3, when assigned as lengths of a leg (width) and hypotenuse (diagonal) of a right triangle, gave rise, by means of the Pythagorean theorem, $a^2 + b^2 = c^2$ (a = height, b = width, c = diagonal) to *integer* values for the length of the missing side a.[1] This is significant, for if one chooses two whole numbers at random, assigning the larger to be the hypotenuse and the smaller to be a leg of a right triangle, the chances are extremely slim that the Pythagorean relationship will produce an integral value for the third side. When a triple of whole numbers does happen to satisfy this relation, we call it a **Pythagorean triple**.

Once this Pythagorean pattern is revealed, it doesn't take much extra work to deduce that the entries in the first column correspond to values of the quantity $\frac{b^2}{a^2}$. For instance, in row 12 of the table, the width of the triangle is $b = 27|59 = 1679$ and its diagonal is $c = 48|49 = 2929$, so by the Pythagorean theorem,

$$a = \sqrt{c^2 - b^2} = 2400 = 40|0,$$

whence

---

[1]There were some exceptions, namely in rows 2, 9, 13, and 15, but these can be explained away as simple computational errors made by the Scribe. For instance, the number 9|1 in row 9 should instead be 8|1, an easy error to make when marking strokes on a tablet, and the number 7|12|1 in row 13 is actually the *square* of the number 2|41 that should have been entered there. Indeed, some speculate that this tablet was a scrap copy, recognized as containing arithmetical errors and discarded in a garbage heap, only to be discovered centuries later in an archaeological dig!

$$\frac{b^2}{a^2} = \frac{1679^2}{2400^2} = \frac{2819041}{5760000}.$$

Long division in sexagesimal (a challenging computation for the student!) produces the quotient 0||29|21|54|2|15, exactly the value we see at the start of row 12. Armed with this knowledge, we can check the Scribe's calculations and, more significantly, we can restore the text from the damaged parts of the tablet. (Decimal equivalents for the entries in columns 2 and 3 are provided in the table in boldface.)

	width	diagonal	name
0||59|00|15	**119** = 01|59	**169** = 02|49	1
0||56|56|58|14|50|06|15	**3367** = 56|07	**4825** = 01|20|25	2
0||55|07|41|15|33|45	**4601** = 01|16|41	**6649** = 01|50|49	3
0||53|10|29|32|52|16	**12701** = 03|31|49	**18541** = 05|09|01	4
0||48|54|01|40	**65** = 01|05	**97** = 01|37	5
0||47|06|41|40	**319** = 05|19	**481** = 08|01	6
0||43|11|56|28|26|40	**2291** = 38|11	**3541** = 59|01	7
0||41|33|45|14|03|45	**799** = 13|19	**1249** = 20|49	8
0||38|33|36|36	**481** = 08|01	**769** = 12|49	9
0||35|10|02|28|27|24|26|40	**4961** = 01|22|41	**8161** = 02|16|01	10
0||33|45	**45** = 45	**75** = 01|15	11
0||29|21|54|02|15	**1679** = 27|59	**2929** = 48|49	12
0||27|00|03|45	**161** = 02|41	**289** = 04|49	13
0||25|48|51|35|06|40	**1771** = 29|31	**3229** = 53|49	14
0||23|13|46|40	**28** = 28	**53** = 53	15

Table 31.2. Plimpton 322 corrected and translated

This confirms that the Scribe must have had some understanding that connected the values of the numbers in columns 2 and 3 of the tablet with measures of sides of a right triangle, and, as some moderately large values are included there, there must have been some systematic procedure for obtaining them; it is hardly likely that they were arrived at by trial and error. Generating such values is not a trivial sort of computational problem, indicating that the Scribe possessed a good deal of sophistication in mathematical technique. Plimpton 322, therefore, provides convincing evidence that the Scribe was part of a mathematical culture that understood the relationship we identify as the Pythagorean Theorem, both as a geometric and an arithmetic relationship, more than 1000 years before Pythagoras was born!

The most thorough analysis of how the tablet entries were produced was suggested by Jöran Friberg [3] and promoted by Eleanor Robson [6]. Robson argues that a natural setting in the context of Babylonian mathematics for generating this table of numbers proceeds from the tradition of *igi-igibi* (number-reciprocal) problems in which the Scribe operated. Such problems have the form: given that a number x and its reciprocal x' have a specified sum $(x + x' = s)$ or difference $(x - x' = d)$, find x. The *igi-igibi* problem is an important special case of a more general problem popular in Babylonian mathematics: given two numbers x and y with given product $xy = p$ and given sum $x + y = s$ (or given difference $x - y = d$), find x and y.

We readily find an algebraic solution to the latter problem once we have the very useful quadratic identity

$$xy + \left(\frac{x-y}{2}\right)^2 = \left(\frac{x+y}{2}\right)^2.$$

As two of the terms in this equation are given data in the problem, the identity provides a means for determining the third. Thus, if $x + y = s$ is given, we obtain $d = x - y = \sqrt{s^2 - 4p}$, and if $x - y = d$ is given, then $s = x + y = \sqrt{4p + d^2}$. In either case, both the sum and difference of the two unknowns are obtained, whence x is half the sum of these ($x = \frac{s+d}{2}$) and y is half their difference ($y = \frac{s-d}{2}$). But since symbolic algebra was unknown to Babylonian culture, the solution just presented could not have been the one employed. How would a Babylonian have solved this problem?

31.5. So What Does It All Mean?

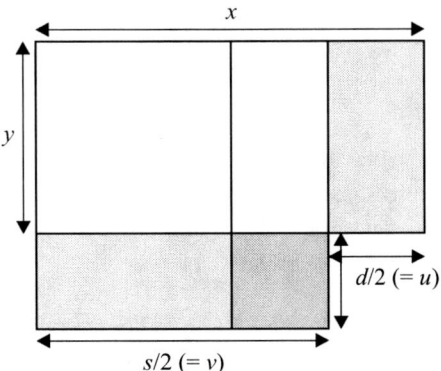

Figure 31.3. Completing the square

Most likely, it was performed by "completing the square," in much the same way as Greek mathematicians did centuries later (as in Euclid's *Elements* ii.5). The rectangle with sides x and y has area p. Mark off the smaller distance y along the longer x, and halve the difference to obtain $d/2$. Cutting this amount off from x leaves half the sum $s/2 = x - d/2$. If we are given d, we can now tear off from one end of the rectangle a portion that is $d/2$ wide and move it underneath the rectangle after giving it a quarter turn (Figure 31.3). We create from this a *gnomon* whose two arms are y long. The gnomon can be completed to a square by adding on the area of the missing corner, a square with side $d/2$. The side of the resulting "completed" square is $s/2$ (that is, $(s/2)^2 = p + (d/2)^2$). From the now known lengths $d/2$ and $s/2$, we easily find $x = \frac{s}{2} + \frac{d}{2}$ and $y = \frac{s}{2} - \frac{d}{2}$. If, on the other hand, we are given s, then marking off $s/2$ from x identifies the width of the arm of the gnomon; tearing off the piece of the rectangle of area p that sticks out from the gnomon and moving it underneath replaces the rectangle p within the gnomon having the same area, since $x - s/2 = y + d/2$. The empty corner of the enclosing square is also a square with side $d/2$ (that is, $(d/2)^2 = p - (s/2)^2$). As before, from the known lengths $d/2$ and $s/2$, we then find x and y.

From a computational perspective, the special case $p = 1$, in which x and $y = x'$ are reciprocals, is significant, for Babylonian computers (human ones, of course) would have had access to extensive prepared tables of reciprocals of numbers as standard arithmetical aids. From such tables of reciprocal pairs x, x', then, our Scribe would be able to determine, by means of this method of completing the square, the values of the halves of their difference $u = \frac{x-x'}{2}$ and their sum $v = \frac{x+x'}{2}$, whence the relation $1 + u^2 = v^2$. Indeed, we uncover the connection with the entries of Plimpton 322 by consulting the following table of reciprocal pairs (as in Table 6 of [6, p. 186]):

x	$x' = 1/x$	$u = (x - x')/2$	$v = (x + x')/2$	$1 + u^2 = v^2$
2||24	0||25	0||59|30	1||24|30	1||59|00|15
2||22|13|20	0||25|18|45	0||58|27|17|30	1||23|46|02|30	1||56|56|14|50|06|15
2||20|37|30	0||25|36	0||57|30|45	1||23|06|45	1||55|07|41|15|33|45
2||18|53|20	0||25|55|12	0||56|29|04	1||22|24|16	1||53|10|29|32|52|16
2||15	0||26|40	0||54|10	1||20|50	1||48|54|01|40
2||13|20	0||27	0||53|10	1||20|10	1||47|06|41|40
2||09|36	0||27|46|40	0||50|54|40	1||18|41|20	1||43|11|56|28|26|40
2||08	0||28|07|30	0||49|56|15	1||18|03|45	1||41|33|45|14|03|45
2||05	0||28|48	0||48|06	1||16|54	1||38|33|36|36
2||01|30	0||29|37|46|40	0||45|56|06|40	1||15|33|53|20	1||35|10|02|28|27|24|26|40
2	0||30	0||45	1||15	1||33|45
1||55|12	0||31|15	0||41|58|30	1||13|13|20	1||29|21|54|02|15
1||52|30	0||32	0||40|15	1||12|15	1||27|00|03|45
1||51|06|40	0||32|24	0||39|21|20	1||11|45|20	1||25|48|51|35|06|40
1||48	0||33|20	0||33|20	1||10|40	1||23|13|46|40

Table 31.3. Reciprocal pairs for Plimpton 322

It should be clear from comparing Table 31.3 with either Table 31.1 or Table 31.2 that the first column entries of Plimpton 322 represent the values of u^2 in the analysis presented above. However, Robson explains that the first column of Plimpton 322 may not be the Scribe's original first column.

> The left-hand side of the tablet is missing, and has been at least since Plimpton acquired it. There is a clean break here, along one of the vertical rulings which divide the surface of the tablet into columns. Traces of glue remain in this break, and it has been implied that the other fraction of the tablet must therefore have been lost in modern times, deliberately or otherwise [6, p. 172].

Another reason for how a break in the clay might have erupted here is possible: the entries may have represented the values of v^2 instead of u^2: since each value of v^2 begins with the digit 1 followed by the digits of u^2, a column of such numbers may have broken along the line of vertical wedges that produced these 1s. Ultimately, however, this ambiguity is inconsequential, since both options lead to very similar interpretations of how the values were computed.

Once a list of reciprocal squares had led to the values $1 + u^2 = v^2$ in column 1, the Scribe would have obtained solutions to the Pythagorean relation and thus, the dimensions of a right triangle with height = 1, width = u and diagonal = v. Columns 2 and 3 of the tablet are then understood as dimensions of a right triangle similar to the one just found, obtained by scaling up the dimensions 1, u and v to dimensions a, $b = au$ and $c = av$, respectively. Here, a is chosen so that all of a, b, c are integers and as small as possible. For instance, in row 1, we have $u = 0||59|30$ and $v = 1||24|30$, and since both end in the digit 30, multiplication by 2 reduces the number of fractional digits: $2u = 1||59$ and $2v = 2||49$. Scaling up further by $60 = 1|00$ produces $120u = 1|59$ and $120v = 2|49$. So with $a = 120 = 2|00$, we find $b = 1|59$ and $c = 2|49$. The other rows in the tablet are handled similarly.

31.6 Why Tabulate These Numbers?

Finally, a natural question arises: what would motivate the Scribe to compute Pythagorean triples? Robson [6] argues that the tablet was devised by the Scribe to generate a "test bank" of right triangle problems. Here is one form such a problem may have taken: given the hypotenuse and one side of a right triangle, find the missing side. (For instance, from row 7: given the right triangle with "width" 2291 and "diagonal" 3541, find the "height." Answer: $\sqrt{3541^2 - 2291^2} = 2700$.)

The Scribe, either a teacher of mathematics or an apprentice teacher preparing materials to use with students in the future, worked to obtain integer dimensions for these triangles so that the required computations would produce integer results. In fact, it could be that the Scribe was a student assigned the sophisticated task of carrying out the analysis we have just laid out, to find the most elementary integer-sided right triangle from each of the reciprocal pairs in a given table of reciprocals. Unfortunately, it appears that the truth of this can never be known with certainty.

31.7 Plimpton 322 in the Classroom

The investigation into understanding what Plimpton 322 says and what purpose it may have served its author can be carried out pretty much entirely by students, under guidance, of course, but with little background. In particular, these activities should be accessible to many high school students.

These explorations can be used to highlight a number of topics:

- a study of the Pythagorean Theorem;

- algorithms for arithmetical computations, both by hand and with hand-held calculators;

- decimal positional notation and representation of numbers in other bases; and

- a deep understanding of what it means to solve a quadratic equation, by considering a nonstandard version of the problem (the *igi-igibi* problem), namely, as a pair of (nonlinear) equations in two unknowns, and a study of its geometric/algebraic solution through completing the square.

31.7. Plimpton 322 in the Classroom

31.7.1 First session (about 45 minutes)

- Share with your students an image of the tablet having sufficient contrast such that the characters can easily be read. Ask them to guess at what it says. Point out the following important features: the kinds of characters that appear in the body of the table are different – and less complicated than – the characters that appear at the tops of the columns; all the writing in the body of the table is made from combinations of just two kinds of strokes (⊤ and ⟨). Focus their attention on the last column: this should convince them that they are looking at representations of the numbers from 1 to 15. This is the key to reading the rest of the tablet. All the other entries are sequences of numbers similarly represented.

- Ask them to identify the largest number that can be read; no number on the tablet larger than 59 can be found (see row 8, column 1). What significance might this have? You might want to display a copy of some tabular array of decimal numbers (exactly what the table of numbers represents is irrelevant) and ask what the largest numerical *symbol* (not the largest *number*) is in that table. This could begin a discussion that compares decimal and sexagesimal positional numeration as in the section "Reading the Tablet" above.

- Once the class is ready to accept that the tablet displays (sequences of) sexagesimal numbers, they are ready to see what these numbers are. Share with them the transliterated version of the tablet in Table 31.1 above. Students can practice what they have learned about sexagesimal numeration by converting entries of columns 2 and 3 of the tablet into decimal form. (Electronic calculators with Degree-Minute-Second features for working with angle measure can be used as aids for performing sexagesimal arithmetic.)

31.7.2 Second session (about 45 minutes)

- Once the entries of columns 2 and 3 have been identified, consider the numbers in column 1. What might these numbers represent? If they are meant to be read as integers, they are VERY large integers indeed. Is this a reasonable interpretation? Note that the numbers in column 1 seem to be ordered from highest to lowest by virtue of the size of the *left*most digit of the number. This leads to a recognition that these numbers are not whole numbers but fractional sexagesimal numbers. There is an "invisible" radix point symbol before the first digit of each of these numbers.

- So what do the numbers in the tablet measure, if anything? At this point, the English translations of the Akkadian words that form the column headings can be revealed to point to a geometric interpretation linked to sides of a right triangle.

- If the numbers in columns 2 and 3 are sides of a right triangle, what are the values of the missing sides (the heights of the corresponding triangles)? This brings in a consideration of the Pythagorean Theorem.

- Determine the missing height values. (The arithmetic required to find the heights could even be performed in sexagesimal notation by those with more computational fortitude. The final form of the translated tablet is in Table 31.2.) Note that most of the rows of the tablet produce integer heights. This drives home a realization that the Scribe must have been well acquainted with the Pythagorean Theorem (1000 years before Pythagoras!). But what methods did the Scribe use to determine such large examples of Pythagorean triples? And why would this have been useful to the Scribe? This continues to support an interpretation that imputes some deeper underlying arithmetical/geometrical organization. Many classroom discussions of the tablet can end at this stage. Others may proceed to investigate further …

31.7.3 Further explorations

Correct the Scribe. It can be mentioned that there are computational errors in the tablet, in rows 2, 8, 9, 13, and 15 of Table 31.1. These errors were found because the rules that govern the computations performed by the Scribe are now understood. Some of the errors are easier to explain than others, and more often than not, they give an indication about how the Scribe performed the calculations necessary to produce the tablet. Working from the last column of Table 31.2, reconstruct the missing digits in column 1 of rows 1, 2 and 3 in Table 31.1, then correct the errors throughout the tablet. Both [2] and [6] discuss possible explanations for each of these errors.

The *igi-igibi* problem and the quadratic equation. Introduce a discussion of the *igi-igibi* problem (in modern notation, $xy = p$, $x + y = s$), an ancient version of the modern problem of solving a quadratic equation. By eliminating one of the variables, have students show that this is equivalent to a single variable quadratic equation. (This discussion requires at least one dedicated hour-long lesson.) Students can then be assigned the following tasks:

- Given values of x from Table 31.3, have students compute their reciprocals x' (whether in sexagesimal or decimal).

- The special case $p = 1$ in the *igi-igibi* problem makes x and $y = x'$ reciprocals. Together with the geometric connection indicated in Figure 31.3, this suggests that the change of variables $u = \frac{x-x'}{2}$ and $v = \frac{x+x'}{2}$ produces the quadratic equation $1 + u^2 = v^2$.

- Determine from the reciprocal pairs in Table 31.3 the corresponding values (in decimal) of u and v. Now the connection with Plimpton 322 is complete, as the final column in Table 31.3 matches the first column in Plimptom 322.

31.8 Conclusion

Making sense of Plimpton 322 illustrates the surprising sophistication of ancient mathematics. It can also bring students into contact with a variety of fundamental mathematical ideas, from arithmetic (positional numeration, pencil-and-paper algorithms for the four arithmetical operations) to geometrical algebra (the *igi-igibi* problem, completing the square as an algebraic and geometric method, similar triangles, scale factors, and the Pythagorean Theorem). The 15 rows of the tablet provide opportunity for a variety of exercises useful for verifying the underlying relations; and the handful of errors scattered across the tablet keep things interesting as well. The ability to decode an unfamiliar numerical notation on an exotic artifact and can engage and intrigue students. In the process, central concepts of arithmetic, algebra and geometry take the stage and bring mathematical order of a high level to what looks at first like meaningless scratches in a shard of clay.

Bibliography

[1] Asger Aaboe, Episodes from the Early History of Mathematics, Washington, DC, Mathematical Association of America, 1964.

[2] R. Creighton Buck, Sherlock Holmes in Babylon. *Amer. Math. Monthly* **87** (1980), no. 5, 335–345.

[3] Jöran Friberg, Methods and traditions of Babylonian mathematics: Plimpton 322, Pythagorean triples and the Babylonian triangle parameter equations. *Historia Math.* **8** (1981), no. 3, 277–318.

[4] George Gheverghese Joseph, *The Crest of the Peacock: non-European roots of mathematics*, Princeton University Press, Princeton, NJ, 2000.

[5] Otto Neugebauer, *The Exact Sciences in Antiquity*, 2nd ed., Brown University Press, Providence, RI, 1957.

[6] Eleanor Robson, Neither Sherlock Holmes nor Babylon: a reassessment of Plimpton 322. *Historia Math.* **28** (2001), no. 3, 167–206.

[7] William Seagle, *Men of Law: from Hammurabi to Holmes*, Hafner, New York, 1971.

32

Thomas Harriot's Pythagorean Triples: Could He List Them All?

Janet L. Beery
University of Redlands

32.1 Introduction

English mathematician and scientist Thomas Harriot (1560–1621) gave the usual formula for Pythagorean triples using his new algebraic notation but he also started to list them in a systematic way. If he could have continued his list indefinitely, would he have listed all of the Pythagorean triples? In exploring this question, students can recognize and describe patterns, write and use algebraic formulas, and construct proofs, including proofs by mathematical induction. Students also have the opportunity to study an historical approach in which a mathematician seemed to believe that tabular presentation of a result was just as valuable, effective, and interesting as symbolic presentation of that result.

This material could be used in an undergraduate number theory course, in a "proofs" or "transition" course, or as enrichment for bright algebra or general education students. At least part of it could be used in college algebra or other general education courses. In lower level courses, the material should be presented in class, either during an interactive lecture or as an in-class exploration. In more advanced courses, it could be presented in class, as homework, or as a project for a small group (or small groups) of students.

32.2 Mathematical Background

Pythagorean triples are ordered triples (x, y, z) of positive integers with the property that $x^2 + y^2 = z^2$. Such a triple can be viewed geometrically as giving the lengths of the sides of a right triangle with hypotenuse of length z and legs of lengths x and y. The Pythagorean triple $(3, 4, 5)$ has been known since antiquity, with the triple $(5, 12, 13)$ making its appearance fairly early on in history as well. We usually do not distinguish between the triples $(5, 12, 13)$ and $(12, 5, 13)$, although we generally prefer the former representation to the latter. A primitive Pythagorean triple (x, y, z) is a Pythagorean triple for which the entries x, y, and z are relatively prime, or, equivalently, for which the greatest common divisor of x, y, and z is 1. Since any divisor of any two of x, y, and z in a Pythagorean triple must divide all three of x, y, and z, then x, y, and z in a primitive Pythagorean triple are in fact pairwise relatively prime. The Pythagorean triples given above are primitive, whereas the Pythagorean triple $(15, 36, 39)$ is not primitive since each of its entries is divisible by 3. Since any Pythagorean triple can be reduced to a primitive Pythagorean triple by dividing each entry by the greatest common divisor of the three entries, we may restrict our search for all Pythagorean triples to a search for all primitive Pythagorean triples.

32.3 Historical Background

The ancient Greek mathematician and philosopher Pythagoras (c. 572–497 BCE) is generally credited with recognizing that triples of the form $(2n + 1, 2n^2 + 2n, 2n^2 + 2n + 1)$, where n is a positive integer, are Pythagorean triples (though not in the modern algebraic form given here). Some attribute the form $(2n, n^2 - 1, n^2 + 1)$, where n is a positive integer such that $n \geq 2$, to Pythagoras as well, while others, including Thomas Harriot, attribute it to the Greek philosopher Plato (c. 429–347 BCE) [1, vol. 2, p. 165; 7, p. 50]. The Greek mathematician Euclid (c. 300 BCE) described geometrically the much more general formula $(2mn, m^2 - n^2, m^2 + n^2)$, where m and n are positive integers with $m > n$, in Book X of his famous *Elements* [6, vol. 3, pp. 63–64]. The Greek mathematician Diophantus (c. 250) certainly knew this formula, too, as he described it via examples in Books II and III of his *Arithmetic* [5, pp. 144–145, 166–167].[1] The *Arithmetic* contained many intriguing problems involving sums and differences of squares.

In sixteenth century Europe there was renewed interest in the mathematics of the ancient Greeks, including that of Diophantus [17, pp. 10, 31–32].[2] Rafael Bombelli (1526–1572) included problems from the *Arithmetic* of Diophantus in his 1572 *Algebra*, printed in Bologna. W. Holzmann (1532–1576), writing under the pen name of Xylander, published the first translation of the *Arithmetic* from Greek into Latin in 1575 in Basel. In 1593, the French mathematician François Viète (1540–1603) based his *Five Books of Zetetica* on the *Arithmetic* of Diophantus.

In England, Thomas Harriot[3] read carefully both the *Arithmetic* of Diophantus (probably in Xylander's translation) and Viète's reworking of it in his *Five Books of Zetetica*. In particular, he rewrote in his own algebraic notation Diophantus' and Viète's derivations of the formula, $(2mn, m^2 - n^2, m^2 + n^2)$, where m and n are positive integers with $m > n$, for generating Pythagorean triples, correcting minor errors [4, folios 204–205]. However, Harriot's list of Pythagorean triples was written in response to the work of yet another mathematician, Michael Stifel of Germany.[4]

32.4 In the Classroom

32.4.1 Harriot's List of Pythagorean Triples

In his 1544 *Arithmetica Integra*, Michael Stifel (1487–1567) discussed "diametrical numbers," which correspond to Pythagorean triples [11, pp. 14v–16]. By a "diametrical number" Stifel meant a product mn of positive integers such that $m^2 + n^2$ is a square. He gave as an example that 12 is a diametrical number with sides 3 and 4 and diameter 5 [11, p. 14v]. He then gave the sequences of diametrical numbers, or Pythagorean triples, traditionally attributed to Pythagoras and to Plato; namely [11, p. 15],

$$\text{Order 1 (Pythagoras):}$$

$$1\frac{1}{3},\ 2\frac{2}{5},\ 3\frac{3}{7},\ 4\frac{4}{9},\ 5\frac{5}{11},\ \ldots$$

$$\text{Order 2 (Plato):}$$

$$1\frac{7}{8},\ 2\frac{11}{12},\ 3\frac{15}{16},\ 4\frac{19}{20},\ 5\frac{23}{24},\ \ldots$$

Here, for example,

$$3\frac{3}{7} = \frac{24}{7}$$

corresponds to the Pythagorean triple $(7, 24, 25)$. Have students compare the first few triples given by these lists with the first few triples given by the formulas above attributed to Pythagoras and Plato. Replacing n by $2n$ in the Platonic formula yields the formula $(4n, 4n^2 - 1, 4n^2 + 1)$, $n \geq 2$, for the entries of Order 2 above. The first several triples for each of orders 1 (Pythagoras) and 2 (Plato) above are listed in Figure 32.1, also in orders 1 and 2. For instance, the

---

[1] Diophantus actually sought and found positive rational solutions of equations of the form $x^2 + y^2 = z^2$ in his examples.

[2] Leonardo of Pisa, or Fibonacci, had included problems of Diophantus in his *Liber Quadratorum* in 1225, but this work was not to resurface in Europe until 1856 [17, p. 11].

[3] For a short biography of Harriot, see the appendix, "Who Was Thomas Harriot?"

[4] Although the Harriot scholar Rosalind Cecilia Tanner (1900–1992) described Harriot's work on Pythagorean triples in [13], it remains little known.

32.4. In the Classroom

```
Order 1)              2)                3)                   4)
     1    0    1           4    3    5        15    8   17         20   21   29
 2 _____       4 _____    12    6 _____    12     8 _____   24
     3    4    5           8   15   17        21   20   29         28   45   53
 2           8    4           20    6            16    8            32
     5   12   13          12   35   37        27   36   45         36   77   85
 2          12    4           28    6            20    8            40
     7   24   25          16   63   65        33   56   65         44  117  125
 2          16    4           36    6            24    8            48
     9   40   41          20   99  101        39   80   89         52  165  173
 2          20    4           44    6            28    8            56
    11   60   61          24  143  145        45  108  117             229
 2          24    4           52    6            32
    13   84   85               197             51  140  149
 2          28                 60    6            36
    15  112  113               257                 185
 2          32                                    40
    17  144  145                                  225
 2          36
    19  180  181
 2          40
    21  220  221

     5)                      6)                   7)
    45   28   53          48   55   73        91   60  109
10 _____   20     12 _____   36    14 _____   28
    55   48   73          60   91  109             137
10          24    12          44                  32
    65   72   97          72  135  153             169
10          28                52                  36
    75  100  125               205                 205
10          32                60                  40
    85  132  157               265                 245
            36
           193
```

And in this way for the rest to infinity.

Here are all the primes,
but not all here are prime.

Note.
First differences of the order
first.   2.   4.   under double
second.   4. 12.   triple
third.   6. 12.   double
fourth.   8. 24.   triple
fifth.   10. 20.   double
sixth.   12. 36.   triple
seventh. 14. 28.   double
eighth. 16. 48.   triple
and so on to infinity

Figure 32.1. Thomas Harriot's orders 1 through 7 of Pythagorean triples

triple $(7, 24, 25)$, corresponding to $3\frac{3}{7}$, appears in the fourth row of Order 1 in Figure 32.1. The triple $(9, 40, 41)$ in the fifth row of Order 1 corresponds to $4\frac{4}{9}$ in Order 1 above, and the triple $(16, 63, 65)$ in the fourth row of Order 2 to $3\frac{15}{16}$ in Order 2 above.

Stifel made the bold claim that his orders 1 and 2 included all primitive Pythagorean triples [11, p. 15]. However, as we have seen, both Euclid and Diophantus knew there were more Pythagorean triples than just these, as did Harriot. Harriot set out to correct Stifel's error not by providing an algebraic formula generating all Pythagorean triples, but rather by providing a complete list of Pythagorean triples. As can be seen in Figure 32.1,[5] Harriot began with Stifel's orders 1 and 2 as his own first and second orders of Pythagorean triples. Each of Harriot's orders contains infinitely many triples and there are infinitely many orders.

To see how Harriot constructed these orders, let us call the triple above the line in each order the "starter" and the first triple below the line the "first triple" (as in [13]). Note that Harriot interchanged the first two entries of the first triple to obtain the starter for the next order. For example, the first triple $(8, 15, 17)$ of the second order becomes the starter $(15, 8, 17)$ of the third order. Notice also that the tables are stepped so that these triples appear side by side.

Within each order, to obtain the next triple, Harriot used rules based on finite differences, described in his table at the bottom of Figure 32.1. Although Harriot did not use the symbol n, or any other symbol, in this table, one can see that the first entries of the triples of the nth order have a constant first difference of $2n$; that is, to obtain the subsequent first entry, add $2n$. For instance, note that in the order 3 table, successive first entries are obtained by adding 6. Note also that the constant difference 6 is recorded in the order 3 table in a column to the left of the column of first entries.

The first differences between second and third entries are not constant, but the second and third entries have a constant second difference of 4 or 8, depending on whether n is odd or even, respectively. Harriot's rule in the table at the bottom of Figure 32.1 is for the *first* first difference in each order—that is, the first difference one adds to the second and third entries of the starter to obtain the second and third entries of the first triple. This first difference is "double" $2n$ or $2(2n) = 4n$ for odd orders and "triple" $2n$ or $3(2n) = 6n$ for even orders. So, for example, in the order 4 table, one adds $3(2 \cdot 4) = 24$ to each of 21 and 29 to obtain 45 and 53. One then adds $24 + 8 = 32$ to these entries to obtain 77 and 85, and so on. The first differences, 24, 32, 40, and so on, are recorded in the order 4 table in a column to the right of the column of third entries.

In four additional folios, or manuscript sheets, Harriot continued the seven orders shown in Figure 32.1 through order 22 with entries up to hypotenuse 1105 [3, ff. 86–89]. Ask students to fill in the missing entries and rows (triples) for the orders shown in Figure 32.1 and to construct tables for orders 8 and 9 using the rules described in the preceding paragraph. Point out that some of the Pythagorean triples in Harriot's tables are primitive and some are not. For instance, $(33, 56, 65)$ is a primitive Pythagorean triple, while the triple right above it in the order 3 table, $(27, 36, 45)$, is not primitive, since each of its entries is divisible by 3 (and 9).

32.4.2 Does Harriot's List Include All Primitive Pythagorean Triples?

In the lower right corner of Figure 32.1, Harriot claimed that his list contained all the primitive ("prime") Pythagorean triples but that not every triple in his list was primitive. Our question, then, is, would Harriot's tables, if extended indefinitely, include all primitive Pythagorean triples?

As we have seen, Harriot knew at least one general formula for Pythagorean triples and derived it elsewhere in his manuscripts [4, ff. 204–205]. Harriot gave this formula as $(rr - ss, \; 2rs, \; rr + ss)$ or, as we would write it, $(r^2 - s^2, \; 2rs, \; r^2 + s^2)$. He was careful to point out that r must be greater than s, but did not specify any other conditions on r and s. We now know that selecting r and s to be relatively prime positive integers with $r > s \geq 1$ and $r - s$ odd always results in a primitive Pythagorean triple and that every primitive Pythagorean triple can be obtained in this way. (Number theory students will know the last condition as r and s are incongruent modulo 2, while general education students should note that one of r or s must be even and the other odd.)

Again, our question is, would Harriot's lists, if continued indefinitely, contain every primitive Pythagorean triple? Equivalently, is every triple of the form described in the preceding paragraph in at least one of Harriot's lists? Since this form is completely characterized by r and s, it would be helpful to identify the values of r and s for the entries in Harriot's lists shown in Figure 32.1. Ask students to compute values of r and s for orders 1 through 4 and to predict those for orders 5 through 8 based on the patterns they see in orders 1 through 4. Figures 32.2 and 32.3 show some of

---

[5]Figure 32.1 is a transcription, and translation from Latin to English, of the manuscript sheet BL Add. MS 6782, folio 85, from [3].

1)

					r	s
	1	0	1		1	0
2				4		
	3	4	5		2	1
				8		
	5	12	13		3	2
				12		
	7	24	25		4	3
				16		
	9	40	41		5	4
				20		
	11	60	61		6	5
				24		
	13	84	85		7	6
				28		
	15	112	113		8	7
				32		
	17	144	145		9	8
				36		
	19	180	181		10	9
				48		
	21	220	221		11	10

2)

					r	s
	4	3	5		2	1
4				12		
	8	15	17		4	1
4				20		
	12	35	37		6	1
4				28		
	16	63	65		8	1
4				36		
	20	99	101		10	1
4				44		
	24	143	145		12	1
				52		
			197		14	1
				60		
			257		16	1

3)

					r	s
	15	8	17		4	1
6				12		
	21	20	29		5	2
6				16		
	27	36	45		6	3
6				20		
	33	56	65		7	4
6				24		
	39	80	89		8	5
6				28		
	45	108	117		9	6
6				32		
	51	140	149		10	7
				36		
			185		11	8

4)

					r	s
	20	21	29		5	2
8				24		
	28	45	53		7	2
8				32		
	36	77	85		9	2
8				40		
	44	117	125		11	2
8				48		
	52	165	173		13	2
				56		
			229		15	2

Figure 32.2. Thomas Harriot's orders 1 through 4 of Pythagorean triples with values for r and s

these values for Harriot's orders 1 through 7. For instance, in the order 3 table in Figure 32.2, $r = 5$ and $s = 2$ yield the triple $(5^2 - 2^2,\ 2 \cdot 5 \cdot 2,\ 5^2 + 2^2)$ or $(21, 20, 29)$.

Students should note that r and s are integers with $r > s \geq 1$ and $r - s$ odd, but that r and s are not always relatively prime. They can now see that there are Pythagorean triples that would not appear in Harriot's list. For instance the triple $(16, 30, 34)$ is not in Harriot's list because it is generated using values of r and s ($r = 5$ and $s = 3$) for which $r - s$ is even. It can be obtained by multiplying by 2 each entry of the primitive Pythagorean triple $(8, 15, 17)$ from Order 2. The triple $(15, 36, 39)$ given above also is not in the list. It can be obtained by multiplying by 3 each entry of the primitive Pythagorean triple $(5, 12, 13)$ from Order 1.

Of course, if we can show that every Pythagorean triple of the form $(r^2 - s^2,\ 2rs,\ r^2 + s^2)$, with r and s integers such that $r > s \geq 1$ and $r - s$ is odd, is in Harriot's list, then we'll have that every triple that also has r and s relatively prime is in the list—that is, every primitive Pythagorean triple is in the list. Have students begin with specific examples and look for patterns. For instance, ask them to determine in which order n, and in which row k of that order, triples with $s = 2$ and $r = 3, 5, 7, 9, 19$, and 119 would occur. (They should see from their completed Figure 32.2 that for

5)

					r	s
	45	28	53		7	2
10				20		
	55	48	73		8	3
10				24		
	65	72	97		9	4
10				28		
	75	100	125		10	5
10				32		
	85	132	157		11	6
				36		
			193		12	7

6)

					r	s
	48	55	73		8	3
12				36		
	60	91	109		10	3
12				44		
	72	135	153		12	3
				52		
			205		14	3
				60		
			265		16	3

7)

					r	s
	91	60	109		10	3
14				28		
			137		11	4
				32		
			169		12	5
				36		
			205		13	6
				40		
			245		14	7

8)

					r	s

Figure 32.3. Thomas Harriot's orders 5 through 7 of Pythagorean triples with values for r and s

$s = 2, r = 3$ is in row 2 of Order 1; $r = 5$ is in row 1 of Order 3; and $r = 7, 9, 19,$ and 119 are in rows 1, 2, 7, and 57 of Order 4.) Do the same for $s = 5$ and $r = 10, 12, 14, 16, 18,$ and 218. Repeat as needed, then have students apply their newfound rule or formula to $s = 17$ and $r = 46, 48, 50, 52, 54,$ and 104.

All students should note that, for a given value of s, the (even) order with that fixed value of s contains all pairs (r, s) with $r > 3s$ (or $r − s > 2s$), and that the remaining pairs (r, s) with $s < r < 3s$ (or $0 < r − s < 2s$) occur in the preceding odd orders. Students capable of writing algebraic formulas can be more specific, noting that, for a given pair (r, s) with $r > s \geq 1$ and $r - s$ odd, either $r < 3s$ or $r > 3s$ (being careful to explain how they know $r \neq 3s$). In the former case, (r, s) is in the order $n = r − s$ table in row $k = \frac{3s-r+1}{2}$ (where the starter is in row 0 and the first triple is in row 1). In the latter case, (r, s) is in the order $n = 2s$ table in row $k = \frac{r-3s+1}{2}$.

Of the many patterns students may detect in Harriot's tables of Pythagorean triples, one of the most interesting is that, in tables of odd order n, $z − y = n^2$, and, in tables of even order n, $z − y = \frac{1}{2}n^2$. Students may confirm these observations using the formulas we have given thus far or using those given in the next paragraph.

32.4.3 Proofs for More Advanced Students

Implicit in the penultimate paragraph of the preceding section are formulas for r and s, which students may have written in order to describe patterns in Harriot's tables and/or to complete the proof above. For odd order n, the entries in row k may be written using $r = \frac{3n-1}{2} + k$ and $s = \frac{n-1}{2} + k$. Here, $r − s = n$. For even order n, the entries in row k are given by $r = \frac{3n-2}{2} + 2k$ and $s = \frac{n}{2}$, and we have $r − s = n − 1 + 2k$. Students may ask how we know our formulas for r and s are correct for Harriot's tables; that is, how we know they generate all entries of Harriot's tables. Students in more advanced courses could be asked to devise induction proofs to show that our formulas for r and s do indeed agree with Harriot's table construction rules. Since every pair (r, s), with r and s integers such that $r > s \geq 1$, generates a Pythagorean triple, such a proof also would establish that every triple in Harriot's list is indeed Pythagorean.

Harriot's table construction rules tell us how to move from one order to the next and, within each order, how to move from one entry to the next. More specifically, they tell us how to "start" each table from the preceding one and, within each table, how to derive each row from the preceding one. Ask students to show separately for odd and even orders n that, according to our formulas for r and s, the first triple ($k = 1$) in the order n table is the same as the starter ($k = 0$) in the order $n + 1$ table. This can be done simply by checking that the formulas for r and s agree. To prove by induction on n, $n \geq 1$, that our formulas give the same starter in each order n as do Harriot's rules, students actually must show that our formulas give the same starter (row $k = 0$) and first triple (row $k = 1$) as do Harriot's rules. After checking that our formulas give Harriot's order $n = 1$ starter and first triple (basis step) and completing the check described above to establish that the order $n + 1$ starter is the same as the order n first triple, what remains for students to show (in order to complete the inductive step) is that the order $n + 1$ first triple is correct. This proof is a special case of the argument described in the next paragraph.

The proof outlined in the preceding paragraph establishes the basis step for a proof by induction on k, $k \geq 0$, that, within each order n, our formulas give the same entries in row k as do Harriot's rules. To prove that our formulas give successive triples within each order according to Harriot's rules, we show that our formulas give Harriot's first differences for row k; namely, $2n$ and $4n + 4k$ for tables of odd order n, and $2n$ and $6n + 8k$ for tables of even order n. For odd orders n, one can check that the differences between corresponding entries of the Pythagorean triples generated by (r, s) and $(r + 1, s + 1)$ are $2(r − s)$, $2(r + s + 1)$, and $2(r + s + 1)$. Substituting the expressions for r and s for n odd from the preceding paragraph results in differences of $2n$, $4n + 4k$, and $4n + 4k$, matching Harriot's first differences. For even orders n, when we compute the differences between entries of the triples generated by (r, s) and $(r + 2, s)$, and substitute the appropriate expressions for r and s from the preceding paragraph, we again obtain Harriot's first differences.

32.4.4 Harriot's Derivation of the Formula $(r^2 − s^2, 2rs, r^2 + s^2)$

Our reasoning above relies on the theorem that the formula $(r^2 − s^2, 2rs, r^2 + s^2)$, with r and s relatively prime integers such that $r > s \geq 1$ and $r − s$ is odd, generates all primitive Pythagorean triples. This theorem is proved in most elementary number theory texts. Harriot's derivation of the formula $(r^2 − s^2, 2rs, r^2 + s^2)$, with r and s integers such that $r > s \geq 1$, followed those of Diophantus (most likely as translated by G. Xylander, the pen name of

32.4. In the Classroom

W. Holzmann) and Viète with minor corrections to their reasoning. He noted both "Diophantus lib. 2.8" and "Zet. lib. 4.1" at the start of his own work [4, f. 204]. What is most notable about Harriot's derivation is the modern algebraic notation he used to convey clearly the details of the argument.

In setting out to solve the problem, "Divide bb into two square numbers" [4, ff. 204–205], Harriot assumed bb (or b^2) is the sum of the squares of a and $b - \frac{sa}{r}$, where r and s were assumed implicitly to be positive integers and explicitly to satisfy $r > s$, and set out to write expressions for a and $b - \frac{sa}{r}$ in terms of r and s. (Here, a and b were assumed to be positive and almost certainly rational.) He then solved the equation

$$aa + bb - \frac{2bsa}{r} + \frac{ssaa}{rr} = bb$$

to obtain

$$a = \frac{2brs}{rr + ss} \quad \text{and} \quad b - \frac{sa}{r} = \frac{brr - bss}{rr + ss}.$$

Multiplication of $(b - \frac{sa}{r}, a, b)$ by $\frac{rr+ss}{b}$ would then yield integers $(rr - ss, 2rs, rr + ss)$ forming a Pythagorean triple, a formula Harriot put to use in nearby manuscript sheets to generate lengths of sides of right triangles [4, ff. 201–203].

None of the aforementioned mathematicians who had what we now know to be the general formula claimed his formula generated all (primitive) Pythagorean triples, although some or all of these mathematicians may have believed or at least suspected so. Diophantus, in Proposition (or Problem) 8 of Book II of the *Arithmetic,* set out "To divide a given square number into two squares" [5, p. 144]. Viète, following Diophantus, wanted "To find numerically two squares equal to a given square." (This was Zetetic I (*Zeteticum* I) in the Fourth Book (*Liber Quartus*) of his *Five Books of Zetetica* (*Zeteticorum Libri Quinque*) [16, p. 124].) Harriot, following both Diophantus and Viète, stated the problem even more succinctly as, "Divide bb into two square numbers" (he actually wrote on one manuscript sheet, "*Dividere bb in duo* □") [4, ff. 204–205]. Note that Diophantus, Viète, and Harriot were interested here in finding at least one way to represent a given rational square number as a sum of rational squares, and not necessarily all ways of doing so. In fact, they went on to solve problems in which they found alternate ways of representing numbers as sums of squares. Both Zetetic II and Zetetic III in Viète's Fourth Book are "To find numerically two squares equal to two other given squares" [16, pp. 125–6]. Euclid sought (in the first of two lemmas preceding Proposition 29 of Book X of *The Elements*) "To find two square numbers such that their sum is also square" [6, vol. 3, p. 63]. This was a slightly different approach from those of Diophantus, Viète, and Harriot, but Euclid did not seem to indicate that he would find all such numbers.

32.4.5 Harriot's Algebraic Notation

Harriot's algebraic notation was very modern and certainly was a great improvement over that of his primary algebraic influence, Viète. Where Harriot wrote the Pythagorean triple formula as $(rr - ss, 2rs, rr + ss)$, Viète wrote it as $(R\, quad - S\, quad,\ S\, in\, R\, 2,\ R\, quad + S\, quad)$, where "in" can be translated as "times" and "quad" as "squared". Using these translations, Harriot scholar Jacqueline Stedall noted that where Viete wrote [10, pp. 8, 10–11],[6]

If to
$$\frac{A\ plane}{B}$$

there should be added
$$\frac{Z\ squared}{G},$$

the sum will be
$$\frac{G\ times\ A\ plane\ +\ B\ times\ Z\ squared}{B\ times\ G},$$

Harriot wrote
$$\frac{ac}{b} + \frac{zz}{g} = \frac{acg + bzz}{bg}.$$

---

[6]Stedall translated Viète's Latin and modernized Harriot's "equals" sign. In particular, she translated Viète's "in" as "times" and his "quad" as "squared".

By "*A plane*" Viète meant a variable representing the area of a planar or rectangular region—that is, the product of two variables—hence Harriot's translation of "*A plane*" to "*ac*". One can see that Harriot's notation was much more succinct and modern-looking than Viète's. Harriot's friend, Nathaniel Torporley, introduced superscript notation for powers as he copied and explained Harriot's work in his own manuscript notes. For instance, he rewrote Harriot's $bbbbaaaa$ as $b^{IV}a^{IV}$ using Roman numerals as exponents. He also wrote this expression as

$$b\hat{4}a\hat{4},$$

making it unclear whether or not he intended $\hat{4}$ to be a superscript [15, p. 69].

32.5 Conclusion

In seventeenth century Europe, tabulations of square roots, logarithms, and trigonometric values were essential to applications such as navigation and astronomy. At the same time, the algebraization of mathematics had begun, with all sorts of mathematical ideas expressed in algebraic language and symbolic notation. Thomas Harriot developed and used the most succinct and modern-looking algebraic notation of his day, yet he also was a master of table construction. Harriot's work with Pythagorean triples shows that he valued both symbolic formulas and numerical tables as means of organizing and communicating mathematical ideas.

It is important and even exciting for our students to see early uses of mathematical ideas and tools that have become commonplace today, and Thomas Harriot's algebraic notation is one such tool. However, it also is important for students to see ways of thinking about and communicating information and ideas that are not as common today, such as Harriot's organization of Pythagorean triples into an infinite sequence of tables, each table containing an infinite sequence of Pythagorean triples. (Students also can use Harriot's tables to show that the set of Pythagorean triples is countably infinite, although there is an easier way to do this!) Of course, students should discuss the advantages and disadvantages of both methods of presentation.

Harriot's organization of the Pythagorean triples into tables also helps make this material accessible to students at all levels, including students in liberal arts or general education mathematics courses who can, at the very least, recognize mathematical patterns in the tables. More advanced students also can use this material to gain practice in writing symbolic formulas and constructing logical arguments or proofs.

If Harriot could have continued his list of Pythagorean triples indefinitely, would he have listed all of the primitive Pythagorean triples? The answer is "yes!" and my hope is that the exploration is as interesting as the answer.

Acknowledgments

I am grateful to Janine Stilt, of the University of Redlands, for her assistance in preparing the figures for this article, and to Sandra Richey, of the University of Redlands Library, for obtaining books and journals for me from libraries near and far. I thank the editors of this volume, Dick Jardine and Amy Shell-Gellasch, along with the anonymous reviewers of this article, for their encouragement and assistance. Finally, I thank the British Library, West Sussex Record Office, University of Delaware Library, Huntington Library, Lambeth Palace Library, and Cambridge University Library for the use of manuscripts, copies of manuscripts, and rare books.

Bibliography

[1] Leonard Eugene Dickson, *History of the Theory of Numbers* (3 vols.), Chelsea, New York, 1952.

[2] Thomas Harriot, *Mathematical Papers of Thomas Harriot*, British Library Additional MSS 6782-6789 (14 vols.) and Petworth MSS 240-241.

[3] ———, "De Triangulis Laterum Rationalium," British Library Additional MS 6782, ff. 84–89, and Petworth MS 241/5, ff. 1–7.

[4] ———, "Diophantus lib. 2.8 and Zet. lib. 4.1," British Library Additional MS 6785, ff. 201–206.

[5] Thomas L. Heath, *Diophantus of Alexandria* (second ed.), Cambridge University Press, 1910; reprinted by Dover Publications, New York, 1964.

[6] ———, *Euclid: The Elements* (3 vols.), St. John's College Press, Annapolis, 1947.

[7] Victor J. Katz, *A History of Mathematics: An Introduction* (second ed.), Addison-Wesley, Reading, Mass., 1998.

[8] John W. Shirley, *Thomas Harriot: A Biography*, Clarendon Press, Oxford, 1983.

[9] Jacqueline Stedall, *A Discourse Concerning Algebra: English Algebra to 1685*, Oxford University Press, 2002.

[10] ———, *The Greate Invention of Algebra: Thomas Harriot's Treatise on Equations*, Oxford University Press, 2003.

[11] Michael Stifel, *Arithmetica Integra*, Nuremberg, 1544.

[12] Rosalind C. H. Tanner, "Thomas Harriot as Mathematician: A Legacy of Hearsay," *Physis*, 9 (1967), 235–247 (Part 1), 257–292 (Part 2).

[13] R. C. H. (Rosalind Cecilia) Tanner, "Nathaniel Torporley's 'Congestor analyticus' and Thomas Harriot's 'De triangulis laterum rationalium'," *Annals of Science*, 34 (1977), 393–428.

[14] Nathaniel Torporley, Lambeth Palace Library Sion College MS Arc L.40.2/L40, ff. 1–34v, 35–54v (two untitled and undated documents concerning Harriot's mathematics).

[15] ———, *Of Differences* (1627), Cambridge University Library MS Add. 9597/17/28.

[16] François Viète, *The Analytic Art: Nine Studies in Algebra, Geometry, and Trigonometry from the Opus Restitutae Mathematicae Analyseos, seu Algebra Nova* (translated by T. Richard Witmer), Kent State University Press, 1983.

[17] André Weil, *Number Theory: An Approach through History from Hammurapi to Legendre*, Birkhauser, Boston, 1984.

[18] D. T. Whiteside, "Review of *Thomas Harriot: Renaissance Scientist* edited by J. W. Shirley," *History of Science*, 13 (1975), 61–70.

Appendix: Who Was Thomas Harriot?

Thomas Harriot (1560–1621) may be best known as the navigator and scientist for Sir Walter Ralegh's[7] 1585–1586 expedition to the Virginia Colony, but he also was the leading English mathematician of his day. Harriot made important discoveries in a wide range of mathematical sciences, including algebra, geometry, navigation, astronomy, and optics. He published only one work during his lifetime, *A Briefe and True Report of the New Found Land of Virginia* (1588), but, at his death, left thousands of manuscript pages of mathematics. Harriot's mathematical work is remarkable both for its content—he obtained many results generally credited to later mathematicians—and for its highly symbolic and visual presentation.

Harriot's 1577 Oxford University matriculation records show that he probably was born in 1560 in Oxford or nearby.[8] After he graduated from Oxford in 1580, Harriot moved to London, where Sir Walter Ralegh employed him to research and teach navigation. Ralegh sent Harriot on a voyage to the New World during 1585–1586, and, upon his return to England, Harriot published *A Briefe and True Report of the Newfound Land of Virginia* (1588), in which he described the flora and fauna of Virginia—North Carolina, actually—and also the customs and language of the people there.

By 1593, Harriot had found a second patron in Henry Percy, the Ninth Earl of Northumberland, known as the "Wizard Earl" for his interest in science. During the 1590s, Harriot continued to work for both of his patrons, Ralegh and Northumberland, on navigation, ballistics, optics, chemistry, and alchemy, and, by the turn of the century, geometry and algebra. In optics, he discovered the sine law of refraction, now known as Snell's Law, before Willebrord Snell

---

[7] Although spelling of names was not consistent during the sixteenth and seventeenth centuries, I have never seen in a document from his era Sir Walter Ralegh's name spelled "Raleigh", as it is in many modern-day history books.

[8] See [8, p. 40] or [9, p. 88]. The biographical information provided here is from [8] and [9].

(1591–1626). For his work in navigation, Harriot obtained the formula for the area of a spherical triangle. He made advances in all of the fields in which he worked,[9] except perhaps for alchemy.

In 1603, the year Queen Elizabeth I died and James I assumed the throne, things started to go very badly for Harriot's patrons. Ralegh was sent to the Tower of London, convicted of treason, and sentenced to death, although he wasn't executed for another 15 years. Then, in 1605, Northumberland and Harriot were sent to the Tower after the Gunpowder Plot (Northumberland's cousin, Thomas Percy, had been involved). Harriot was released almost immediately, but Northumberland was to serve another 16 years. Although both of Harriot's patrons were in prison, they continued to support Harriot, and he kept working on the mathematical and scientific topics listed above and also making astronomical observations. He observed what later would become known as Halley's comet in 1607, the satellites of Jupiter at about the same time as Galileo in 1610, and sunspots from 1611 to 1613. By 1618, when Ralegh was executed, Harriot himself was in very poor health. He was suffering from cancer of the nose, probably brought on by the smoking habit he had picked up in Virginia.

Three days before he died in 1621, Harriot prepared a will,[10] in which he put his friend, Nathaniel Torporley (1564–1632), in charge of sorting through his mathematical papers and publishing the good stuff. Although Torporley's handwritten explanations of Harriot's work on Pythagorean triples and on finite difference interpolation survive,[11] he ended up publishing none of Harriot's work. Walter Warner (1557–1643), another Harriot associate, did publish some of Harriot's algebra in the *Artis Analyticae Praxis* (*Practice of the Analytic Art*) in 1631, ten years after Harriot died.

The history of the Harriot manuscripts is a story in itself.[12] There currently are over 4000 manuscript sheets in the British Library in London and almost 900 of them at Petworth House in West Sussex, which was Northumberland's country home [2]. The manuscripts were thought to be lost, then were discovered under the stable accounts at Petworth House in 1784, then not studied again until the 1830s, then not again until the 1880s. In the meantime, in 1810, most of the manuscript sheets were transferred to the British Museum, but the split was not made carefully: one finds some papers on Pythagorean triples, for instance, at Petworth House and others at the British Library. The manuscripts contain astronomical observations, including drawings of the sun and moon; navigational tables; results of scientific experiments; mathematical scratchwork; studies of various mathematicians' works, most notably François Viète (1540–1603); and a few more polished pieces, including a short treatise on finite difference interpolation and a lengthy treatise on algebra.[13] Study of Harriot's mathematical manuscripts was revived in the 1950s and 1960s by scholars E.G.R. Taylor, R.C.H. Tanner, J.A. Lohne, J.V. Pepper, and others, and continues today with work of J.A. Stedall, the present author, and others.

---

[9] The Newton scholar, D. T. Whiteside, wrote that Harriot had a "profound grasp and creative understanding of the whole field of the exact sciences of his day.... Harriot in fact possessed a depth and variety of technical expertise which gives him good title to have been England's—Britain's—greatest mathematical scientist before Newton." (See [18], p. 61.)

[10] Harriot's will is reproduced in [12, pp. 244–247].

[11] See [14, ff. 26–34v] and [15, pp. 1–98], respectively.

[12] The brief history of the manuscripts given here is from [8].

[13] Harriot's treatise on finite difference interpolation is in [2] at BL Add. MS 6782, ff. 107–146v. For Harriot's treatise on algebra, see [10].

33

Amo, Amas, Amat! What's the sum of that?

Bernoulli's Account of the the Divergent Harmonic Series in Latin

Clemency Montelle
University of Canterbury
New Zealand

33.1 Introduction

Every course in undergraduate calculus contains some component of the examination of series and the various tests to establish their convergence. One of the most important series is the Harmonic series, which is not only mathematically interesting per se, but also appears frequently as an ideal 'comparison' series to determine the convergence or divergence of other series. At some point, the formal proof of its divergence must be covered. This paper provides a quirky alternative to the format and the content of the standard proof usually offered; a capsule based on an examination of the actual primary source of the proof, as it originally appeared, in Latin.

This capsule should ideally be offered before covering the various convergence tests, and just after examining geometric series. It could be particularly fitting to include it as part of your coverage of the divergence test as the Harmonic series is often *the* example cited to demonstrate that the convergence of terms in a series that tend to zero is not sufficient to guarantee the convergence of the actual series.

Given the richness of historical insight, the relevance of the mathematics, and indeed the novelty for the students, the presentation of this primary source is ideal for the undergraduate mathematics classroom. Grabbing the attention of the students by presenting something completely different, yet utterly relevant, may very well renew their enthusiasm as well as stimulating curiosity and assisting their grasp of this topic.

Observing that multiple proofs exist for a given result can be used to great effect in the classroom. It emphasizes to students the multifarious ways in which mathematical results can be proven and analyzed, reminding them that there is no 'right' way to mathematize, rather there is much room for creativity and ingenuity. Presenting this proof may also be an opportunity to touch upon some of the more philosophical issues in mathematics, such as quality and forms of proof.

This example will also somewhat painlessly introduce them to the '*reductio ad absurdum*' form of argumentation, a technique commonly employed in higher mathematics.

This capsule is intended to take 50 minutes (one lecture). A follow-up mini-assignment may be completed at the teacher's discretion.

33.2 The Harmonic Series

Mathematical intuition is a strange thing! Almost always we can count on it to give us a reliable impression of a given mathematical scenario, but there exist instances in which it spectacularly lets us down. For instance, one of the more important series in calculus is the so-called Harmonic series:

$$\sum_{k=1}^{\infty} \frac{1}{k} = 1 + \frac{1}{2} + \frac{1}{3} + \frac{1}{4} + \frac{1}{5} + \cdots$$

This series is not only distinctive because of its elementary formulation, but because it behaves quite contrary to our initial expectations. This infinite series has terms that decrease but actually diverges, which seems counterintuitive!

33.2.1 The Harmonic Series Diverges

The proof of this that most often appears in the undergraduate classroom is due to the 14th century French mathematician Nicole Oresme (1323–1382) [3], which establishes the divergence by selective groupings and comparisons. The proof, as it might appear nowadays, considers the partial sums. For simplicity's sake, we can simply show that a subset of partial sums (in this case S_{2^n}) diverges, which is sufficient. Consider the following partial sums:

$$\begin{aligned}
S_1 &= 1 \\
S_2 &= 1 + \frac{1}{2} \\
S_4 &= 1 + \frac{1}{2} + \left(\frac{1}{3} + \frac{1}{4}\right) & > 1 + \frac{1}{2} + \left(\frac{1}{4} + \frac{1}{4}\right) & = 1 + \left(2 \times \frac{1}{2}\right) \\
S_8 &= 1 + \frac{1}{2} + \left(\frac{1}{3} + \frac{1}{4}\right) + \left(\frac{1}{5} + \cdots + \frac{1}{8}\right) & > 1 + \frac{1}{2} + \left(\frac{1}{4} + \frac{1}{4}\right) + \left(\frac{1}{8} + \cdots + \frac{1}{8}\right) & = 1 + \left(3 \times \frac{1}{2}\right) \\
&\vdots & \vdots & \vdots \\
S_{2^n} &= \cdots & & = 1 + \left(n \times \frac{1}{2}\right)
\end{aligned}$$

so that, by comparison, it follows that this subset of partial sums gets arbitrarily large as n tends to infinity and thus the series diverges.[1]

33.3 Bernoulli and the Harmonic Series

33.3.1 Historical Background

The Harmonic series was still attracting the attention of mathematicians some 350 years after Oresme, when it caught the curiosity of members from one of the most renowned mathematical clans, the Bernoullis. In a span of just three generations, the Bernoulli family produced no less than eight famous mathematicians. Arguably the two most prominent were siblings, Jakob (b. 1654) and his younger brother Johann[2] (b. 1667). These two brothers were highly successful and prolific mathematicians; both had close associations with other notable mathematicians such as L'Hôpital and Leibniz, and frequented the same mathematical circles. Jakob was, in fact, directly responsible for most of his younger brother's schooling in mathematics, after the latter became disenchanted with the medical profession, and the two were close colleagues and collaborators, keeping up regular correspondence and applying themselves to similar problems. However, far from being supportive, their fraternal proximity proved to be only a hindrance to their collegiality, as many historians observe that their relationship was fraught with tension and rivalry [2, 5, 6]. Extant letters and communications reveal frequent open antagonism and hostile jealousy.[3] However, it may be that this very competitiveness was responsible for their incredible productiveness and prolific successes in mathematics.

---

[1] For various other proofs of this see [7] and [9].

[2] These two individuals are also known by the anglicized forms of their names James and John, or their French equivalents Jacques and Jean. For a nice overview of the family and a genealogical chart, see Boyer ([2]) p. 415–6.

[3] In a rather theatrical portrayal of the relationship, historian Hal Hellman attributes this "sibling rivalry of the highest order" to a "kid brother" syndrome—that is, Jakob continuously struggled to accept the fact that his much younger brother had become his equal, if not his superior. See Hellman in his delightfully dramatic chapter "Bernoulli versus Bernoulli" pp. 73–93.

33.3. Bernoulli and the Harmonic Series

With respect to the Harmonic series, both Jakob and Johann devoted their attention to the various details and related results. Jakob published a proof first in 1689 in his work *Tractatus de Seriebus Infinitis*, which is generally included as an appendix to his larger work, the *Ars Conjectandi*, which was published poshumously in 1713. His account is entirely different from Oresme's proof and for that reason is to be examined here.

33.3.2 Examining Bernoulli's Proof in Latin

If you thought mathematical proofs were bad enough to penetrate, the prospect of examining them in Latin might be downright disagreeable! Indeed it is true that Latin is a language that is no longer used, and although it forms the basis of modern day English, it is an inflected language with strict grammatical rules and intricacies. But the beauty of reading ancient *mathematical* texts is quite precisely the reason that mathematics, unlike literature, is for the most part universal; its logical structure and predictable results mean that as long as you have a good analytical mind, can read numbers, and have a sense of adventure, in many cases you can unexpectedly go some way to gaining insight into the content.

Bernoulli's exposition on the Harmonic Series in Latin is one such case. You will find it surprisingly easy to decipher and determine the details of this proof, given that the bulk of the demonstration is largely 'numerical'. The description in prose, where it appears, is also straightforward to grasp, as the Latin technical terminology used to describe the mathematical features is almost identical to the English terms we use today.

You will notice terms, such as , 'series', 'harmonic', 'infinite', 'fraction', 'numerator', are literally the same words (except for small modifications to their endings which give an indication of their grammatical function within the sentence). Notice the unusual way of depicting a lower-case 's' as 'f'-like symbol, without the cross-bar. You and your students will find that your mathematical background will be sufficient to allow you to piece together the initial statement of the theorem (see Figure 33.1):

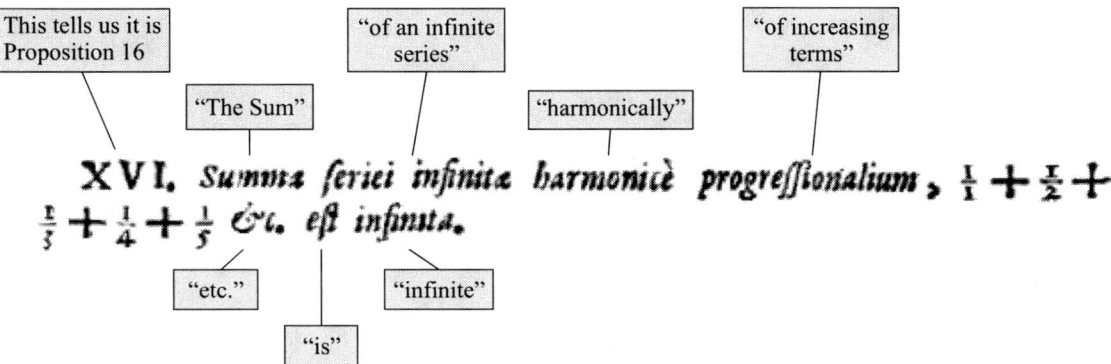

Figure 33.1. Title excerpt with explanation

33.3.3 Mathematical Treasure Hunt

Further exploration will reveal the contents of the mathematical proof. Have a glance over the image of Bernoulli's proof straight from the original manuscript (Figure 33.22) and see what you can recognize. Use the translation on the facing page to help you. Consider it as a mathematical treasure hunt to seek out all the basics you need to make out the proof! Consider:

- How many different series can you find (eight in total)? How are they related to one another?
- In what ways does Bernoulli's depiction of series differ from our modern one?
- Bernoulli bases his proof on an earlier result he proved. Can you find his reference to an earlier proposition? Can you derive what it might have covered?
- The proof is based on a *reductio ad absurdum* type argument, that is, assuming the opposite of a claim or theorem to be proven, then, reasoning logically deriving a contradiction. From here, the claim is 'disproved' and thus its opposite (your original claim) must be true, by the law of the excluded middle. Just looking at the letters and ways in which the series are related, can you figure out how it is used in this case?

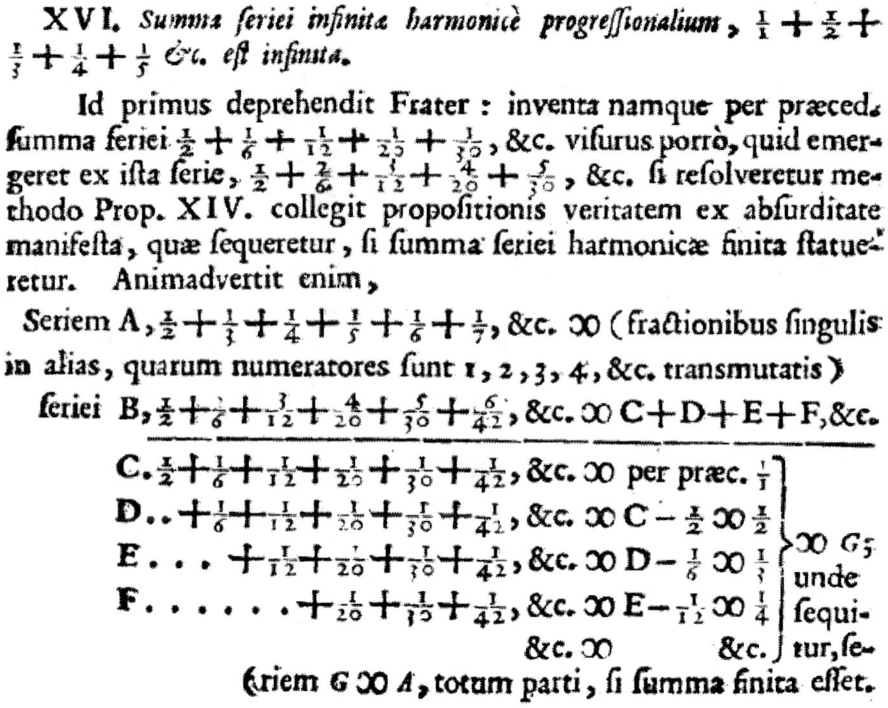

XVI. *Summa seriei infinita harmonicè progressionalium*, $\frac{1}{1} + \frac{1}{2} + \frac{1}{3} + \frac{1}{4} + \frac{1}{5}$ &c. *est infinita.*

Id primus deprehendit Frater: inventa namque per præced. summa seriei $\frac{1}{2} + \frac{1}{6} + \frac{1}{12} + \frac{1}{20} + \frac{1}{30}$, &c. visurus porro, quid emergeret ex ista serie, $\frac{1}{2} + \frac{2}{6} + \frac{3}{12} + \frac{4}{20} + \frac{5}{30}$, &c. si resolveretur methodo Prop. XIV. collegit propositionis veritatem ex absurditate manifesta, quæ sequeretur, si summa seriei harmonicæ finita statueretur. Animadvertit enim,

Seriem A, $\frac{1}{2} + \frac{1}{3} + \frac{1}{4} + \frac{1}{5} + \frac{1}{6} + \frac{1}{7}$, &c. ∞ (fractionibus singulis in alias, quarum numeratores sunt 1, 2, 3, 4, &c. transmutatis)

seriei B, $\frac{1}{2} + \frac{2}{6} + \frac{3}{12} + \frac{4}{20} + \frac{5}{30} + \frac{6}{42}$, &c. ∞ C+D+E+F,&c.

C. $\frac{1}{2} + \frac{1}{6} + \frac{1}{12} + \frac{1}{20} + \frac{1}{30} + \frac{1}{42}$, &c. ∞ per præc. $\frac{1}{1}$

D. . $+ \frac{1}{6} + \frac{1}{12} + \frac{1}{20} + \frac{1}{30} + \frac{1}{42}$, &c. ∞ C $- \frac{1}{2}$ ∞ $\frac{1}{2}$

E. . . $+ \frac{1}{12} + \frac{1}{20} + \frac{1}{30} + \frac{1}{42}$, &c. ∞ D $- \frac{1}{6}$ ∞ $\frac{1}{3}$

F. . . . . $+ \frac{1}{20} + \frac{1}{30} + \frac{1}{42}$, &c. ∞ E $- \frac{1}{12}$ ∞ $\frac{1}{4}$

&c. ∞ &c.

∞ G; unde sequitur, seriem G ∞ A, totum parti, si summa finita esset.

Figure 33.2. Bernoulli's proof from *Tractatus*

33.3.4 The Translation

XVI The sum of an infinite series of harmonically increasing terms $\frac{1}{1} + \frac{1}{2} + \frac{1}{3} + \frac{1}{4} + \frac{1}{5}$ etc. is infinite.

[My] brother discovered this first. Indeed, with the sum of the series $\frac{1}{2} + \frac{1}{6} + \frac{1}{12} + \frac{1}{20} + \frac{1}{30}$, etc. having been found previously, he wanted, in turn, to see what emerged from the following series $\frac{1}{2} + \frac{2}{6} + \frac{3}{12} + \frac{4}{20} + \frac{5}{30}$, etc. if it was reduced by the method [set out] in Proposition XIV. He checked the truth of the proposition from a clear contradiction, which follows if the sum of an harmonic series is supposed to be finite.

For, he observed that

Series A, $\quad \frac{1}{2} + \frac{1}{3} + \frac{1}{4} + \frac{1}{5} + \frac{1}{6} + \frac{1}{7}$, etc. is equal (from the individual fractions, whose numerators are 1, 2, 3, 4, etc. rewritten as others)

to series $\quad B, \quad \frac{1}{2} + \frac{2}{6} + \frac{3}{12} + \frac{4}{20} + \frac{5}{30} + \frac{6}{42}$, etc. equal to $\quad C + D + E + F, etc.$

$C \quad . \quad \frac{1}{2} + \frac{1}{6} + \frac{1}{12} + \frac{1}{20} + \frac{1}{30} + \frac{1}{42}$, etc. equal to as prev. $\quad \frac{1}{1}$

$D \quad . . \quad + \frac{1}{6} + \frac{1}{12} + \frac{1}{20} + \frac{1}{30} + \frac{1}{42}$, etc. equal to $\quad C - \frac{1}{2}$ equal to $\frac{1}{2}$

$E \quad . . . \quad + \frac{1}{12} + \frac{1}{20} + \frac{1}{30} + \frac{1}{42}$, etc. equal to $\quad D - \frac{1}{6}$ equal to $\frac{1}{3}$

$F \quad . . . . . \quad + \frac{1}{20} + \frac{1}{30} + \frac{1}{42}$, etc. equal to $\quad E - \frac{1}{12}$ equal to $\frac{1}{4}$

etc. equal to etc.

equal to G; from which it follows that the series G is equal to A, the whole to the part, if the sum were finite.

33.4 The Mathematical Explanation

After stating the result of the theorem, Bernoulli immediately credits his brother, Johann, 'id primus deprehendit Frater' with the initial discovery of this result. There are two important mathematical points which should be raised before we examine his proof. Firstly, the Bernoullian account does not invoke the notion of 'convergence' or 'divergence', for these concepts were yet to be given mathematical expression. Rather, he articulates the problem in terms of the sum being finite or infinite. Secondly, Bernoulli expresses himself mathematically a little differently from what we are used to. For example, he never describes a series by means of a general term, nor uses the modern 'sigma' notation. Rather he represents an infinite series by giving the first few terms followed by "*etc.*". In keeping with this, the following mathematical commentary has been careful to avoid unnecessary modern references as well. It's not always easy!!

Bernoulli's proof, ironically enough, is based on a *convergent* series and the structure of his argument is based on a *reductio ad absurdum* style of proof. That is, he uses the relations between various series to derive a contradiction to prove his result.

He begins with the series,

$$\frac{1}{2} + \frac{1}{6} + \frac{1}{12} + \frac{1}{20} + \frac{1}{30} \cdots$$

which he showed earlier in the work (Proposition XIV) to have a finite sum equal to 1.

He then describes another series, related to the above series, in that the denominators are the same, but the numerators are now 1, 2, 3, 4, etc... He calls this series B:

$$B = \frac{1}{2} + \frac{2}{6} + \frac{3}{12} + \frac{4}{20} + \frac{5}{30} + \frac{6}{42} \cdots$$

The individual fractions that make up this series are of course reducible, and indeed this series is in fact the Harmonic series he stated in his introduction, missing its first term. He notes this equality, and calls this equivalent series, series A.

$$A = \frac{1}{2} + \frac{1}{3} + \frac{1}{4} + \frac{1}{5} + \frac{1}{6} + \frac{1}{7} \cdots$$

The advantage of expressing series B as a series of reducible fractions is that it allows Bernoulli to be able to split up the terms and regroup them. Teasing apart the numbers, we can imagine series B rewritten as:

$$B = \frac{1}{2} + \frac{1}{6} + \frac{1}{6} + \frac{1}{12} + \frac{1}{12} + \frac{1}{12} + \frac{1}{20} + \frac{1}{20} + \frac{1}{20} + \frac{1}{20} \cdots$$

which Bernoulli regroups into smaller series, calling them C, D, E, F, and so on. Series C is:

$$C = \frac{1}{2} + \frac{1}{6} + \frac{1}{12} + \frac{1}{20} + \frac{1}{30} \cdots$$

the sum of which is 1; series D begins with the term $\frac{1}{6}$ etc., series E with the term $\frac{1}{12}$ etc. so that:

$$C = \frac{1}{2} + \frac{1}{6} + \frac{1}{12} + \frac{1}{20} + \frac{1}{30} + \frac{1}{42} + \cdots$$

$$D = \frac{1}{6} + \frac{1}{12} + \frac{1}{20} + \frac{1}{30} + \frac{1}{42} + \cdots$$

$$E = \frac{1}{12} + \frac{1}{20} + \frac{1}{30} + \frac{1}{42} + \cdots$$

$$F = \frac{1}{20} + \frac{1}{30} + \frac{1}{42} + \cdots$$

$$\vdots$$

Diagram 1

Using the already established result that series C has a sum of 1,[4] Bernoulli determines the sums of the subsequent series:

Series	Sum		
C			$= 1$
D	$= C - \frac{1}{2}$	$= 1 - \frac{1}{2}$	$= \frac{1}{2}$
E	$= D - \frac{1}{6}$	$= \frac{1}{2} - \frac{1}{6}$	$= \frac{1}{3}$
F	$= E - \frac{1}{12}$	$= \frac{1}{3} - \frac{1}{12}$	$= \frac{1}{4}$
$\vdots$			

Diagram 2

He now considers this array columnwise, just as he did in Proposition XIV. Adding the sub-series as shown in diagram 1, he gets

$$C + D + E + F + \cdots = \frac{1}{2} + \left(\frac{1}{6} + \frac{1}{6}\right) + \left(\frac{1}{12} + \frac{1}{12} + \frac{1}{12}\right) + \left(\frac{1}{20} + \frac{1}{20} + \frac{1}{20} + \frac{1}{20}\right)\cdots$$

which is equal to series B, which was the series of unreduced fractions equivalent to series A, the Harmonic series missing its first term.

But, adding the terms which are contained in diagram 2, he gets:

$$C + D + E + F + \cdots = 1 + \frac{1}{2} + \frac{1}{3} + \frac{1}{4} + \cdots$$

---

[4]In the previous proposition (XV) he considers the infinite series:

$$N = \frac{a}{c} + \frac{a}{2c} + \frac{a}{3c} + \frac{a}{4c} + \frac{a}{5c} \text{ etc.}$$

and

$$P = \frac{a}{2c} + \frac{a}{3c} + \frac{a}{4c} + \frac{a}{5c} + \frac{a}{6c} \text{ etc.} = N - \frac{a}{c}$$

He subtracts the second from the first, term by term, to get the series:

$$N - P = \left(\frac{a}{c} - \frac{a}{2c}\right) + \left(\frac{a}{2c} - \frac{a}{3c}\right) + \left(\frac{a}{3c} - \frac{a}{4c}\right) + \left(\frac{a}{4c} - \frac{a}{5c}\right) + \cdots = \frac{a}{c}$$

so that:

$$Q = \frac{a}{c} = \frac{a}{2c} + \frac{a}{6c} + \frac{a}{12c} + \frac{a}{20c} + \frac{a}{30c} \text{ etc.}$$

Setting $a = c$, he concludes that the sum of this infinite series is 1. There are some inherent problems in this derivation, which Bernoulli recognizes. In calculus classes today, you might show that the series converges in a number of different ways. One can recognize that the series can be rewritten as:

$$\frac{1}{1\cdot 2} + \frac{1}{2\cdot 3} + \frac{1}{3\cdot 4} + \frac{1}{4\cdot 5} + \cdots$$

Rewriting this in closed form, consider the related finite series:

$$\sum_{k=1}^{n} \frac{1}{k(k+1)} = \sum_{k=1}^{n} \left(\frac{1}{k} - \frac{1}{k+1}\right)$$

using a partial fraction decomposition. Noticing that this is a telescoping sum, the expression for the nth partial sum S_n of this series is:

$$S_n = 1 - \frac{1}{n+1}$$

and

$$\lim_{n\to\infty} S_n = 1$$

so that the series indeed converges and the sum of that series is 1.

which he states is equal to G. Comparing these two series, there results:

$$A = C + D + E + F + \cdots = G = 1 + \frac{1}{2} + \frac{1}{3} + \frac{1}{4} + \cdots = 1 + A$$

or

$$A = 1 + A$$

an obvious contradiction! Bernoulli phrases this as "the whole [is equal] to the part" (*totum parti*). In other words, the Harmonic series, which is $1 + A$ would be equal to 'its part', that is, A (the Harmonic series missing its first term). This would be a clear contradiction (*absurditate manifesta*) if we assume the sum is finite. Therefore, Bernoulli concludes the only alternative, namely that the sum of the harmonic series *must* be infinite.[5]

33.5 Conclusion

This proof reveals a time at which the mathematical manipulation of series as well as analytically rigorous conceptions of infinity were at their infancy. Indeed the proof relies on very simple techniques, it is a proof that appeared 150 years before the mathematically rigorous treatment of series. On its own it can be an alternative or complementary form of proving divergence without invoking the notion of partial sums.

Bibliography

[1] Howard Anton, Irl Bivens, and Stephen Davis, *Calculus* 8th ed., John Wiley and Sons, 2005.

[2] Carl B. Boyer, *A History of Mathematics*, 2nd Ed., John Wiley and Sons, 1991.

[3] H.L.L. Busard, *Quaestiones Super Geometriam Euclidis*, Brill, Leiden Netherlands, 1961.

[4] Florian Cajori, *A History of Mathematical Notations*, 2 Vols., Dover, 1993, reprinted from the work first published as two volumes by the Open Court Publishing Company, La Salle Illinois, in 1928 and 1929.

[5] William Dunham, "The Bernoullis and the Harmonic Series", *The College Mathematics Journal* Vol. 18, No. 1 (Jan., 1987), pp. 18–23.

[6] Hal Hellman, *Great Feuds in Mathematics Ten of the Liveliest Disputes Ever* John Wiley and Sons, Inc., Hoboken, New Jersey, 2006.

[7] Steven J. Kifowit and Terra A. Stamps, "The harmonic series diverges again and again", *AMATYC Review* Vol. 27 No. 2, Spring 2006, pp. 31–43.

[8] Morris Kline, *Mathematical Thought from Ancient to Modern Times*, Oxford University Press, New York, 1972.

[9] David E. Kullman, "What's Harmonic about the Harmonic Series?" *The College Mathematics Journal*, Vol. 32, No. 3. (May, 2001), pp. 201-203.

[10] D. J. Struik, (ed), *A Source book in Mathematics*, (1200–1800), Harvard University Press, 1969.

Teachers Note

The following worksheet may be useful as a follow-up to the above capsule. After introducing and covering the capsule in class, and in particular, interactively working through the Latin text with and without its translation as suggested, this mini-assignment forms a useful follow-up activity for students to carry out either individually or in small groups in class, or as a 'take-home'. The students then engage in mathematics by considering its history, as well as making a historical investigation connected to the mathematics they are learning. The result is a product in which students demonstrate historical awareness and connections as well as consolidating the relevant mathematical properties.

---

[5] For further details see also [4, 5, 8 and 10].

MINI-ASSIGNMENT

In class, you considered Bernoulli's Latin account with its translation that the sum of the Harmonic series is infinite.

1. From the steps Bernoulli takes, confirm the mathematical function of the ∞ symbol. Given that the Latin word for 'equal' is *aequalis*, speculate on the origin of this symbol.

2. Briefly summarize how Bernoulli proves that the sum of the Harmonic series is infinite, recasting his account into modern notation.

3. (a) Using modern notation, state the general term of the series Bernoulli calls B:
$$\frac{1}{2} + \frac{1}{6} + \frac{1}{12} + \frac{1}{20} + \frac{1}{30} + \cdots$$

   (b) Find exact values for the first four partial sums.

   (c) Find a closed form for the nth partial sum (Hint: A partial fraction decomposition may be helpful).

   (d) Using your result in (c) show that this series converges by calculating the limit of the nth partial sum and confirm Bernoulli's reckoning of its sum.

4. Often, examining an idea in another context can highlight some of the features of one's own present circumstances. Write a reflective paragraph on the process and outcomes of examining a mathematical idea in its original context. What does it clarify about your own mathematical ways of doing things and what features of modern mathematics does it reveal? Did this historical activity help you to understand the harmonic series, and more general series, better?

XVI. *Summa seriei infinitæ harmonicè progressionalium*, $\frac{1}{1} + \frac{1}{2} + \frac{1}{3} + \frac{1}{4} + \frac{1}{5}$ &c. *est infinita.*

Id primus deprehendit Frater : inventa namque per præced. summa seriei $\frac{1}{2} + \frac{1}{6} + \frac{1}{12} + \frac{1}{20} + \frac{1}{30}$, &c. visurus porrò, quid emergeret ex ista serie, $\frac{1}{2} + \frac{2}{6} + \frac{3}{12} + \frac{4}{20} + \frac{5}{30}$, &c. si resolveretur methodo Prop. XIV. collegit propositionis veritatem ex absurditate manifesta, quæ sequeretur, si summa seriei harmonicæ finita statueretur. Animadvertit enim,

Seriem A, $\frac{1}{2} + \frac{1}{3} + \frac{1}{4} + \frac{1}{5} + \frac{1}{6} + \frac{1}{7}$, &c. ∞ (fractionibus singulis in alias, quarum numeratores sunt 1, 2, 3, 4, &c. transmutatis)

seriei B, $\frac{1}{2} + \frac{2}{6} + \frac{3}{12} + \frac{4}{20} + \frac{5}{30} + \frac{6}{42}$, &c. ∞ C+D+E+F, &c.

C. $\frac{1}{2} + \frac{1}{6} + \frac{1}{12} + \frac{1}{20} + \frac{1}{30} + \frac{1}{42}$, &c. ∞ per præc. $\frac{1}{1}$

D. . . $+ \frac{1}{6} + \frac{1}{12} + \frac{1}{20} + \frac{1}{30} + \frac{1}{42}$, &c. ∞ C $- \frac{1}{2} \infty \frac{1}{2}$

E. . . . $+ \frac{1}{12} + \frac{1}{20} + \frac{1}{30} + \frac{1}{42}$, &c. ∞ D $- \frac{1}{6} \infty \frac{1}{3}$

F. . . . . . $+ \frac{1}{20} + \frac{1}{30} + \frac{1}{42}$, &c. ∞ E $- \frac{1}{12} \infty \frac{1}{4}$

&c. ∞ &c. ∞ G; unde sequitur, seriem G ∞ A, totum parti, si summa finita esset.

34

The Harmonic Series: A Primer

Adrian Rice
Randolph-Macon College

34.1 Introduction

Students in a first course of real analysis are often bewildered by many things, but perhaps the main difficulties they encounter are centered around three fundamental concepts: the notion of infinity and infinite processes; the phenomena of convergence and divergence; and the construction of rigorous proofs. It is in just such a course that historical information can be used to good effect. After all, the fact that these are issues with which mathematicians have grappled for centuries will doubtless be of some reassurance to the student struggling to master them.

What may be less comforting (but important, nevertheless, for a student to know) is that while mathematical results, once proved, are permanent, the arguments on which they are based are sometimes less so. In other words, as mathematics has developed, so has the concept of mathematical rigor. The result is that, although a theorem first proved 250 years ago is still true, the proof originally given for it might not be regarded as fully satisfactory by today's mathematicians. Given that the result still holds, however, it is not normally too difficult to formulate an alternative proof that attains modern standards of mathematical rigor.

A good example is the subject of this chapter, the harmonic series, which provides both a simple and an important introduction to issues surrounding the convergence and divergence of infinite series. Moreover, by virtue of its rich history and the related mathematical topics that arise from its study, it is a particularly appropriate tool to use when introducing students to the rudiments of classical analysis. There are many ways that the material contained in this chapter can be conveyed to students, depending on how much time the instructor is willing to devote to the topic in class. Thus, while they have been written to form the basis of a lecture on the harmonic series, with related activities to be undertaken either in class or for homework, the following suggestions should be in no way regarded as definitive, but should rather be used as a guide, illustrating one possible way in which students may be introduced to this fascinating subject.

34.2 Historical preliminaries

One of the earliest mathematical proofs to involve the concept of infinity concerns one of the most fundamental results in number theory, namely, the infinitude of the prime numbers. This was Proposition 20 in Book IX of Euclid's *Elements* (*c.* 300 B.C.), possibly the most influential and widely read mathematical text of all time. Although Euclid's proof was phrased in a way that differs from modern presentations, its logic and structure have definitely stood the test of time. The method used is *reductio ad absurdum*, or proof by contradiction, one of the most basic and crucial proof techniques available to the mathematician. When introducing students to the concept of proofs that involve infinite quantities, this is often a good place to start.

Theorem 1. *There are infinitely many prime numbers.*

Proof. Suppose that there are only finitely many primes, arranged in order of magnitude $2, 3, 5, \ldots, p$. Consider the quantity $q = (2 \times 3 \times 5 \times \cdots \times p) + 1$. This new number must either be prime itself or divisible by a prime greater than p. In either case there must be a larger prime number than p, which contradicts our initial assumption that p was the largest prime. Hence the number of primes is infinite.

To the non-mathematician, this result is somewhat surprising and the method used to prove it disarmingly brief. To the pure mathematician, it is one of the most beautiful theorems in the whole subject. But to both, its power lies in its amazing simplicity. Rather than tackle the (impossible) task of directly verifying the existence of infinitely many positive integers whose only proper divisors are themselves and 1, the proof approaches the issue from the opposite direction, considering only finite possibilities, before refuting the initial supposition. In such a way, it avoids a direct confrontation with infinity, but obtains the desired result nonetheless. As G. H. Hardy once wrote in praise of this method of proof [8, p. 34]: "a chess player may offer the sacrifice of a pawn or even a piece, but a mathematician offers *the game*."

When introducing students to Theorem 1, it is usually best to state and prove the result in class; then, as a homework exercise, ask them to find an alternative demonstration.[1] Incidentally, when presenting Euclid's proof, it is often interesting to give J. E. Littlewood's one-line version, "condensed for the professional" mathematician [1, p. 40], just to show how short it really can be:

"If $p_1, p_2, \ldots, p_n$, $1 + p_1 p_2 \ldots p_n$ is not divisible by any p_m."

After Euclid, matters concerning either infinite quantities or involving infinite processes continued to crop up in mathematics from time to time: for example, in the method of exhaustion used by Archimedes around 250 B.C., and in the work of some medieval Indian mathematicians. But during the 17th century, questions involving infinity moved to the very forefront of mathematics, thanks to the creation of one of the most powerful and versatile of all mathematical subjects: the calculus.

By the late 1600s, due to the groundbreaking work of the likes of Wallis, Mercator, Gregory, Newton and Leibniz, the closely related study of infinite series was at the cutting edge of mathematical research. For mathematicians of this time, the idea of a process that continued forever but could still have a finite result was both remarkable and counter-intuitive. Nevertheless, certain basic facts would have been known. For example, the sum

$$1 + \frac{1}{2} + \frac{1}{4} + \frac{1}{8} + \cdots \frac{1}{2^{n-1}} + \cdots = 2 \tag{34.1}$$

was a simple consequence of the fact that (for $|r| < 1$) the geometric series

$$a + ar + ar^2 + ar^3 + \cdots ar^{n-1} + \cdots = \frac{a}{1-r}.$$

Via the use of new calculus-related methods, knowledge of infinite series increased as further interesting results were obtained, such as

$$1 - \frac{1}{2} + \frac{1}{3} - \frac{1}{4} + \cdots + \frac{(-1)^{n-1}}{n} + \cdots = \log 2 \tag{34.2}$$

and

$$1 - \frac{1}{3} + \frac{1}{5} - \frac{1}{7} + \cdots + \frac{(-1)^{n-1}}{2n-1} + \cdots = \frac{\pi}{4}. \tag{34.3}$$

Although by the time students take a first course in real analysis, they will probably have seen equations (34.1), (34.2) and (34.3), it is still instructive for them to be reminded of these results, and perhaps asked to re-derive them. At the same time, distinction should be made between these convergent infinite series and the different kinds of non-convergent series, which either diverge to infinity, like $1 + 2 + 3 + 4 + 5 + \cdots$, or have no sum at all, like $1 - 2 + 3 - 4 + 5 - \cdots$. Students will then be ready to be introduced to the most famous divergent series of all.

---

[1] For example, in [9], Hardy and Wright give three different proofs of this theorem.

34.3 Introducing the harmonic series

At first sight, it would appear that the sum of the reciprocals of the positive integers should not be too hard to find. It does, after all, bear considerable resemblance to series (34.1), (34.2), and (34.3) above, all of which turn out to be nicely convergent:

$$\sum_{n=1}^{\infty} \frac{1}{n} = 1 + \frac{1}{2} + \frac{1}{3} + \frac{1}{4} + \frac{1}{5} + \cdots + \frac{1}{n} + \cdots$$

A common feature of all of these series is that, in every case, their terms all tend towards zero, a fact which often leads the beginner into error. While it is true that for an infinite series to be convergent the limit of its terms must be zero, the converse is not true; and the harmonic series is the most famous example of this. Although every successive term in this series gets closer and closer to zero, its sum is still infinite. Not that one would necessarily know this from numerical experimentation, however. The sum of its first 10 terms comes to just under 3, the first 100 to a little over 5, and the first 1000 terms to nearly $7\frac{1}{2}$. Indeed, it is a worthwhile exercise to have the students find $\sum \frac{1}{n}$ for values of n such as 20, 50, and 100 in their own time, and to form a conjecture regarding the convergence of the harmonic series on this evidence. They could obviously be forgiven for thinking that the sum is gradually converging to a finite upper bound, due to the extreme slowness of its divergence. Nevertheless, the upshot is that, while the harmonic series initially looks as though it should converge to a single number, it will *eventually* exceed any finite value and diverge to infinity.

The name of this fascinating series is derived from the close relationship that exists between the disciplines of mathematics and music. Musical notes are usually produced by strings or columns of air, called harmonic oscillators, due to the vibrations they emit. Each harmonic oscillator will have a lowest possible frequency, f, which determines the note created when oscillations occur along the whole length l of the string or air column. This fundamental frequency f almost always has accompanying faster sound waves called overtones which, ideally, have frequencies $f, 2f, 3f, 4f, 5f$, etc. Consequently, the wavelengths of these harmonic overtones will be $l, \frac{1}{2}l, \frac{1}{3}l, \frac{1}{4}l, \frac{1}{5}l$, and so on. It is from this sequence of "harmonic" wavelengths that the harmonic series gets its name.

Despite its lack of a finite value, the harmonic series is far from useless, having a variety of real-world applications including the study of traffic flow, structural integrity, and even weather prediction. Suppose that all weather is random and that the average global temperature in any year has no effect on subsequent temperatures. How many years in the 21st century will have record-breaking temperatures? The first year of the century (2001) was a record breaker. In 2002, there was a probability of $\frac{1}{2}$ that the average temperature would be higher than the previous year, so the expected number of record-breaking years in the first two years of the century is $1 + \frac{1}{2}$. In the third year, we have a probability of $\frac{1}{3}$ that the temperature will be higher than both 2001 and 2002, giving an expected value of $1 + \frac{1}{2} + \frac{1}{3}$ record-breaking years. Continuing on, we find that, in a period of n years, we should expect to have $1 + \frac{1}{2} + \frac{1}{3} + \frac{1}{4} + \frac{1}{5} + \cdots + \frac{1}{n}$ years where global temperature records are broken. And since the sum of the first 100 terms of the harmonic series comes to just over 5, that is how many record-breaking years we should expect this century.

Obviously, given an infinite number of years, we would expect an infinite number of records to be broken — an intuitive, and highly non-rigorous, demonstration of the series' divergence. For homework, the students could be asked to look for a formal proof of its divergence that would be considered rigorous today and discuss why such a proof satisfies modern standards of rigor. They will probably find either a demonstration involving an application of the Integral Test or a proof, now standard in most calculus textbooks, that is a modern presentation of an argument first given by the medieval mathematician, Nicole Oresme, around 1350 [10, p. 76]:

> Add to a magnitude of 1 foot: $\frac{1}{2}, \frac{1}{3}, \frac{1}{4}$ foot, etc.; the sum of which is infinite. In fact, it is possible to form an infinite number of groups of terms with a sum greater than $\frac{1}{2}$. Thus: $\frac{1}{3} + \frac{1}{4}$ is greater than $\frac{1}{2}$; $\frac{1}{5} + \frac{1}{6} + \frac{1}{7} + \frac{1}{8}$ is greater than $\frac{1}{2}$; $\frac{1}{9} + \frac{1}{10} + \cdots + \frac{1}{16}$ is greater than $\frac{1}{2}$, etc.

The standard modern presentation of Oresme's argument is as follows:

Theorem 2 (Modern proof). *The harmonic series $\sum_{n=1}^{\infty} \frac{1}{n}$ diverges.*

Proof. Let $s_{2^k} = 1 + \frac{1}{2} + \frac{1}{3} + \frac{1}{4} + \cdots + \frac{1}{2^k}$ be the 2^kth partial sum of the series. It can clearly be written as

$$s_{2^k} = 1 + \left(\frac{1}{2}\right) + \left(\frac{1}{3} + \frac{1}{4}\right) + \left(\frac{1}{5} + \frac{1}{6} + \frac{1}{7} + \frac{1}{8}\right) + \cdots + \left(\frac{1}{2^{k-1}+1} + \cdots + \frac{1}{2^k-1} + \frac{1}{2^k}\right)$$

$$> 1 + \left(\frac{1}{2}\right) + \left(\frac{1}{4} + \frac{1}{4}\right) + \left(\frac{1}{8} + \frac{1}{8} + \frac{1}{8} + \frac{1}{8}\right) + \cdots + \left(\frac{1}{2^k} + \cdots + \frac{1}{2^k} + \frac{1}{2^k}\right)$$

$$= 1 + \frac{1}{2} + \frac{1}{2} + \frac{1}{2} + \cdots + \frac{1}{2}$$

$$= 1 + k\left(\frac{1}{2}\right).$$

So $s_{2^k} = \sum_{n=1}^{2^k} \frac{1}{n} > 1 + k\left(\frac{1}{2}\right)$.

Now, $\sum_{n=1}^{\infty} \frac{1}{n} = \lim_{k\to\infty} s_{2^k} > \lim_{k\to\infty} 1 + k\left(\frac{1}{2}\right)$,

and since this final expression has no limit, the harmonic series diverges.

When presenting the standard modern version of the proof, it is always instructive to provide the students with alternatives, particularly ones that are not entirely satisfactory today. And with the divergence of the harmonic series, history provides us with several of these to choose from. There is a proof by Pietro Mengoli from 1647 [5, pp. 204-5], one each by both Jakob and Johann Bernoulli from 1689 [4, p. 30; 5, pp. 196-8], as well as one by Leonhard Euler from 1748 [4, p. 31]. We present a later proof, apparently given in 1828 by the 19th-century British mathematician J. J. Sylvester at the age of 14 when he was a student of Augustus De Morgan [2, p. 246]. It takes for granted the fact (see (34.2) above) that the alternating series $1 - \frac{1}{2} + \frac{1}{3} - \frac{1}{4} + \cdots$ converges to $\log 2$.

Theorem 2 (Sylvester's proof). *The harmonic series $\sum_{n=1}^{\infty} \frac{1}{n}$ diverges.*

Proof. Let $S = 1 + \frac{1}{2} + \frac{1}{3} + \frac{1}{4} + \frac{1}{5} + \cdots$. The series can then be written as

$$S = 1 + \frac{1}{2} + \frac{1}{3} + \frac{1}{4} + \frac{1}{5} + \frac{1}{6} + \cdots$$

$$= 1 - \frac{1}{2} + \frac{1}{3} - \frac{1}{4} + \frac{1}{5} - \frac{1}{6} + \cdots$$

$$+ 1 + \frac{1}{2} + \frac{1}{3} + \cdots$$

$$= \log 2 + S.$$

Now, clearly $\log 2$ is finite. But no finite quantity S could equal itself plus another non-zero finite quantity. Hence $S = 1 + \frac{1}{2} + \frac{1}{3} + \frac{1}{4} + \frac{1}{5} + \cdots$ is infinitely large and the series diverges.

To the beginner in analysis, this proof may seem convincing enough, and there is no doubt that it is ingenious. However, modern mathematicians would disapprove of the algebraic manipulation of infinite series as if they were finite quantities. Moreover, deducing that the sum S is infinite simply because $S = \log 2 + S$ differs considerably from the modern mathematical approach. Nevertheless, it does not take too much effort to convert Sylvester's proof into a watertight demonstration of the theorem and, for the student, this would be a good exercise in mathematical reasoning. Therefore, in class, either Mengoli's, the Bernoullis', Euler's, or Sylvester's proof of Theorem 2 should be presented, followed by a discussion of its various merits and/or limitations. In such a way, students may be guided to construct their own, more rigorous, proofs.

34.4 A "prime" piece of mathematics

Some of the most brilliant results involving infinite series were obtained by the 18th-century genius, Leonhard Euler. Euler worked in so many different areas of pure and applied mathematics, and wrote so voluminously on each, that it is impossible to form an impression of which area was his forte. Nevertheless, there is no doubt that he had a special fondness for infinite series, contributing numerous results throughout his career, earning himself the nickname of "analysis incarnate" [4, p. 54]. One of the earliest, from 1735, is perhaps the most famous, concerning not the harmonic series itself, but its convergent cousin, the sum of the reciprocals of the squares [7, pp. 73-86, 178-81]:

$$\sum_{n=1}^{\infty} \frac{1}{n^2} = \frac{\pi^2}{6}.$$

This was followed by related results concerning reciprocals of higher powers, such as

$$\sum_{n=1}^{\infty} \frac{1}{n^4} = \frac{\pi^4}{90} \quad \text{and} \quad \sum_{n=1}^{\infty} \frac{1}{n^6} = \frac{\pi^6}{945},$$

as well as variations on the theme involving, for example, odd squares

$$\sum_{n=1}^{\infty} \frac{1}{(2n-1)^2} = \frac{\pi^2}{8}$$

and alternating series of odd cubic reciprocals

$$\sum_{n=1}^{\infty} \frac{(-1)^{n+1}}{(2n-1)^3} = \frac{\pi^3}{32}.$$

All of these results essentially concerned infinite series of the form $\sum_{n=1}^{\infty} 1/n^s$, where $s \geq 1$ is an integer. But in a paper of 1737, Euler proved a startling connection between infinite *sums* involving reciprocals of integers and infinite *products* involving powers of primes [7, p. 230].

Theorem 3 (Euler's proof). $\displaystyle\sum_{n=1}^{\infty} \frac{1}{n^s} = \prod_{p \text{ prime}} \frac{p^s}{p^s - 1}.$

Proof. Euler began by letting $x = 1 + \frac{1}{2^s} + \frac{1}{3^s} + \frac{1}{4^s} + \cdots$. Dividing this by 2^s gave him

$$\frac{1}{2^s} \cdot x = \frac{1}{2^s} + \frac{1}{4^s} + \frac{1}{6^s} + \frac{1}{8^s} + \cdots.$$

Subtracting this from his expression for x resulted in

$$\frac{2^s - 1}{2^s} \cdot x = 1 + \frac{1}{3^s} + \frac{1}{5^s} + \frac{1}{7^s} + \cdots. \tag{34.4}$$

He continued by dividing this equation by 3^s:

$$\frac{2^s - 1}{2^s} \cdot \frac{1}{3^s} \cdot x = \frac{1}{3^s} + \frac{1}{9^s} + \frac{1}{15^s} + \frac{1}{21^s} + \cdots.$$

Subtracting this from (34.4) gave

$$\frac{2^s - 1}{2^s} \cdot \frac{3^s - 1}{3^s} \cdot x = 1 + \frac{1}{5^s} + \frac{1}{7^s} + \frac{1}{11^s} + \cdots.$$

In like manner, he obtained

$$\frac{2^s - 1}{2^s} \cdot \frac{3^s - 1}{3^s} \cdot \frac{5^s - 1}{5^s} \cdot x = 1 + \frac{1}{7^s} + \frac{1}{11^s} + \frac{1}{13^s} + \cdots$$

and so on. Given that this process could be continued ad infinitum, Euler obtained the equation

$$\left(\frac{2^s-1}{2^s} \cdot \frac{3^s-1}{3^s} \cdot \frac{5^s-1}{5^s} \cdot \frac{7^s-1}{7^s} \cdots\cdots\right) x = 1,$$

and thereby concluded that

$$x = 1 + \frac{1}{2^s} + \frac{1}{3^s} + \frac{1}{4^s} + \cdots = \frac{2^s}{2^s-1} \cdot \frac{3^s}{3^s-1} \cdot \frac{5^s}{5^s-1} \cdot \frac{7^s}{7^s-1} \cdots,$$

or, in modern notation,

$$\sum_{n=1}^{\infty} \frac{1}{n^s} = \prod_{p \text{ prime}} \frac{p^s}{p^s-1}.$$

As with Sylvester's proof that the harmonic series diverges, Euler's proof of this famous identity, while highly ingenious, has some serious drawbacks. In particular, the wholesale manipulation of the infinite expression $1 + \frac{1}{2^s} + \frac{1}{3^s} + \frac{1}{4^s} + \cdots$ as if it were a single finite quantity, x, is highly dubious, and there is also no justification given for the fact that $\left(\frac{2^s-1}{2^s} \cdot \frac{3^s-1}{3^s} \cdot \frac{5^s-1}{5^s} \cdot \frac{7^s-1}{7^s} \cdots\cdots\right) x$ converges to 1. Once again, though, it is useful to show students this proof in class, and have them locate and attempt to "fix" the logical gaps for homework, before giving them a fully rigorous demonstration. This modern version, adapted from a proof given in [11, pp. 271–2] is particularly useful as it is structurally identical to Euler's, but with all of the logical gaps filled in.

Theorem 3 (Modern proof). $\sum_{n=1}^{\infty} \frac{1}{n^s} = \prod_{p \text{ prime}} \frac{p^s}{p^s-1}.$

Proof. Let $s \geq 1 + \delta$, where $\delta > 0$. Then, just as in Euler's proof, subtracting the series $2^{-s} \sum_{n=1}^{\infty} \frac{1}{n^s}$ from $\sum_{n=1}^{\infty} \frac{1}{n^s}$ gives us

$$(1 - 2^{-s}) \sum_{n=1}^{\infty} \frac{1}{n^s} = 1 + \frac{1}{3^s} + \frac{1}{5^s} + \frac{1}{7^s} + \cdots.$$

In like manner, subtracting from this the series $3^{-s} \sum_{n=1}^{\infty} \frac{1}{n^s}$ gives

$$(1 - 2^{-s})(1 - 3^{-s}) \sum_{n=1}^{\infty} \frac{1}{n^s} = 1 + \frac{1}{5^s} + \frac{1}{7^s} + \frac{1}{11^s} + \cdots,$$

and so on, until eventually

$$(1 - 2^{-s})(1 - 3^{-s}) \cdots (1 - p^{-s}) \sum_{n=1}^{\infty} \frac{1}{n^s} = 1 + \sum{}' n^{-s},$$

where $\sum'$ indicates that the summation is only over those integers $n > p$ which are not divisible by the primes $2, 3, \ldots, p$. Now, since $s \geq 1 + \delta$,

$$\left|\sum{}' n^{-s}\right| \leq \sum{}' n^{-1-\delta} \leq \sum_{n=p+1}^{\infty} n^{-1-\delta},$$

and, since $\lim_{p \to \infty} \sum_{n=p+1}^{\infty} n^{-1-\delta} = 0$, the product

$$(1 - 2^{-s})(1 - 3^{-s}) \cdots (1 - p^{-s}) \sum_{n=1}^{\infty} \frac{1}{n^s}$$

converges to 1 as $p \to \infty$. Therefore,

$$\sum_{n=1}^{\infty} \frac{1}{n^s} = \frac{1}{(1-2^{-s})(1-3^{-s})\cdots(1-p^{-s})\cdots},$$

or, more concisely,

$$\sum_{n=1}^{\infty} \frac{1}{n^s} = \prod_{p \text{ prime}} \frac{1}{1-p^{-s}} = \prod_{p \text{ prime}} \frac{p^s}{p^s - 1}.$$

The function $\zeta(s) = \sum_{n=1}^{\infty} 1/n^s$ is now known as the Riemann zeta function, after Bernard Riemann who, defining it for complex values of s, discovered many of its crucial properties. Students often find it interesting to find out (perhaps for a project or expository paper) about one of the most prized unsolved problems in the whole of mathematics, the famous Riemann Hypothesis, concerning the non-trivial zeros of $\zeta(s)$, and how it arises precisely from the relationship proved in Theorem 3.[2] At a more basic level, this identity also provides some fascinating numerical results, which it is useful to have the students derive for themselves, for example:

$$\zeta(2) = 1 + \frac{1}{4} + \frac{1}{9} + \frac{1}{16} + \cdots = \frac{4}{3} \cdot \frac{9}{8} \cdot \frac{25}{24} \cdot \frac{49}{48} \cdots \cdots = \frac{\pi^2}{6}$$

$$\zeta(4) = 1 + \frac{1}{16} + \frac{1}{81} + \frac{1}{256} + \cdots = \frac{16}{15} \cdot \frac{81}{80} \cdot \frac{625}{624} \cdot \frac{2401}{2400} \cdots \cdots = \frac{\pi^4}{90}$$

$$\zeta(6) = 1 + \frac{1}{64} + \frac{1}{729} + \frac{1}{4096} + \cdots = \frac{64}{63} \cdot \frac{729}{728} \cdot \frac{15625}{15624} \cdot \frac{117649}{117648} \cdots \cdots = \frac{\pi^6}{945}.$$

But the definition of the Riemann zeta function and Theorem 3 hold only for $s > 1$. Thus, while the harmonic series would appear to be the special case of $\zeta(s)$ when $s = 1$, in practice it is undefined. However, it is true to say that

$$\sum_{n=1}^{\infty} \frac{1}{n} = \lim_{s \to 1+} \prod_{p \text{ prime}} \frac{1}{1-p^{-s}},$$

an identity which helps the students solve one final homework problem; namely, to derive an alternative proof of Theorem 1.

Corollary 4. *There are infinitely many prime numbers.*

Proof. By Theorem 2,

$$\sum_{n=1}^{\infty} \frac{1}{n} = \infty.$$

Therefore, by Theorem 3, the expression

$$\lim_{s \to 1+} \prod_{p \text{ prime}} \frac{1}{1-p^{-s}}$$

must also equal infinity. But each component in this product is finite. The total product could only reach infinity if there were an infinite number of components, and for that to happen, there must be an infinite number of primes.

34.5 Conclusion

We have thus come full circle, ending with a completely different proof of Theorem 1 and illuminating not only a connection between analysis and number theory, but also a link between the seemingly distinct mathematical worlds of the discrete and the continuous. As we said at the beginning, the ideas contained in this chapter should be regarded

---

[2] A good popular account is given in [3]. For more advanced students, [6] is an excellent study.

merely as suggestions and not as the last word on the subject. Individual teachers will be able to determine which topics and activities are within the capabilities of their students, and may omit what they feel is unnecessary. Some, for example, may believe that an introduction to the Riemann zeta function is too advanced for a first course in real analysis, while others might regard having the students find $\sum \frac{1}{n}$ for certain values of n as too elementary. Indeed, no matter how the instructor chooses to present the material contained in this chapter, the only major mistake he or she can make is to fail to tailor that presentation to fit the abilities of the students.

Although this chapter contains ideas for a class in real analysis, it also provides tantalizing glimpses of topics from other areas, such as number theory and complex analysis, thus hopefully whetting students' appetites for further mathematical subjects of higher sophistication and scope. Perhaps more significantly, this material shows a connection between two different but related forms of infinity: the divergence of the harmonic series and the infinitude of the prime numbers. Finally, and for the pure mathematician this is essential, it shows students that the same result can be proved in a variety of contrasting ways, giving them practice at comparing the structure and validity of different proofs, and providing them with an idea of what constitutes rigorous mathematics and what does not. And all of this is achieved with the careful use of appropriate examples from the history of mathematics.

Bibliography

[1] Béla Bollobás, ed., *Littlewood's Miscellany*, Cambridge University Press, Cambridge, 1986.

[2] Augustus De Morgan, "On the Summation of Divergent Series," *Journal of the Institute of Actuaries and Assurance Magazine*, **12** (1865), 245-252.

[3] John Derbyshire, *Prime Obsession: Bernhard Riemann and the Greatest Unsolved Problem in Mathematics*, Joseph Henry Press, Washington D.C., 2003.

[4] William Dunham, *Euler: The Master of Us All*, The Mathematical Association of America, Washington D.C., 1999.

[5] William Dunham, *Journey Through Genius: The Great Theorems of Mathematics*, John Wiley and Sons, New York, 1990.

[6] H. M. Edwards, *Riemann's Zeta Function*, Academic Press, New York, 1974. Dover reprint, 2001.

[7] Leonhard Euler, *Opera Omnia*, 1st series, vol. 14, B. G. Teubner, Leipzig and Berlin, 1925.

[8] G. H. Hardy, *A Mathematician's Apology*, Cambridge University Press, Cambridge, 1940.

[9] G. H. Hardy and E. M. Wright, *An Introduction to the Theory of Numbers*, 4th edition, Clarendon Press, Oxford, 1960.

[10] Nicole Oresme, *Quaestiones super Geometriam Euclidis*, ed. H. L. L. Busard, E. J. Brill, Leiden, 1961.

[11] E. T. Whittaker and G. N. Watson, *A Course of Modern Analysis*, 4th edition, Cambridge University Press, Cambridge, 1962.

35

Learning to Move with Dedekind

Fernando Q. Gouvêa
Colby College

> *Es steht schon bei Dedekind.*
> — Emmy Noether

The history of mathematics sometimes calls our attention to intellectual hurdles that our students must face, showing that ideas and conceptual moves that have become second nature to us are in fact quite daring and difficult to learn. This article focuses on a particular conceptual move, which we call "the Dedekind move" because it was so characteristic of Richard Dedekind's work. Briefly put, the idea is *to define a mathematical object as a set of other mathematical objects*. We then treat the whole set as a single thing, and do our best to forget its original plural nature.

Students typically meet this idea for the first time in an "introduction to abstraction" class, when they learn about equivalence classes. It really comes into its own, however, when the quotient construction is introduced in abstract algebra. This is a notorious stumbling block for students. A little history can help us understand why, and suggests some ideas for helping students over the hump.

35.1 Historical Background: What Dedekind Did

Suppose we are confronted with the need to come up with a definition of some mathematical entity. There are many ways to go about this. Some mathematical definitions, for example, are entirely functional: we explain what it *does*, and ignore completely the issue of what it *is*. But this is rarely completely satisfying. How do we even know that the object in question exists? Some construction is usually wanted.

Richard Dedekind[1] was faced with this situation more than once. His approach was fairly consistent. First, working with objects already known, he found a *set* of objects (usually an *infinite* set) that completely determined the entity he was trying to construct. Then he defined the new entity to *be* that set.

What is a real number?

Our first example is probably the best known: Dedekind's construction of the real numbers. As Dedekind himself tells the story in [5] (we cite from the translation in [3]), the problem presented itself to him when he was first teaching calculus, in 1858. He explained to his students that bounded increasing sequences must converge, but found that he

---
[1] For a sketch of Dedekind's life and work, see [1].

could only justify this claim by drawing pictures and appealing to his students' geometric intuition. It frustrated him that he was not able to give a real proof.

Dedekind was a good enough teacher to realize that for his students the geometric approach was probably sufficient. He was a good enough mathematician to see that this approach "can make no claim to being scientific." The reason he could make no progress, he saw, was that he had no clear notion of what a real number *was*. Since to prove that the sequence converges requires one to find, construct, or prove the existence of a real number, there was no way to prove the theorem without such a notion.

Of course, Dedekind was perfectly aware that one could take the convergence of bounded increasing sequences as an *axiom*; in fact, he says that this is what one should do in a calculus class "if one does not wish to lose too much time." But this approach did not satisfy him.

The calculus class had started in early Fall, 1858. By November, he had a solution. First of all, he took the rational numbers as known. Then he noted that to pin down a real number α, all one has to do is to specify two sets of rational numbers: the set A of all rational numbers less than or equal to α and the set B of all rational numbers larger than α. He called the pair of sets (A, B) a *cut*.

This, of course, is a routine observation. Noticing it was not the new thing about Dedekind's work. The unexpected move was to turn the logic around: rather than think that the real number α determined the two sets (A, B), he proposed that the cut *defined* the real number α.

In order for this not to be circular, it is necessary to describe which pairs of sets we want without mentioning α at all. But that is fairly easy. We want three things: every element of A should be smaller than every element of B, B should not have a smallest element, and A and B together should contain all rational numbers. Once we see that, it's easy: define a real number to be any such pair of sets, and show that we can obtain all of the usual structure (algebraic operations, order, limiting processes) in terms of this definition.

This move is so much a part of the mental toolkit of mathematicians that we no longer see how strange it is. A real number has just been defined as a pair of sets with certain properties. The set of all real numbers is then the set of all such pairs of sets. Every one of the sets in the game is infinite, and every definition involves this infinity directly.

Consider, for example, the sum $\alpha_1 + \alpha_2$ of two real numbers $\alpha_1 = (A_1, B_1)$ and $\alpha_2 = (A_2, B_2)$. It must be some cut (A, B). To construct A we take every possible element $x_1 \in A_1$ and every possible element $x_2 \in A_2$, and define A to be the set of all sums $x_1 + x_2$ (this is a sum of rational numbers, so we know what that means). Can we actually find A? Doesn't doing so require an infinite number of additions? Does this make sense? Many mathematicians found this a little disturbing.

In fact, as Dedekind noted in his introduction to [8], several of his contemporaries missed the point. They thought that all Dedekind was saying was that any real number determined and was determined by a pair of sets (A, B). As Dedekind pointed out, that observation goes back all the way to Eudoxus' definition of equality of ratios in Euclid's *Elements*, Book V. The new idea was that one could *create* a real number from the rationals by taking such a pair of sets, and that by taking all the cuts one would get (exactly) all the real numbers.

What is an ideal prime factor?

The Dedekind move appears again in his theory of algebraic numbers. The issue was to understand the arithmetic of rings such as $\mathbf{Z}[i]$, $\mathbf{Z}[\sqrt{2}]$, and $\mathbf{Z}[\sqrt{-5}]$. These are integral domains in which we can define "primes" in the usual way, as numbers that do not factor non-trivially, and then look for "prime factorizations." Everyone's first assumption is that things would work as usual, and in some cases they do. As Ernst Eduard Kummer discovered, however, those turn out to be the exceptional cases.

The simplest example of the problem (and one that Dedekind himself used) is in the ring $\mathbf{Z}[\sqrt{-5}]$. It is easy to check, first, that

$$(1 + 2\sqrt{-5})(1 - 2\sqrt{-5}) = 21 = 3 \cdot 7.$$

With a little bit more work, one can check that each of the four numbers $1 + 2\sqrt{-5}$, $1 - 2\sqrt{-5}$, 3, and 7 are "prime" in the sense that they cannot be factored. Thus, the number 21 factors in two completely different and unrelated ways in $\mathbf{Z}[\sqrt{-5}]$. So while "prime factorizations" exist, they are *not unique*. Since uniqueness plays a crucial role in much of elementary number theory, this is a problem.

35.1. Historical Background: What Dedekind Did

Faced with this, one could just throw up one's hands in despair, or perhaps focus on determining for which rings it actually would work "correctly," so that prime factorizations are unique. Kummer had another idea. He argued that one could rescue uniqueness by postulating that invisible divisors are in play. Specifically, he suggested something like this. We would see the weird factorization of 21 above if in fact there were hidden primes p, q, p', q' (the latter two would be the complex conjugates of the first two) such that

$$1 + 2\sqrt{-5} = p \cdot q$$
$$1 - 2\sqrt{-5} = p' \cdot q'$$
$$3 = p \cdot p'$$
$$7 = q \cdot q'$$

and so

$$21 = p \cdot q \cdot p' \cdot q'$$

Kummer argued that the apparent non-uniqueness should be understood as *demonstrating* that such undetectable prime factors were actually there. He called on chemistry for an analogy, pointing out that chemists postulated the existence of atoms and molecules for similar reasons, without being able to actually observe them. He called these invisible numbers "ideal prime factors."

This would remain only a fiction if we couldn't say more; Kummer's amazing insight was to see that in fact, we can. What, after all, does one want from a prime number p? Not to look at it! Instead, what one wants is to be able to decide what numbers are divisible by p, and, for those that are, to decide what power of p appears in their factorization. For ideal primes such as p and q above, Kummer did exactly that. In other words, he showed how to take numbers $a + b\sqrt{-5}$ and determine whether they are divisible by p, and how many times. Then he showed that one could use these ideal prime divisors (and ideal divisors in general) to construct a unique factorization theory for any ring of algebraic numbers. The ideal primes remained just symbols; there is no element of $\mathbf{Z}[\sqrt{-5}]$ that is a common divisor of 3 and $1 + 2\sqrt{-5}$, but we still name that common divisor p and work with it.

We might be satisfied with this. Kummer's "ideal primes" are really functions from the non-zero elements of our ring to the positive integers; the value of the function at an (algebraic) integer α is the number of times α is "divisible" by p. There are some problems, of course. When we write $3 = pp'$ it's not really clear what the product signifies. And it's not clear what a product like pq^2 might be, since it does not seem to be an element of $\mathbf{Z}[\sqrt{-5}]$.

Dedekind found this situation not to his liking. He seems to have wanted to be able to point to something and say "that's what an ideal prime divisor is." And he wanted to define the product of these things, preferably using a method that actually had something to do with multiplying. That's where the move comes in.

If we look back at the integers, we see that a prime number in $\mathbf{Z}$ is completely determined by the set of integers that are divisible by it, which we might write down as $p\mathbf{Z}$. The prime p is simply the smallest positive element in that set. It turns out that we can completely describe such sets *without specifying that they are sets of multiples*. As every abstract algebra teacher knows, they are non-empty sets of integers that are closed under addition and under multiplication by an arbitrary integer. In other words, they are what we now call "ideals" in $\mathbf{Z}$.

Now, the subset of $\mathbf{Z}[\sqrt{-5}]$ consisting of "multiples" of our mysterious "ideal prime divisor" p has the same properties: it is an ideal in that ring of algebraic integers. In fact, it is a non-zero prime ideal, just as the ideal generated by a prime number in $\mathbf{Z}$ is. It just doesn't have a generator. So the Dedekind move is to define the ideal prime divisor to *be* this set. In modern terms, we might describe it as replacing ideal primes with prime ideals.

Given that, it is possible to re-do Kummer's theory, dispensing with the mysterious ideal primes and replacing them, instead, with nice sets of algebraic numbers, the ideals. That is what Dedekind did. One can read a very clear account of this, including a detailed analysis of the example of factoring 21 in $\mathbf{Z}[\sqrt{-5}]$, in [7].

As before, not everyone liked (and some mathematicians[2] still don't like) this move. Leopold Kronecker didn't think working with an infinite set of numbers was any clearer than working with an algorithm for finding how many times a number is divisible by p, and proposed a completely different foundation for the theory. It was only as mathematicians

---

[2] For example, see [9].

got more and more used to working with sets and thinking of a whole set as in some sense *one* thing, in the early 20th century, that Dedekind's point of view came to dominate.

Other instances

Even earlier in his career, in [2, §7], first published in 1857, Dedekind had already used his move, though admittedly in a minor way. He was studying congruence classes modulo p of polynomials with integer coefficients. In his time it was still common to think of the elements of, say, $\mathbf{Z}/n\mathbf{Z}$ via a canonical choice of representatives. Dedekind was one of the first to note that it would be easier to "just use the entire congruence class," i.e., to think of the elements of $\mathbf{Z}/n\mathbf{Z}$ as *sets of integers*, as we teach students to do today. (In his paper, of course, the rings in question were $\mathbf{Z}[x]/p\mathbf{Z}[x]$ and its quotients.) It seems to have taken some time for this idea to really "take."

Another situation in which Dedekind used his "move," this time in quite an audacious manner, was to answer the question "what is an integer?" His answer, explained in [8], was basically that the integers "are" any set[3] that satisfies a certain list of axioms (almost the same as the Peano axioms). He then showed that any infinite set contains many such subsets, any of which will do as a representation of the integers. He gave a philosophical argument to "prove" that infinite sets exist; given that, his approach gives a construction of the integers. The "any such subset" part of this means that the integers are constructed up to a certain kind of equivalence (order-preserving bijections). In effect, Dedekind defines the integers as an equivalence class of infinite sets.[4]

35.2   In the Classroom 1: Transition to Proofs

Sometimes the most important pedagogical contribution of the history of a idea is to alert us that an idea is neither obvious nor "natural." In fact, the Dedekind move is a choice, one that has become so dominant in modern treatments that we no longer feel that any alternatives are available. The quotient group *is* the set of all cosets, and that's it.

Thinking historically, we see that alternative routes are available. And, in the light of the difficulty even great mathematicians had with the Dedekind move, we see that our students' difficulties are quite natural.

A second observation is that the Dedekind move has more than one use. Dedekind used it to *construct* well known objects, to *clarify* the nature of objects that were not yet well understood, and to eliminate the need to choose a canonical representative for an equivalence class.

The history also highlights one of the things that makes the move feel unnatural: we are proposing to *deal with a set of things as itself a single thing*. For most mathematicians, this is a completely familiar conceptual move, but for our students it can be quite difficult. How can we make it easier for them?

The minimal thing to do, of course, is simply to notice it. Call your students' attention to what a daring move it is, discuss alternatives, and generally clue them in that something new and beautiful has just entered the scene. The natural place to do this is in a "transition to proofs" or discrete mathematics course.

What is a fraction?

Sporting a scraggly gray beard, Prof. Niemand prepares to teach one more class in his *Introduction to Abstract Mathematical Thought*. Today he wants to introduce the notion of an equivalence relation and of equivalence classes. He walks into class and asks his students, "What is $\frac{3}{5}$?"

The students are used to this kind of shenanigan. "It's three fifths," says a bored-looking young woman, Amy, in the front row.

"Ah, but what does that mean?" asks Prof. Niemand, trying and failing to make his eyes sparkle. The students knew this was coming, of course.

"It's what you get when you divide three by five," says Tim.

---

[3]This description of Dedekind's concept is more than a little anachronistic, at least in the language used, and may well give historians heartburn; still, it captures the main idea.

[4]The currently fashionable set-theoretical construction of the integers goes the other way and chooses a canonical representative for each equivalence class, namely the (finite) von Neumann ordinals.

"How do you know you would get anything?" shoots back the professor. That gets total silence, so it is time to rile them up a bit more. "OK, consider this," he says, writing on the board:

$$\frac{3}{5} = \frac{6}{10}.$$

"I know you learned that in grade school, but those things do not look equal to me. Equal means *the same*, and they're not the same."

"Well, they are both solutions to the equation $5x = 3$," says Emily, an algebraist at heart.

"So that equation has two solutions?" asks the professor.

"No," says Emily, "that's why they're equal. They both solve the same equation, so they're the same thing."

"Really? But quadratic equations do have two solutions, don't they? So you're claiming there is a solution and that there's only one. How do you know?"

Emily frowns. "I guess I mean that we just create a number 3/5, and define it to be the solution to that equation."

"But what allows you to 'just create' a number?" asks the professor. "Don't you have to show that this number exists? I don't even know what kind of thing it is!"

"Well, it's a thing that has a numerator and a denominator. Fractions are just things like that, and you just decree that some of them are equal to others," says Ben the formalist.

"I thought I knew what 'equal' means," says the professor, "but maybe not. I thought equal meant 'exactly the same', but you're saying you're allowed to say that two things that are clearly not the same are still equal."

"But of course! Isn't $0.5 = 1/2$?" says Amy, and we're back on the spiral again.

From fractions to equivalence relations

Most of our students have some sort of concept of what a fraction is. But fractions are inherently objects with multiple representations. "Clearly" 3/5 and 6/10 are two different symbols for the same *thing*. What is that thing?

There are, of course, many answers to this, as the dialogue above tries to show. One other possibility, for example, would be that $\frac{a}{b}$ is something that acts on numbers divisible by b in a certain specified way (we would then call that action "multiplication," of course). In that case, the equality would mean that $\frac{3}{5}$ and $\frac{6}{10}$ act in exactly the same way on any number on which both can act, i.e., on any multiple of 10. Or one could follow Kummer and argue that both fractions are names of some mysterious entity out there.

Students may well come up with various different concepts of a fraction. A particularly tempting one is to short-circuit the problem by saying that a fraction *is* its representation in lowest terms (and with a positive denominator). This is analogous to choosing unique representatives for integers modulo n.

This is tempting, but it forces us to say that $\frac{6}{10}$ is not actually a fraction, but rather some sort of "pre-fraction." It also makes explaining how to add fractions a little bit harder, because the formula

$$\frac{a}{b} + \frac{c}{d} = \frac{ad + bc}{bd}$$

is not strictly true unless we add "and then you replace the pre-fraction at the end by its underlying fraction." It might be fun to ask students to prove that this operation is associative.

With luck, we might find that one of our students is a radical nominalist, willing to argue that a fraction is simply *the list of all of its names*. That is the Dedekind move in this situation: a fraction is an equivalence class of ordered pairs of numbers.

The "what is a fraction" discussion will probably fit in about 20 minutes of class time, leaving time for the teacher to investigate what is necessary for the notion of "defined equality" to work properly. This leads to the definition of an equivalence relation and the recognition that the set of all names for a fraction is precisely an equivalence class.

Is it a set or is it a thing?

So now we have a notion of a fraction as an equivalence class of pairs of numbers. Should we be satisfied? Probably not! Who really believes that a real number *is* a pair of sets of rational numbers? After all, there are alternative constructions, such as the one involving equivalence classes of Cauchy sequences. The partisans of one construction do not have arguments with the partisans of another about the true nature of the real numbers. They argue about which construction is better, clearer, or pedagogically most effective.

It is the same for fractions: while it is useful to think of a fraction as an equivalence class of pairs of integers, it seems a little weird to claim that "one half" really means an infinite set of pairs of integers. So once the basic move is made, it is time to discuss with our students what it is that we have achieved. In other words, what is the point?

For rational and real numbers alike, the point seems to be this: we *know* what we want to construct, at least on an intuitive level. The question becomes, then, whether this intuitive notion is sufficiently precise and free of contradictions. This is what the Dedekind move achieves: it shows that if we believe in fractions and in sets of fractions, then we can believe in real numbers as well.

We can then keep digging: do we believe in fractions? Well, if we believe in integers and sets of integers, then we are allowed to believe in fractions. And so on, for as long as we like. The current fashion is to get down to set theory itself, axiomatize that, and stop. And then, of course, to forget the construction! It does not help, when thinking about the integer 2, to think of it as the set $\{\emptyset, \{\emptyset\}\}$.

The more philosophical discussion of the goal of such constructions can be postponed, but it should happen at some point, since otherwise students will never quite get the point of formal constructions. For example, if a construction of the real numbers is part of the syllabus, the philosophical discussion could well belong there.

35.3 In the Classroom 2: Abstract Algebra

In the abstract algebra classroom, the Dedekind move achieves its full power. The most natural place to highlight it is when we teach the quotient construction. Here the main point is to obtain something that has certain properties. The nature of the elements of that something (i.e., that they are equivalence classes, cosets, whatever) is not that important. Being familiar and comfortable with the Dedekind move allows us to create the objects we want with minimal fuss.

Congruence classes

To teach the quotient construction in this spirit, one might start by trying a bare-hands approach first. For example, we could define $\mathbf{Z}/n\mathbf{Z}$ as a set consisting of n symbols:

$$\{\overline{0}, \overline{1}, \overline{2}, \ldots, \overline{n-1}\}.$$

Then we define operations as follows: to add (or multiply) $\overline{a}$ and $\overline{b}$, add (or multiply) a and b, and then check if the answer is greater than or equal to n. If not, just put a bar on top of the answer; if it is, divide by n, take the remainder, and put a bar on top.

Students will probably be happy with this. Now challenge them to prove that the resulting object is a ring. This is doable, but annoying. Suppose we want to prove addition is associative. Look at

$$\overline{a} + (\overline{b} + \overline{c}) \qquad \text{versus} \qquad (\overline{a} + \overline{b}) + \overline{c}.$$

If we could just remove the bars, this would follow at once from associativity in $\mathbf{Z}$. But with the definition above, each of the parentheses may require a division with remainder step, and things get complicated.

If your students already know modular arithmetic, this exercise can be done in 10 or 15 minutes of class time, and the challenge of proving that the resulting object is indeed a ring can be assigned as homework. In that case, the next class can open with a discussion of the students' proofs.

Analyzing the proofs will show that everything becomes much easier if we decide to allow ourselves to write $\overline{a}$ for *any* integer a. (To put it in fancy terms, we want to have not only the quotient $\mathbf{Z}/n\mathbf{Z}$ but also the canonical homomorphism $\mathbf{Z} \longrightarrow \mathbf{Z}/n\mathbf{Z}$.) As in the case of fractions, once we do this we have objects with multiple names:

$$\overline{0} = \overline{n} = \overline{2n} = \ldots$$

What are all those names actually naming? Again, the Dedekind move takes the easy way out of defining the object in question to be the set of all of its names. Discussing this in detail, and highlighting the analogy with fractions, can help students get a better grip on what is actually going on.

Something a bit harder

The special thing about $\mathbf{Z}/n\mathbf{Z}$, of course, is that the numbers from 0 to $n - 1$ provide a natural set of representatives. The next activity should push further, looking at a quotient where such a natural set is not so easily available. For example, one might look at $\mathbf{Z}[i]$ and consider the quotient by (the ideal generated by) $1 + 2i$. This is still fairly small (five elements), so it can be done "by hand," but there is no natural set of representatives. Describing it as "arithmetic modulo $1 + 2i$" highlights the analogy to the case of the integers.

Doing this in detail will take a good deal of class time, probably a full class period. In particular, one needs to establish a way to decide whether an element $a + bi$ is divisible by $1 + 2i$. (We are suddenly back to Kummer!) Once this example is worked out, however, it becomes easier to see that in the general case we do not want to try to locate a set of representatives, and hence will be much better off by using the cosets themselves.

Quotients in general

Having done these exercises, we are ready to construct quotients in general. This is similar for rings or groups, of course. Let's stick with rings so that we can be specific.

Given a ring R and a subset I, we want to construct a new ring R/I *and a function* $R \longrightarrow R/I$ which we will denote by $r \mapsto \overline{r}$. We want $\overline{r}$ to be one of many names of an element of R/I, and the rule is that if $r_1 - r_2 \in I$, then $\overline{r_1}$ and $\overline{r_2}$ are names for *the same* element.

Now we deploy the Dedekind move: since that's what we want, we just define *an element in R/I is the set of all its names*, and check that (as long as I is an ideal) this makes sense. We now have the standard definition of the quotient ring, and, with any luck, our students have it too.

The philosophical discussion has a crucial role here as well. Yes, we will define an element of R/I to be a particular kind of subset of R, but the goal of doing that is just to get an easy construction. In the end, we want our students to treat a coset as an object. They need to be told that explicitly. To further that goal, we should avoid notations that highlight the set-theoretical construction. So while it is certainly "legal" to write $n \in \overline{0}$, we would do well to avoid doing it, just as we don't say 3 is an element of the left hand subset of π.

35.4 Moving with Dedekind

Finally, the question poses itself: should we bother to teach our students the Dedekind move? After all, most linear algebra books have decided that mentioning the quotient space in that course is a bad idea. Is the move important enough to spend time on it in other courses?

The answer is, of course, yes. Though in its initial conception the Dedekind move was mostly a way to give a construction of known objects, it has become a standard tool in mathematics, used over and over to create ever more complex objects, from L^p spaces to the algebraic closure of a general field.

The history of Dedekind's move, however, does leave us with two pedagogical lessons, which we have tried to highlight. The first is that the move needs to be learned; it does not "come naturally." The second is that it is actually a double move: we define an object to be a set of simpler objects, yes, but then we forget the "internal structure" of our object, treating it once again as a single thing.

Bibliography

[1] Kurt-R. Biermann. "Dedekind." In *Dictionary of Scientific Biography*, ed. by C. C. Gillispie. Scribners, 1970–1981.

[2] Richard Dedekind. *Abriß einer Theorie der höheren Kongruenzen in bezug auf einem reelen Primzahl-Modulus*, 1857. In [4], vol. 1, item V.

[3] Richard Dedekind. *Essays in the Theory of Numbers*, containing translations of [5] and [8]. Translated by Wooster Woodruff Beman. Open Court, 1901. Reprinted by Dover Publications, 1963.

[4] Richard Dedekind. *Gesammelte Mathematische Werke*, ed. by R. Fricke, E. Noether, and O. Ore. Brunswick, 1930–1932. Reprinted by Chelsea Publishing, 1969.

[5] Richard Dedekind. *Stetigkeit und Irrationalzahlen*, 1872. In [4], vol. 3, item L.

[6] Richard Dedekind. *Sur la Théorie des Nombres Entiers Algébriques*. Gauthier-Villars, 1877.

[7] Richard Dedekind. *Theory of Algebraic Integers*. Translated and introduced by John Stillwell. Cambridge University Press, 1996. (Translation of [6] with an extended introduction.)

[8] Richard Dedekind. *Was sind und was sollen die Zahlen?*, 1888. In [4], vol. 3, item LI.

[9] Harold M. Edwards, "Dedekind's Invention of Ideals," in *Studies in the History of Mathematics*, ed. by Esther R. Phillips. MAA, 1987.

About the Editors

Dick Jardine is a US citizen born in Vienna, Austria. He graduated with a BS from the United States Military Academy, West Point, NY and earned his Ph.D. in Mathematics at Rensselaer Polytechnic Institute. His last dozen publications (journal articles, book chapters, and the MAA Notes volume *From Calculus to Computers*) have focused on undergraduate mathematics teaching and learning, to include the use of the history of mathematics in teaching mathematics. Dick has served on the MAA Committee on the Teaching of Undergraduate Mathematics, the Subcommittee on the History of Mathematics, and the Committee on Assessment.

Amy Shell-Gellasch is a US citizen born in Royal Oak, MI. She is currently Visiting Assistant Professor of Mathematics at Beloit College, Beloit, WI. Amy graduated from the University of Michigan with a B.S. Ed. in math and physics. She earned her M.A. (math) from Oakland University and her D.A. from the University of Illinois at Chicago (2000). Amy's thesis was an historical piece on mathematician Mina Rees. She was named an R. L. Moore Project NeXT Fellow in 2000 and spent three years as a post-doctorate at the United States Military Academy where she conducted historical research with Fred Rickey. Her focus is the history of mathematics and its uses in teaching. For example, her most recent publications are *From Calculus to Computers* with Dick Jardine and *Hands On History*, both in the MAA Notes series, as well as several historical posters for the MAA.

She is very involved in the MAA and the HOM SIGMAA. She currently Chairs the MAA Committee on SIGMAAs. She co-founded and is current Programs Chair of the HOM SIGMMA and founded the HOM SIGMAA Student Paper contest. She has organized numerous meetings and sessions to include the 2009 JMM MAA short course, Exploring the Great Books of Mathematics.